This set of abstract patterns represents the seventeen possible diperiodic space groups discussed in Section 1-6. An explicit indication of the point and translational symmetries is shown in Figure 1-25 and a different type of pattern will be found in Figure 1-26. It is interesting to identify the lattice and the point operations associated with each of these figures by comparing them with the corresponding ones in Figure 1-25. These patterns were created and reproduced with the permission of Arthur L. Loeb, Ledgemont Laboratory, Kennecott Copper Company.

PRENTICE-HALL SOLID STATE PHYSICAL ELECTRONICS SERIES

NICK HOLONYAK, Jr, Editor

ANKRUM *Semiconductor Electronics*

BURGER AND DONOVAN, Editors *Fundamentals of Silicon Integrated Device Technology: Vol. I, Oxidation, Diffusion, and Epitaxy. Vol. II, Bipolar and Unipolar Transistors*

GENTRY et al. *Semiconductor Controlled Rectifiers: Principles and Applications of p-n-p-n Devices*

LAUDISE *The Growth of Single Crystals*

NUSSBAUM *Applied Group Theory for Chemists, Physicists and Engineers*

NUSSBAUM *Electromagnetic and Quantum Properties of Materials*

NUSSBAUM *Semiconductor Device Physics*

PANKOVE *Optical Processes in Semiconductors*

ROBERTS AND VANDERSLICE *Ultrahigh Vacuum and Its Applications*

VAN DER ZIEL *Solid State Physical Electronics, 2nd ed.*

WALLMARK AND JOHNSON, Editors *Field-Effect Transistors: Physics, Technology and Applications*

WESTINGHOUSE ELECTRIC CORP. *Integrated Electronic Systems*

APPLIED GROUP THEORY FOR CHEMISTS, PHYSICISTS AND ENGINEERS

PRENTICE-HALL INTERNATIONAL, INC., *London*
PRENTICE-HALL OF AUSTRALIA, PTY. LTD., *Sydney*
PRENTICE-HALL OF CANADA, LTD., *Toronto*
PRENTICE-HALL OF INDIA PRIVATE LIMITED, *New Delhi*
PRENTICE-HALL OF JAPAN, INC., *Tokyo*

APPLIED GROUP THEORY FOR CHEMISTS, PHYSICISTS AND ENGINEERS

ALLEN NUSSBAUM

PROFESSOR OF ELECTRICAL ENGINEERING
UNIVERSITY OF MINNESOTA

PRENTICE-HALL, INC.

ENGLEWOOD CLIFFS, NEW JERSEY

Current printing (last digit):

10 9 8 7 6 5 4 3 2 1

13-040832-8

Library of Congress Catalog Card Number: 77-146961

Printed in the United States of America

To

Understanding

Belief

Peace

PREFACE

There are probably just two acceptable reasons for writing another book on group theory: either the author has something new to add to our knowledge of the subject or he has a better way of presenting what is already known. It is my hope that the contents of this book meet the second criterion. As an experimental solid-state physicist who has struggled to understand the language of the theorist and the mathematician, I have long felt that a really simple explanation of how group theory is applied would be welcomed not only by people in this field but also by engineers and physical chemists as well. This feeling has been reinforced by the demand for reprints of three earlier attempts along these lines (*Proc. IRE* **50**, 1762; *Am J. Phys.* **36**, 529; and *Advances in Solid State Physics*, Vol. 18, edited by Seitz and Turnbull). The approach in these three review articles emphasized the physical significance of symmetry and its use in determining normal modes and energy band structures of crystalline materials. Any proofs given were either intuitive or based on demonstration via examples.

This present book is an extension of the same kind of treatment to other areas such as bonding problems, infrared spectroscopy, and crystal structure. In choosing topics I tried to select those which I felt would be of greatest interest. Hence, I have not covered as many applications as are to be found in the books of Heine and Tinkham nor have I been as thorough in treating any given topic as have Hamermesh or Wilson, Decius, and Cross. In fact, my objective in part has been to provide an easy transition to these classics on applied group theory. Another simplification that has been made is to idealize physical situations insofar as possible. For example, in showing how group theory helps us interpret infrared and Raman spectra, I have assumed that these are pure vibrational phenomena; rotational and interaction bands are ignored.

By way of background, it would be helpful for the student to be familiar with classical and quantum mechanics. However, this is not absolutely necessary since the Lagrange and Schroedinger equations are generated from physical arguments. The amount of mathematics required is also very small; some previous contact with partial differential equations would be helpful.

Because these requirements are rather low, this book should be usable by advanced undergraduates in science or engineering. In my own teaching, I have presented it primarily to first- and second-year graduate students.

It is quite obvious that a book covering the topics considered here must depend heavily on work done by others. I have tried to acknowledge all published sources from which I have borrowed ideas. I have also had a great deal of help from my colleagues, Ray Vaitkus, Peter Stone, Ronald Soderstrom, John Rayside, and Alan deMonchy. In addition, the lectures given by Professor Morton Hamermesh during the summer of 1969 have been an unforgettable source of inspiration. Finally, I would like to reiterate my debt to the many friends who helped with the review articles which serve as the nucleus of this book.

A.N.

CONTENTS

CHAPTER ONE GROUP THEORY: THE MATHEMATICS OF SYMMETRY

CHAPTER TWO REPRESENTATIONS AND VIBRATIONS

CHAPTER THREE **INFRARED AND RAMAN VIBRATIONAL SPECTRA**

CHAPTER FOUR **MOLECULAR BINDING**

CHAPTER FIVE **ELECTRONS IN CRYSTALS**

CHAPTER ONE GROUP THEORY: THE MATHEMATICS OF SYMMETRY

1-1 The concept of symmetry

The ordinary meaning of *symmetry* implies balance, as seen in a well-organized painting or a pleasing flower arrangement. On the other hand, we may be more precise by describing the chair of Fig. 1-1 as symmetrical with respect to a plane bisecting the seat and the back; this is known as *bilateral symmetry*. A spoked wheel (Fig. 1-2) may have *six-fold rotational symmetry* about its axle; it also possesses *reflection symmetry* in any plane determined by the axle and a spoke.

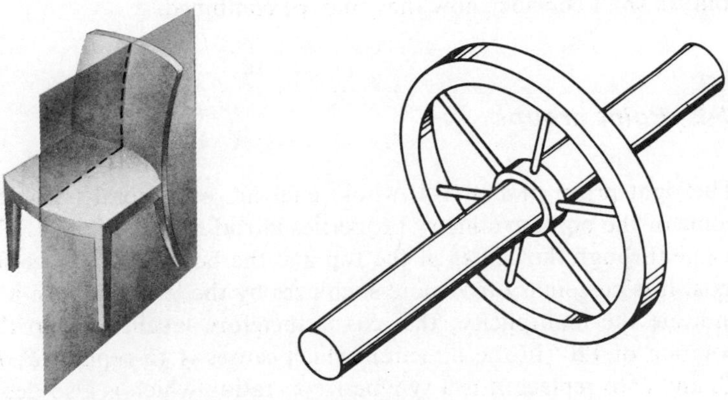

1-1 An example of bilateral symmetry **1-2** Six-fold rotational symmetry

PROBLEM 1-1

What other reflection symmetry planes does the wheel of Fig. 1-2 have?

The most symmetrical object we can visualize readily is a sphere. Every plane through the center is a reflection plane, and every axis through the

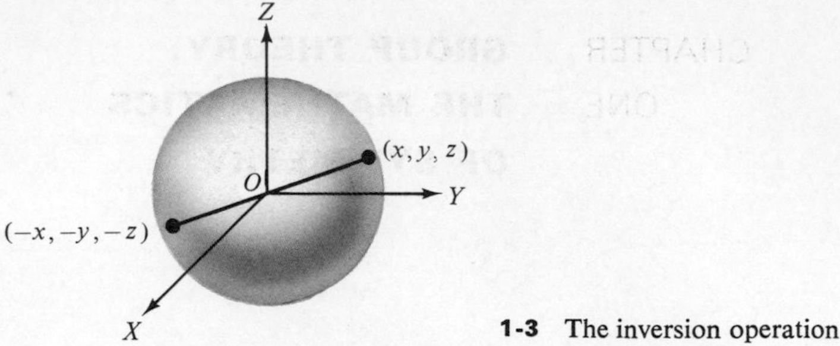

1-3 The inversion operation

center has infinite-fold rotational symmetry. Another way of expressing the nature of a sphere is in terms of *inversion symmetry*. Placing the center at the origin O of a coordinate system $OXYZ$ (Fig. 1-3), any point on the surface has coordinates (x, y, z), which may be positive or negative numbers. If we draw a line from this point through the center of the sphere, it will intersect the surface at the diametrically opposite point $(-x, -y, -z)$. These two points are said to be related by inversion symmetry. We note that the inversion of a sphere through its center, its rotation about an axis, and its reflection in a plane all have one feature in common: the center of the sphere is unaltered. These processes are therefore called *point symmetry operations*, and we shall consider how they may be combined.

1-2 Point groups

The right prism of Fig. 1-4, whose ends are equilateral triangles, possesses some of the point symmetry properties introduced in Sec. 1-1. For example, a line through the center of the top and the bottom is a three-fold rotation axis. It is customary to denote such axes by the letter C, with a subscript to indicate the multiplicity; the axis is therefore labelled C_3 in the figure. A rotation of 120° in the direction which causes A to replace B, B to replace C, and C to replace A is a symmetry operation which is also denoted by C_3. (This double use of a symbol should cause no confusion, since it is helpful to remember that an operation C_3 is a rotation about an axis C_3.) A second such rotation in the same direction —equivalent to 240° in all—is denoted by $(C_3)^2$, or C_3^2 for short, and the third application of this operation returns the prism to its original position. Three consecutive 120° rotations are therefore equivalent to leaving the prism stationary; we call this the *identity operation* E (from the German *einheit*, for unity).

The figure also shows that there is a reflection plane parallel to the top and bottom that divides the prism horizontally. This plane is denoted by σ_h, where σ is the customary symbol for any reflection plane and the subscript

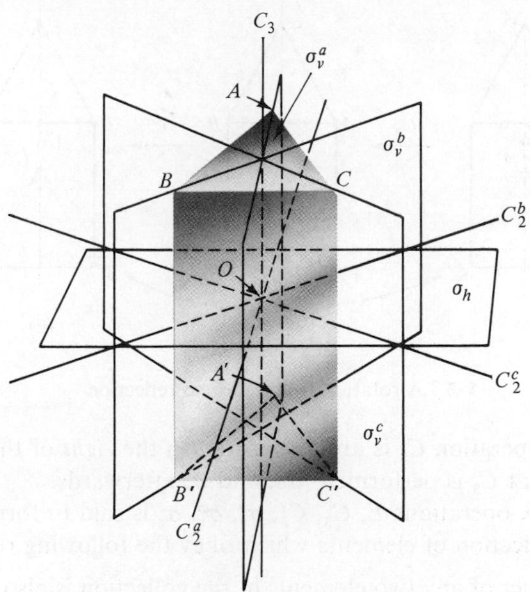

1-4 An illustration of a point group

h denotes that it is horizontal (a principal rotation axis such as C_3 is always oriented vertically). Other symmetry elements shown in the figure are three vertical reflection planes $\sigma_v^a, \sigma_v^b, \sigma_v^c$ and three two-fold axes C_2^a, C_2^b, C_2^c, where the superscripts *a*, *b*, and *c* distinguish each element in the set of three.

PROBLEM 1-2

(a) Does the triangular prism of Fig. 1-4 possess inversion symmetry?
(b) What kinds of right prisms have inversion symmetry?

Let us consider now the set of six operations $E, C_3, C_3^2, \sigma_v^a, \sigma_v^b, \sigma_v^c$, and let us combine any two. For example, a 120° rotation C_3 has the effect indicated in Fig. 1-5(b). If this is followed by a reflection σ_v^a, which is a reflection in the original plane σ_v^a and *not* in the plane *AA'O* of Fig. 1-5(b). The result shown in Fig. 1-5(c) will be recognized as equivalent to the single operation σ_v^b. We symbolize this by writing

$$\sigma_v^a C_3 = \sigma_v^b \tag{1-1}$$

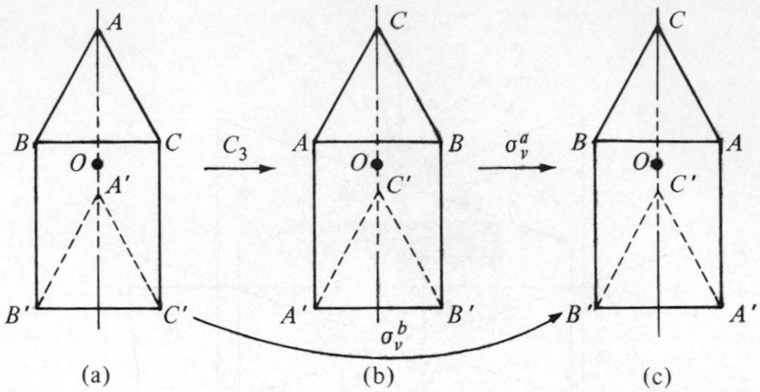

1-5 A rotation followed by a reflection

where the first operation C_3 is always written to the *right* of the second one σ_v^a, indicating that C_3 is performed first and σ_v^a afterwards.

The set of six operations E, C_3, C_3^2, σ_v^a, σ_v^b, σ_v^c is said to form a *group*. A group is any collection of elements which obey the following rules:

1. The product of any two elements in the collection is also a member of the collection.
2. The collection must contain an identity or unit operation E.
3. The associative law of multiplication holds.
4. Every element has a unique inverse which is also a member of the collection.

Rule 1 is illustrated in Eq. (1-1). Rule 2 means that

$$EC_3 = C_3E = C_3, \qquad EC_3^2 = C_3^2E = C_3^2, \qquad \text{etc.} \tag{1-2}$$

Rule 3 states that

$$(C_3C_3^2)\sigma_v^a = C_3(C_3^2\sigma_v^a) \tag{1-3}$$

for these three elements, and rule 4 defines the *inverse* C_3^{-1} of a typical element C_3 as

$$C_3C_3^{-1} = C_3^{-1}C_3 = E \tag{1-4}$$

We can establish a *multiplication*, or *Cayley, table* for this six-element group (Table 1-1) by working out all the products of the form (1-1).

PROBLEM 1-3

(a) Verify the fourth row of Table 1-1.
(b) Verify that the group of Table 1-1 obeys rules 1 through 4.

(c) Prove that an element appears once only in any row (or column) of the table.

The Cayley table shows an interesting feature of this group: the three elements E, C_3, and C_3^2 also form a group and this is indicated by the heavy lines. This group E, C_3, C_3^2 is called a *subgroup* of the six-element group, a subgroup being any set of elements in a group which also obey the group rules. In addition, this subgroup is a *cyclic group*; it may be completely generated with a single element C_3, since $C_3^3 = E$, $C_3^4 = C_3$, and so on.

TABLE 1-1 ***Cayley Table for the Group C_{3v}***

First Operation (columns); *Second Operation* (rows)

	E	C_3	C_3^2	σ_v^a	σ_v^b	σ_v^c
E	E	C_3	C_3^2	σ_v^a	σ_v^b	σ_v^c
C_3	C_3	C_3^2	E	σ_v^c	σ_v^a	σ_v^b
C_3^2	C_3^2	E	C_3	σ_v^b	σ_v^c	σ_v^a
σ_v^a	σ_v^a	σ_v^b	σ_v^c	E	C_3	C_3^2
σ_v^b	σ_v^b	σ_v^c	σ_v^a	C_3^2	E	C_3
σ_v^c	σ_v^c	σ_v^a	σ_v^b	C_3	C_3^2	E

PROBLEM 1-4

(a) Does the group of Table 1-1 have any other subgroups?

(b) Show that the integers (positive, negative, and zero) are a group with respect to addition, where 0 is the unit element.

The group E, C_3, C_3^2 is customarily designated as C_3, so that this symbol denotes the following three concepts:

1. A three-fold axis of rotational symmetry.
2. A 120° rotation about a three-fold axis.
3. The cyclic group consisting of C_3, C_3^2, and $C_3^3 = E$.

However, no confusion is caused by this multiple use of the symbol, since the particular meaning is obvious from the context.

For the six-element group of Table 1-1, we use the symbol C_{3v}, which indicates that we add a single reflection plane (σ_v^a, for example) to the group

C_3. Adding a single plane automatically generates the other two, since the three-fold symmetry would otherwise be destroyed. We can verify this from the Cayley table by realizing that the operations $C_3\sigma_v^a$ and $\sigma_v^a C_3$ would not be members of the group if σ_v^c and σ_v^b were undefined.

Let us now extend the group C_{3v} by incorporating into it the symmetry elements σ_h and C_2^a, C_2^b, and C_2^c of Fig. 1-4. By diagrams like that of Fig. 1-5, we can establish that

$$\sigma_v^a \sigma_h = C_2^a, \qquad C_2^c C_3^2 = C_2^b, \qquad \text{etc.}$$

However, we also find that $\sigma_h C_3$ does *not* correspond to any symmetry operation we have encountered (Fig. 1-6). In order for σ_h to be included in a larger group containing the six members of C_{3v}, we must define a new symmetry

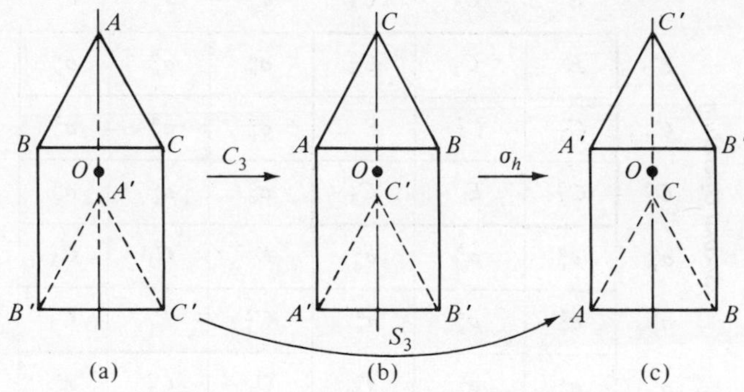

1-6 Generation of the operation S_3

operation called an *improper rotation* S_3. This operation consists of a 120° rotation combined with a reflection in the horizontal plane σ_h. The reason for this name is as follows: if we choose the geometrical center O of the prism as the origin of a right-handed coordinate system $OXYZ$, with OZ along C_3 and OX and OY anywhere in the plane σ_h, then the operation S_3 will merely rotate OX and OY through 120°, but the direction of OZ will be reversed. This transforms the right-handed system into a left-handed one. Having defined S_3 as $\sigma_h C_3$, we must also define S_3' as $\sigma_h C_3^2$; this is a 240° rotation combined with a horizontal reflection.

PROBLEM 1-5

Verify that

$$S_3^2 = C_3^2, \qquad S_3'^2 = C_3$$

TABLE 1-2 ***Cayley Table for the Group D_{3h}***

First Operation (columns); *Second Operation* (rows)

	E	C_3	C_3^2	σ_v^a	σ_v^b	σ_v^c	σ_h	S_3	S_3'	C_2^a	C_2^b	C_2^c
E	E	C_3	C_3^2	σ_v^a	σ_v^b	σ_v^c	σ_h	S_3	S_3'	C_2^a	C_2^b	C_2^c
C_3	C_3	C_3^2	E	σ_v^c	σ_v^a	σ_v^b	S_3	S_3'	σ_h	C_2^c	C_2^a	C_2^b
C_3^2	C_3^2	E	C_3	σ_v^b	σ_v^c	σ_a^a	S_3'	σ_h	S_3	C_2^b	C_2^c	C_2^a
σ_v^a	σ_v^a	σ_v^b	σ_v^c	E	C_3	C_3^2	C_2^a	C_2^b	C_2^c	σ_h	S_3	S_3'
σ_v^b	σ_v^b	σ_v^c	σ_v^a	C_3^2	E	C_3	C_2^b	C_2^c	C_2^a	S_3'	σ_h	S_3
σ_v^c	σ_v^c	σ_v^a	σ_v^b	C_3	C_3^2	E	C_2^c	C_2^a	C_2^b	S_3	S_3'	σ_h
σ_h	σ_h	S_3	S_3'	C_2^a	C_2^b	C_2^c	E	C_3	C_3^2	σ_v^a	σ_v^b	σ_v^c
S_3	S_3	S_3'	σ_h	C_2^c	C_2^a	C_2^b	C_3	C_3^2	E	σ_v^c	σ	σ_v^a
S_3'	S_3'	σ_h	S_3	C_2^b	C_2^c	C_2^a	C_3^2	E	C_3	σ_v^b	σ_v^c	σ_v^a
C_2^a	C_2^a	C_2^b	C_2^c	σ_h	S_3	S_3'	σ_v^a	σ_v^b	σ_v^c	E	C_3	C_3^2
C_2^b	C_2^b	C_2^c	C_2^a	S_3'	σ_h	S_3	σ_v^b	σ_v^c	σ_v^a	C_3^2	E	C_3
C_2^c	C_2^c	C_2^a	C_2^b	S_3	S_3'	σ_h	σ_v^c	σ_v^a	σ_v^b	C_3	C_3^2	E

Classes: K_1 (E); K_2 (C_3, C_3^2); K_3 (σ_v^a, σ_v^b, σ_v^c); K_4 (σ_h); K_5 (S_3, S_3'); K_6 (C_2^a, C_2^b, C_2^c)

Subgroups: C_{3v}; C_3

Using the definitions of S_3 and S_3' combined with relations such as those in Problem 1-5, we can establish the Cayley table (Table 1-2) for this twelve-element group.

The symbol for this large group is D_{3h}, where D denotes the existence of *dihedral axes.* A dihedral axis is a two-fold axis normal to a principal axis, C_3 in this case. As before, we need add only a single dihedral axis such as C_2^a, and the other two are automatically generated.

PROBLEM 1-6

Show that the symbol D_{3h} completely specifies the symmetry of the twelve-element group of Table 1-2. In particular, show that a subscript v is unnecessary.

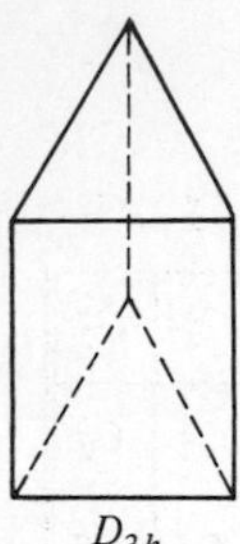

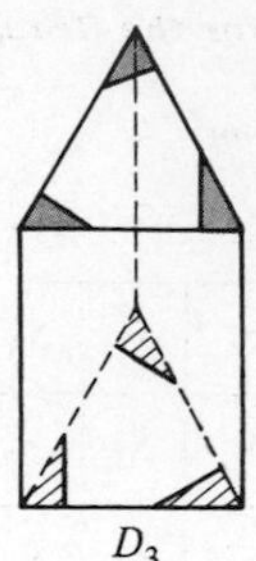

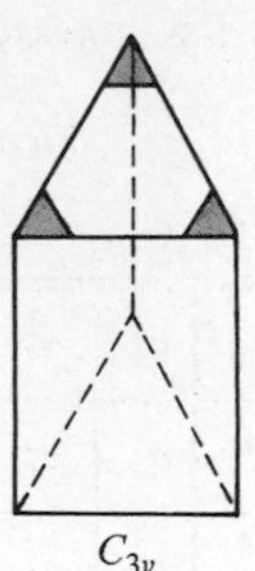

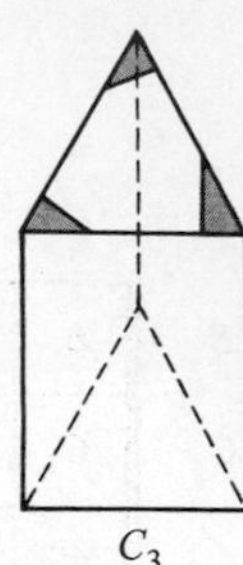

1-7 Some subgroups of D_{3h}

Again, we note that the group D_{3h} contains the groups C_{3v} and C_3 as subgroups (and others as well). Also, we see that all the symmetry axes and planes intersect at the point O. This point is stationary after any of the twelve operations; that is, it is *invariant*. A group containing an invariant point is called a *point group*. Actually, the point group associated with a real triangular prism is the dihedral group D_3, containing only the axes C_3, C_2^a, C_2^b, C_2^c, rather than the group D_{3h}. The reason is that reflection operations cannot be performed upon a solid prism. The full group of twelve operations represents all the possible ways of simultaneously rearranging the points A, B, C, A', B', C'. It is the group of highest symmetry associated with these points and is called the *holohedral group* of the prism. If we paint the corners of the prism in the fashion indicated in Fig. 1-7, we can lower its inherent symmetry and illustrate a number of the subgroups.

PROBLEM 1-7

What is the smallest subgroup which contains C_3 and σ_h? What symbol would describe it? Draw a sketch to illustrate this group.

1-3 Classes and cosets

Let P and Q be two members of a group. They are said to be members of the same *class* if they are related by the equation

$$Q = X^{-1}PX \tag{1-5}$$

where X is any member of the group, including P and Q themselves. For example, let

$$P = C_3, \qquad X = S_3$$

Then

$$X^{-1} = S_3^{-1} = S_3'$$

and $$Q = S_3'C_3S_3 = S_3'S_3' = C_3$$

If X is chosen as σ_v^a, then

$$Q = \sigma_v^a C_3 \sigma_v^a = \sigma_v^a \sigma_v^c = C_3^2$$

Continuing in this fashion with the other members of the group, we find that Q is either C_3 or C_3^2, and these two elements form a class, which we denote as K_2 at the bottom of Table 1-2.

PROBLEM 1-8

(a) Verify that the six classes of D_{3h} are

$$K_1 = E, \qquad K_4 = \sigma_h$$
$$K_2 = C_3, C_3^2, \qquad K_5 = S_3, S_3'$$
$$K_3 = \sigma_v^a, \sigma_v^b, \sigma_v^c, \qquad K_6 = C_2^a, C_2^b, C_2^c$$

(b) Show that E is in a class by itself in any group.

Problem 1-8 reveals the significance of the concept of a class: the members of a given class are sets of symmetry operations which are themselves related by operations of the group. For example, K_3 is the class of vertical reflection planes. These planes σ_v are related to one another by the rotations C_3 about the three-fold axis, so that they should form a single class. On the other hand, there is no symmetry operation of the group which converts a plane σ_v into

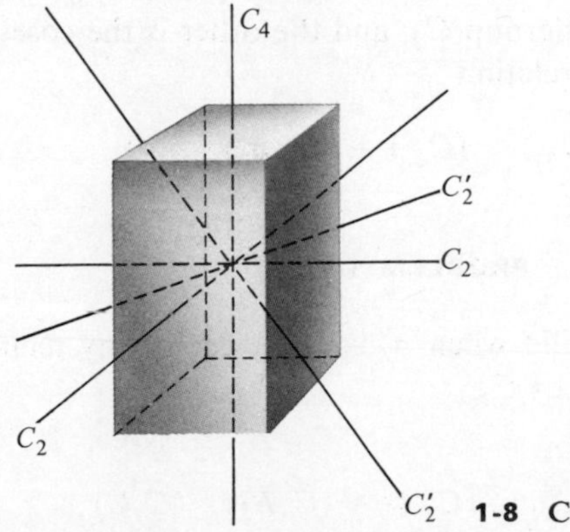

1-8 Classes of two-fold axes

the horizontal reflection plane σ_h. Hence, σ_h is in a class by itself. The same sort of argument holds for the two kinds of rotation axes, C_3 and the dihedral axes. A more subtle application of this principle is found in the right prism with a square base of Fig. 1-8. This solid has two-fold axes through the centers of opposite rectangular faces and it has another set of dihedral axes, denoted C_2', through the centers of opposite edges. The prism also has a vertical four-fold axis C_4 which will transform either one of the C_2 or C_2' axes into the other but will not change an axis through the centers of the faces into one through the edges or vice versa. The C_2 axes are therefore in a different class from the C_2' axes, as may be verified by working out the Cayley table, even though both kinds are dihedral operations.

PROBLEM 1-9

(a) Verify that the classes of the group C_{3v} are K_1, K_2, K_3 as listed in Problem 1-8(a).

(b) Verify that each member of the group C_3 is in a class by itself.

(c) Explain physically, using the argument following Problem 1-8, why C_3 and C_3^2 are in the same class in the group C_{3v} but in separate classes in the group C_3.

Another way of classifying the elements of a group is in terms of a *coset*. Consider for example the subgroup C_{3v} and the operation σ_h. The *right coset* of C_{3v} and σ_h, denoted by $(C_{3v}\,\sigma_h)$, is the set of elements defined by the equation

$$\begin{aligned}(C_{3v}\,\sigma_h) &= (E\sigma_h,\, C_3\sigma_h,\, C_3^2\sigma_h,\, \sigma_v^a\sigma_h,\, \sigma_v^b\sigma_h,\, \sigma_v^c\sigma_h)\\ &= (\sigma_h,\, S_3,\, S_3',\, C_2^a,\, C_2^b,\, C_2^c)\end{aligned} \tag{1-6}$$

Left cosets are defined in an analogous fashion, but we shall generally work with right cosets. Equation (1-6) shows that the group D_{3h} can be broken into two parts; one is its subgroup C_{3v} and the other is the coset of C_{3v} and σ_h. We indicate this by the relation

$$(D_{3h}) = (C_{3v}) + (C_{3v}\sigma_h) \tag{1-7}$$

PROBLEM 1-10

(a) Show that (1-7) is valid when σ_h is replaced by any member of D_{3h} which is not also a member of C_{3v}.

(b) Show that

$$(D_{3h}) = (C_3) + (C_3A) + (C_3B) + (C_3C) \tag{1-8}$$

where A is any member of the class K_3 of Table 1-2, and B and C are to be determined.

Problem 1-10 reveals a very important aspect of the relation between a group and its subgroups. If we define the *order* g of a group as the number of elements it contains, then it may be shown that the order h of any subgroup is a factor of g. That is,

$$g = mh \tag{1-9}$$

where $m = 1, 2, 3, \ldots, g$. For example, the order of D_{3h} is 12, so that $h = 3$, $m = 4$ for C_3 and $h = 6$, $m = 2$ for C_{3v}. Equation (1-9) is known as *Lagrange's theorem*; our approach to relations of this kind will be to establish their validity through examples rather than give general proofs. Equation (1-7) shows that the group D_{3h} can be broken into two parts, one of which is the subgroup C_{3v} and the other the coset $(C_{3v}\sigma_h)$ of (1-6). This coset is not a group, since it does not contain the identity element E. Further, it contains no members of C_{3v}; σ_v is not a member of this subgroup, so that none of the products $(C_{3v}\sigma_h)$ can be either. These products $(C_{3v}\sigma_h)$ are all different, since $A\sigma_h \neq B\sigma_h$ if A and B are different members of C_{3v}. Also, these products must generate the remainder of the group, since any element not generated could not be a member of the group. Hence, the group D_{3h} of order g is broken up into two equal parts, one a subgroup C_{3v} of order h. This means that h must be equal to $g/2$. In Problem 1-10, we have a subgroup for which $h = 3$. Actually, m could have the values

$$m = 1, 2, 3, 4, 6, 12$$

where $m = 1$, or $g = h = 12$, means that a group is one of its subgroups. In formal group theory, a subgroup for which $h < g$ is called a *proper* subgroup; we shall not bother here with this distinction.

1-4 A survey of the point groups

We may enumerate the possible point groups involving rotations by using a method of Coxeter,[1] originally due to Felix Klein.[2] Since these groups leave one point invariant, we may regard them as operations on the surface of a sphere with a stationary center. Any rotation will also leave two points on the surface unaltered. These points, called *poles*, are the intersections of

[1]H. S. M. Coxeter, *Introduction to Geometry*, John Wiley & Sons, Inc., New York, 1961.

[2]F. Klein, *Lectures on the Icosahedron* (2nd ed.), Kegan Paul, London, 1913.

the rotation axis with the surface. If some other operation of a group converts this axis into a different one, the new poles are said to be *equivalent* to the old ones. For example, the three dihedral axes of a triangular prism generate equivalent poles. We shall show that the number of equivalent poles for each n-fold axis C_n is g/n, where g is the order of the group.

To do this, we realize that some point on the surface fairly close to a pole will generate a polygon of n sides under the n rotations of the axis. The other rotations of the group convert this polygon into identical ones about the other poles, where we are assuming temporarily that the group is composed solely of rotations. The g operations of the group (including E) when done in any order generate a total of g points arranged in polygons around the poles. Since the polygons each have n sides, these must be g/n polygons or poles. Also, the $g - 1$ rotations of the group are divided into $n - 1$ rotations about each n-fold axis, or $(n - 1)/2$ for each pole. Hence, there are a total of $[(n - 1)/2](g/n)$ rotations other than E for each set of equivalent poles. Summing over nonequivalent poles will therefore give the total number of rotations, which we know to be $g - 1$, so that

$$g - 1 = \frac{g}{2} \sum \frac{n - 1}{n}$$

or

$$2 - \frac{2}{g} = \sum \left(1 - \frac{1}{n}\right) \tag{1-10}$$

The quantities g and n which satisfy (1-10) must be integers. In particular, $g = 1$ means that the group only has one member (which must be E), so that we need consider only the values $g = 2, 3, \ldots$. Using this fact ($g \geqq 2$) imposes the inequality

$$1 \leqq 2 - \frac{2}{g} < 2 \tag{1-11}$$

PROBLEM 1-11

Verify (1-11).

This in turn means that there must be at least two, but not more than three, nonequivalent sets of poles. For in Eq. (1-10) we are summing over nonequivalent sets and the single term $[1 - (1/n)]$ in the sum would be less than 1, which violates (1-11). On the other hand, four or more terms on the right of (1-10) would sum to at least

$$4(1 - \tfrac{1}{2}) = 2$$

since $n = 2$ is the smallest value that can be used.

The first alternative to consider, then, is two sets of nonequivalent poles. This gives, from (1-10), the relation

$$2 - \frac{2}{g} = 1 - \frac{1}{n_1} + 1 - \frac{1}{n_2}$$

or

$$\frac{g}{n_1} + \frac{g}{n_2} = 2 \tag{1-12}$$

We have already seen that n and g must be integers. Thus, the only solution of (1-12) is $n_1 = n_2 = g$. This is the situation corresponding to the cyclic groups C_n, where $n = n_1 = n_2 = g$. There is a single axis of n-fold symmetry with a pole at each end, and the poles are nonequivalent (i.e., not related by a symmetry operation of the group). We have already encountered C_3 as an example and we now see that operations

$$C_n, C_n^2, \ldots, C_n^{n-1}, C_n^n = E$$

form a group for any integer n.

Turning now to the case of three sets of nonequivalent poles, Eq. (1-10) becomes

$$2 - \frac{2}{g} = 1 - \frac{1}{n_1} + 1 - \frac{1}{n_2} + 1 - \frac{1}{n_3}$$

or

$$\frac{1}{n_1} + \frac{1}{n_2} + \frac{1}{n_3} = 1 + \frac{2}{g} \tag{1-13}$$

Without loss of generality, we may take $n_1 \leqq n_2 \leqq n_3$. Eq. (1-13) has no solution if n_1, n_2, and n_3 are all equal to or greater than 3; at least one of them, say n_1, must be 2. Hence, (1-13) becomes

$$\frac{1}{n_2} + \frac{1}{n_3} = \frac{1}{2} + \frac{2}{g} \tag{1-14}$$

PROBLEM 1-12

Show that n_2 and n_3 cannot both be as large as 4.

To agree with the results of Problem 1-12, we shall let n_2 be 2 or 3, and n_3 be unrestricted. For the combination $n_1 = 2$, $n_2 - 2$, Eq. (1-14) gives

$$n_3 = \frac{g}{2}$$

The resulting groups, defined by $n_1 = 2$, $n_2 = 2$, $n_3 = g/2$, are the dihedral groups D_n. They contain an n-fold axis ($n = n_3 = g/2$) and $g/2$ two-fold axes normal to this principal axis.

PROBLEM 1-13

Verify (1-14) for the group D_3.

When $n_1 = 2$ and $n_2 = 3$, only three values of n_3 satisfy (1-14).

PROBLEM 1-14

Prove that $n_1 = 2$, $n_2 = 3$ leads to $n_3 = 3$, 4, or 5.

These values of n_1, n_2, and n_3 are summarized in Table 1-3. No other combinations are possible, since both n_2 and n_3 cannot be as large as 4. The entry in the table following D_n, for which we find that $g = 12$, leads to the *tetrahedral group T*. It is the symmetry group of the regular four-sided pyramid formed from equilateral triangles (Fig. 1-9). This group is actually

TABLE 1-3 ***Enumeration of the Proper Point Groups***

n_1	n_2	n_3	g	*Group designation*
1	1		1	C_1
2	2		2	C_2
3	3		3	C_3
⋮	⋮		⋮	⋮
n	n		n	C_n
2	2	2	4	D_2
2	2	3	6	D_3
2	2	4	8	D_4
⋮	⋮	⋮	⋮	⋮
2	2	n	$2n$	D_n
2	3	3	12	T
2	3	4	24	O
2	3	5	60	W

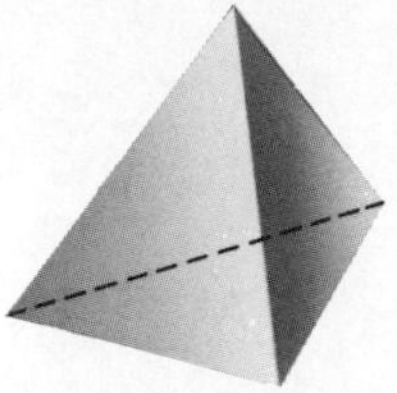

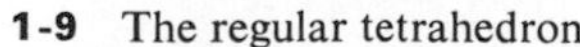

1-9 The regular tetrahedron

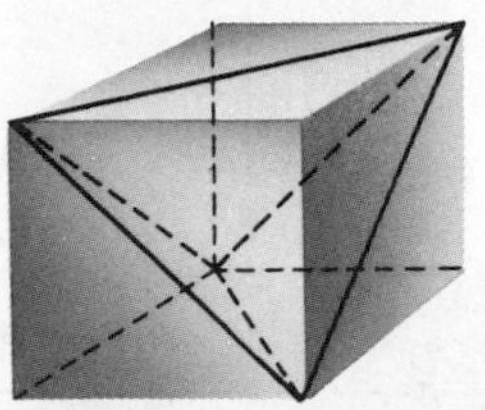

1-10 Relation of the tetrahedron to the cube

a subgroup of the symmetry group of a cube; the relation of the two structures is shown in Fig. 1-10.

The next line in the table corresponds to the *octahedral group* O, which is also a subgroup of the group of the cube. Figure 1-11 shows an octahedron and its relation to the cube. For the last value of n_2 ($n_2 = 5$), we obtain the sixty-element group W of the *dodecahedron*, a figure composed of twelve regular pentagons (Fig. 1-12).

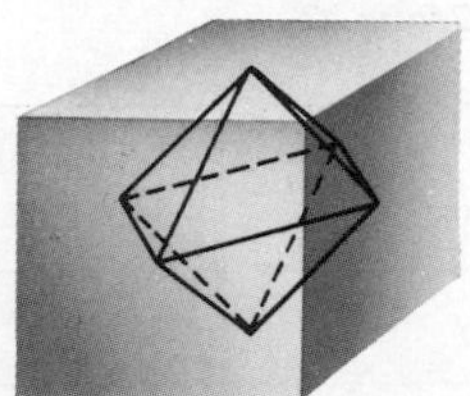

1-11 Relation of the octahedron to the cube

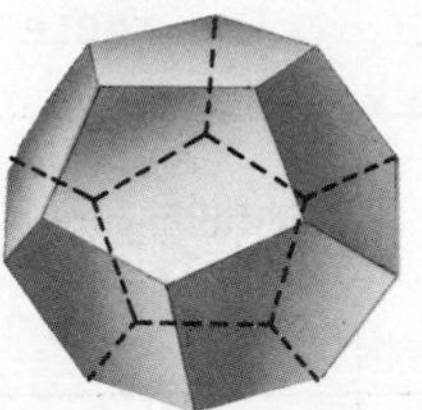

1-12 The regular dodecahedron

The groups listed in Table 1-3 contain only rotations. Points on the surfaces of solids such as the octahedron or the cube are also related by inversion and reflection symmetry. Hence, we may obtain new groups by combining the inversion operation with the groups already developed. Inversion may be specified in terms of an *inversion operator* J, which converts every point (x, y, z) into the point $(-x, -y, -z)$. Incorporating J into the octahedral group O gives the *full cubic group* O_h. This is the holohedral group (as defined at the end of Sec. 1-3) of the cube, and it has 48 members.

Although we may systematically go through the list in Table 1-3 and enumerate the new groups obtained by the inclusion of a reflection or an inversion operator, we shall consider here only the last three entries; the groups derived from C_n and D_n will be examined later in connection with stereographic projections. The expansion of T and O leads to three more groups (T_d, T_h, and O_h), the five groups together forming the *cubic lattice system*. This name comes from the fact that these groups specify the arrange-

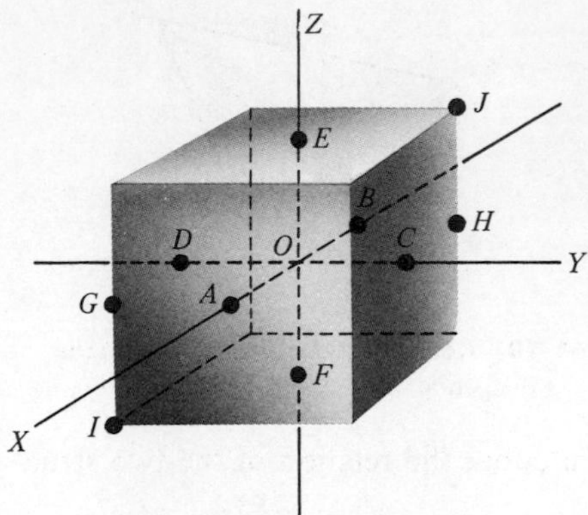

1-13 Symmetry axes of a cube

ment of atoms in cubic crystals. The 24 operations which can actually be performed on a cube—the *covering operations*—are listed in Table 1-4; Fig. 1-13 illustrates these operations.

TABLE 1-4 ***Octahedral Group O***

Symbol	Multiplicity	Definition	Typical axis in Fig. 1-13
E	1	The identity	
C_4	6	$\pm 90°$ rotation about cube axes	*AB*
C_4^2	3	180° rotation about cube axes	*AB*
C_2	6	180° rotation about axes parallel to face diagonals	*GH*
C_3	8	$\pm 120°$ rotation about body diagonals	*IJ*

PROBLEM 1-15

Show that the octahedron has the symmetry elements listed in Table 1-4.

This problem indicates that the cube and the octahedron have identical symmetries. A similar relation exists between the dodecahedron of Fig. 1-12 and the *icosahedron*, a solid composed of twenty equilateral triangles (Fig. 1-14); both have the rotational symmetries indicated by the last line of Table 1-3. Thus, group theory has brought us to the five regular solids known to the Greek geometers.

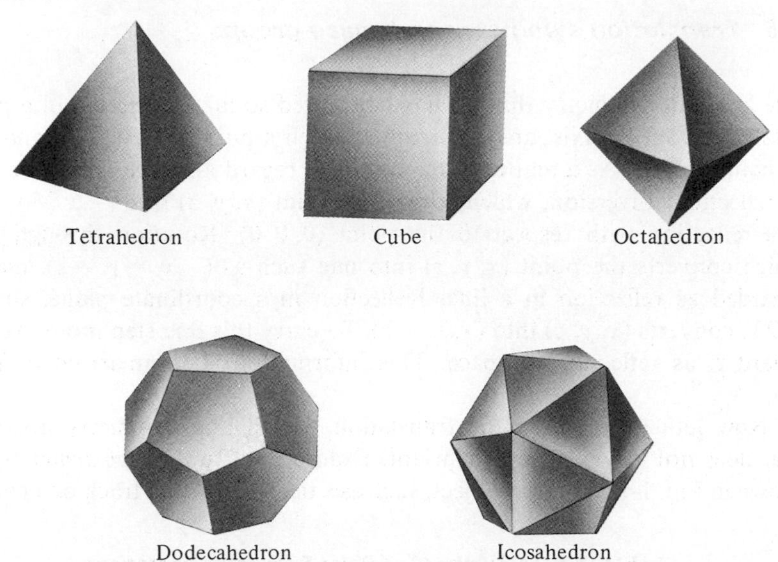

1-14 The five regular solids

To find the remaining cubic groups, let us start with the group T of order 12 and add a *diagonal reflection plane* σ_d. This is a plane passing through the points F, G, E, H, and O of Fig. 1-13; as Fig. 1-10 shows, it is an improper symmetry element of the tetrahedron.

PROBLEM 1-16

Show that there are six operations of the form σ_d.

The combination of σ_d with T gives a 24-element group denoted T_d. We may also combine T with reflections σ_h in the coordinate planes to obtain the group T_h. Then T_h plus operations consisting of two-fold rotations about axes like GH followed by a reflection σ_h produce the full cubic group O_h (which accounts for its symbol).

Molecules may possess symmetries specified by any of these point groups, but we shall see that crystals are restricted to a total of 32 such groups. They are known, of course, as the *crystallographic point groups*, and these groups have symbols such as those of Table 1-3. This set of symbols is known as the *Schoenflies* notation.

1-5 Translation symmetry and space groups

The kinds of symmetry that we have discussed so far—reflection in a plane, rotation about an axis, and inversion through a point—are the components of point groups. As a matter of fact, we may regard all three types as a form of reflection. Inversion, which converts a point (x, y, z) into $(-x, -y, -z)$. is a reflection with respect to the point $(0, 0, 0)$. Rotation through 180°, which converts the point (x, y, z) into one such as $(-x, -y, -z)$, may be regarded as reflection in a line. Reflection in a coordinate plane, such as XOY, converts (x, y, z) into $(x, y, -z)$. To carry this one step more, we may regard E as reflection in space. This information is summarized in Table 1-5.

Now let us introduce the translation, which is a symmetry operation that does not leave a point invariant. Examples of *translation symmetry* are shown in Fig. 1-15. A given object, such as a tie in a railroad track or a column

TABLE 1-5 ***A Summary of Point Symmetry Operations***

Reflection in	Operation	Symbol	Typical sign change
Point	Inversion	J	$- - -$
Line	Rotation	C_2	$- - +$
Plane	Reflection	σ_h or σ_v	$- + +$
Space	Identity	E	$+ + +$

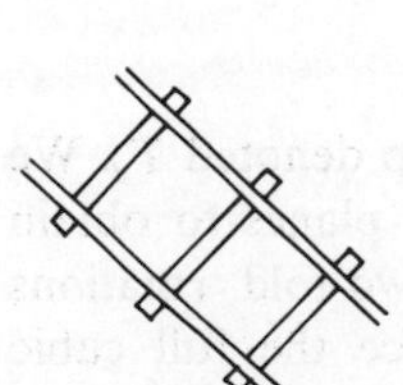

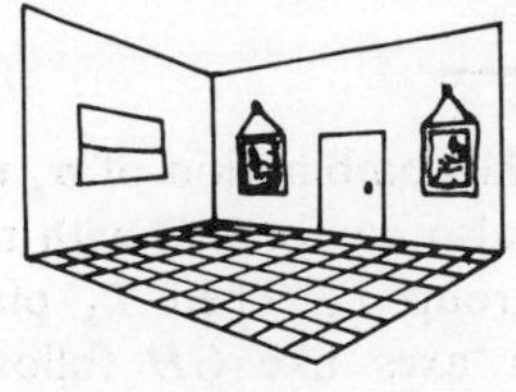

1-15 Examples of translational symmetry

on the front of a building, repeats itself at regular intervals. In the case of floor tiles, the translation symmetry appears in two different directions.

PROBLEM 1-17

(a) Give an example of a two-dimensional system for which the directions of translational symmetry are not at right angles.
(b) Give an example of a three-dimensional system.

If translational symmetry bears any relation to the point symmetries that we have previously considered, then we should be able to incorporate it into groups which combine both kinds. Let us then investigate the way in which this may be done and, also, the interaction of translation and point symmetries. The simplest type of translation symmetry consists of a row of equally spaced points (Fig. 1-16). This arrangement is called a *one-dimensional lattice*,

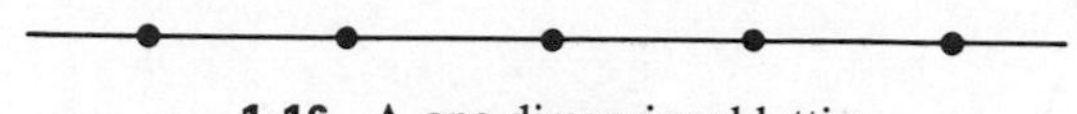

1-16 A one-dimensional lattice

and it possesses inherent symmetry. For example, it may be rotated about, or inverted through, any point.

PROBLEM 1-18

What other symmetry does the lattice of Fig. 1-16 have?

Let us place at each point of this lattice a completely asymmetric pattern: the half-arrow of Fig. 1-17(a). Putting this arrow at an arbitrary angle with respect to the line of points gives the repetitive pattern of Fig. 1-17(b). We may obtain other linear patterns in an intuitive way by combining translation and point operations. In Fig. 1-17(c), a second arrow is derived from the initial one at some arbitrary point by a 180° rotation C_2 about an axis through this point and normal to it. This pair of arrows is then repeated by the translation to produce Fig. 1-17(d).

PROBLEM 1-19

Show that Fig. 1-17(d) may be obtained by reversing the order of the two operations which generated it.

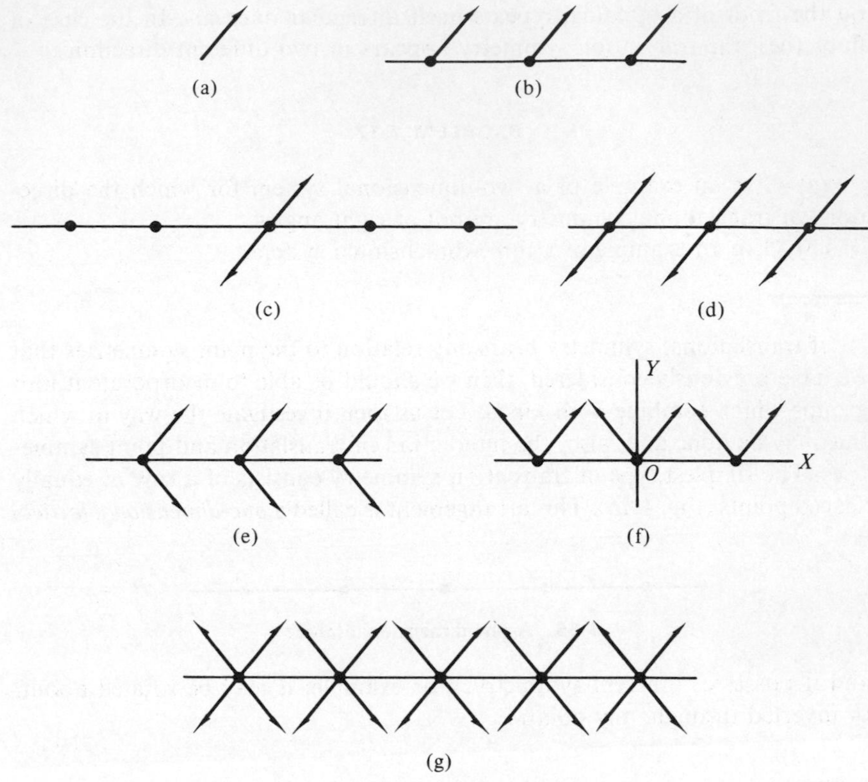

1-17 Combination of point and translational symmetry operations

Further arrangements are obtained by combining a translation with σ_x, the reflection in the line of translation [Fig. 1-17(e)], with σ_y, the reflection in the perpendicular direction [Fig. 1-17(f)], and with both σ_x and σ_y, which implies C_2 as well [Fig. 1-17(g)].

Our approach so far has been rather empirical. Let us return to group theory by introducing a *translation operator* T, which moves any given pattern one lattice point to the right. Its inverse, denoted by T^{-1}, is then the corresponding motion to the left. These operations can be repeated an indefinite number of times, since one-dimensional lattices must be infinitely long in either direction.

PROBLEM 1-20

Explain why the statement just above is true.

To avoid difficulties with infinite lengths, we shall imagine that one-dimensional lattices are bent round on themselves, making their effective length finite. For example, Fig. 1-18(a) shows a lattice containing just three points. Joining the ends so that point 1 is coincident with point 3 gives Fig. 1-18(b). With this choice of length, the two operations T and T^{-1} are equivalent, since they have the same effect on the lattice. That is,

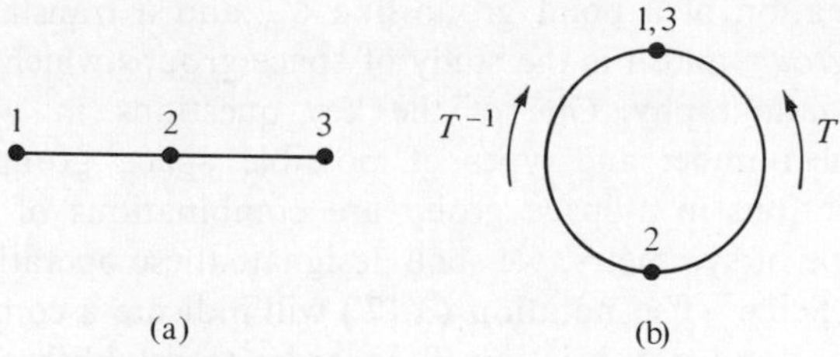

1-18 (a) A finite one-dimensional lattice; (b) use of periodic boundary conditions

$$T = T^{-1} \quad \text{or} \quad T^2 = E$$

Hence, we have a two-element *translation group* consisting of E and T.

PROBLEM 1-21

Assume that the lattice of Fig. 1-18(a) has four points and that the first and last point coincide. Find the elements of the translation group and their inverses, and write down the Cayley table.

There is also a group associated with each point of the lattice. In Fig. 1-17(g), the symmetry operations are σ_x, σ_y, and C_2, so that we have the four-element group whose multiplication table is Table 1-6. This point group is known as C_{2v}.

TABLE 1-6 ***Cayley Table for the Group C_{2v}***

	E	σ_x	σ_y	C_2
E	E	σ_x	σ_y	C_2
σ_x	σ_x	E	C_2	σ_y
σ_y	σ_y	C_2	E	σ_x
C_2	C_2	σ_y	σ_x	E

PROBLEM 1-22

Based on what has been stated about C_{3v}, explain the notation C_{2v}.

The combination of a point group like C_{2v} and a translation group is called a *space group*, and it is the study of space groups which underlies the science of crystallography. One of the key questions in which we are interested is the number and types of possible space groups. Since the symmetry operations in a space group are combinations of a translation and a point type of symmetry, we shall designate these operations by a dual symbol due to Seitz.[3] The notation $(X|T)$ will indicate a combination of a point operation X and a translation T. In order to establish an algebra for these Seitz symbols which is internally consistent, it is necessary to abandon the algebra tor translations used with Fig. 1-18(b); that is, rather than write

$$T^2 = E$$

we shall use the fact that T denotes a translation vector $\mathbf{t}$, so that

$$\mathbf{t} + \mathbf{t} = 0$$

Hence, a pure point symmetry operation is indicated by $(X|0)$ and a pure translation by $(E|T)$. Then the compound unit element is denoted by $(E|0)$.

Let us generate a space group multiplication table by using the ring-lattice of Fig. 1-18. If we place an asymmetric arrow at the point labelled O, choose the x-axis along the ring in the direction of T, and the y-axis normal to the plane of the ring, we obtain the situation of Fig. 1-19(a). The operation $(\sigma_x|0)$ followed by $(E|T)$ brings the arrow to the position shown in Fig. 1-19(b), and since this result cannot be obtained by either a pure translation or a pure point operation, we must introduce the compound operation $(\sigma_x|T)$ into the space group. In the same way, we find that the other compound operations of Table 1-7 are produced, and the products indicated may be verified by using the procedure of Fig. 1-19(b).

This table contains a number of subgroups, each of which corresponds to a linear pattern. For example, the two elements $(E|0)$, $(E|T)$ comprise the translation subgroup of Fig. 1-17(b), reproduced as Fig. 1-20(a), where the patterns are shown in linear rather than circular form. A more complex subgroup has the four members $(E|0)$, $(E|T)$, $(C_2|0)$, and $(C_2|T)$, with the corresponding pattern shown in Fig. 1-20(b). The next three patterns in this figure we have also encountered before and the associated subgroups are indicated. Figure 1-20(f), however, is new, and it illustrates what is known as

[3] F. Seitz, *Ann. Math.*, **37**, 17 (1936).

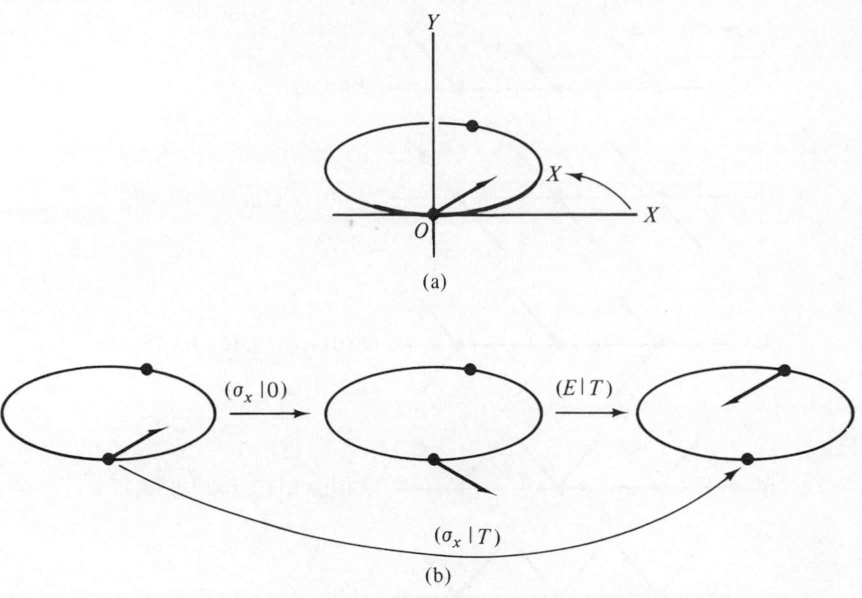

1-19 Symmetry operations in a periodic lattice

TABLE 1-7 ***Cayley Table for a Monoperiodic Space Group***

	$(E\|0)$	$(E\|T)$	$(\sigma_x\|0)$	$(\sigma_x\|T)$	$(\sigma_y\|0)$	$(\sigma_y\|T)$	$(C_2\|0)$	$(C_2\|T)$
$(E\|0)$	$(E\|0)$	$(E\|T)$	$(\sigma_x\|0)$	$(\sigma_x\|T)$	$(\sigma_y\|0)$	$(\sigma_y\|T)$	$(C_2\|0)$	$(C_2\|T)$
$(E\|T)$	$(E\|T)$	$(E\|0)$	$(\sigma_x\|T)$	$(\sigma_x\|0)$	$(\sigma_y\|T)$	$(\sigma_y\|0)$	$(C_2\|T)$	$(C_2\|0)$
$(\sigma_x\|0)$	$(\sigma_x\|0)$	$(\sigma_x\|T)$	$(E\|0)$	$(E\|T)$	$(C_2\|0)$	$(C_2\|T)$	$(\sigma_y\|0)$	$(\sigma_y\|T)$
$(\sigma_x\|T)$	$(\sigma_x\|T)$	$(\sigma_x\|0)$	$(E\|T)$	$(E\|0)$	$(C_2\|T)$	$(C_2\|0)$	$(\sigma_y\|T)$	$(\sigma_y\|0)$
$(\sigma_y\|0)$	$(\sigma_y\|0)$	$(\sigma_y\|T)$	$(C_2\|0)$	$(C_2\|T)$	$(E\|0)$	$(E\|T)$	$(\sigma_x\|0)$	$(\sigma_x\|T)$
$(\sigma_y\|T)$	$(\sigma_y\|T)$	$(\sigma_y\|0)$	$(C_2\|T)$	$(C_2\|0)$	$(E\|T)$	$(E\|0)$	$(\sigma_x\|T)$	$(\sigma_x\|0)$
$(C_2\|0)$	$(C_2\|0)$	$(C_2\|T)$	$(\sigma_y\|0)$	$(\sigma_y\|T)$	$(\sigma_x\|0)$	$(\sigma_x\|T)$	$(E\|0)$	$(E\|T)$
$(C_2\|T)$	$(C_2\|T)$	$(C_2\|0)$	$(\sigma_y\|T)$	$(\sigma_y\|0)$	$(\sigma_x\|T)$	$(\sigma_x\|0)$	$(E\|T)$	$(E\|0)$

a *glide* symmetry operation. A glide is a single operation composed of a translation and a reflection with respect to the translation direction. Unlike the patterns of Fig. 1-20(a) through (e), the design (one, two, or four arrows) does not repeat itself for each translation operation, but for every other one.

It should now occur to the reader to raise the following question: why is there no glide pattern corresponding to the subgroup $(E\,|\,0)$, $(\sigma_y\,|\,T)$? One way

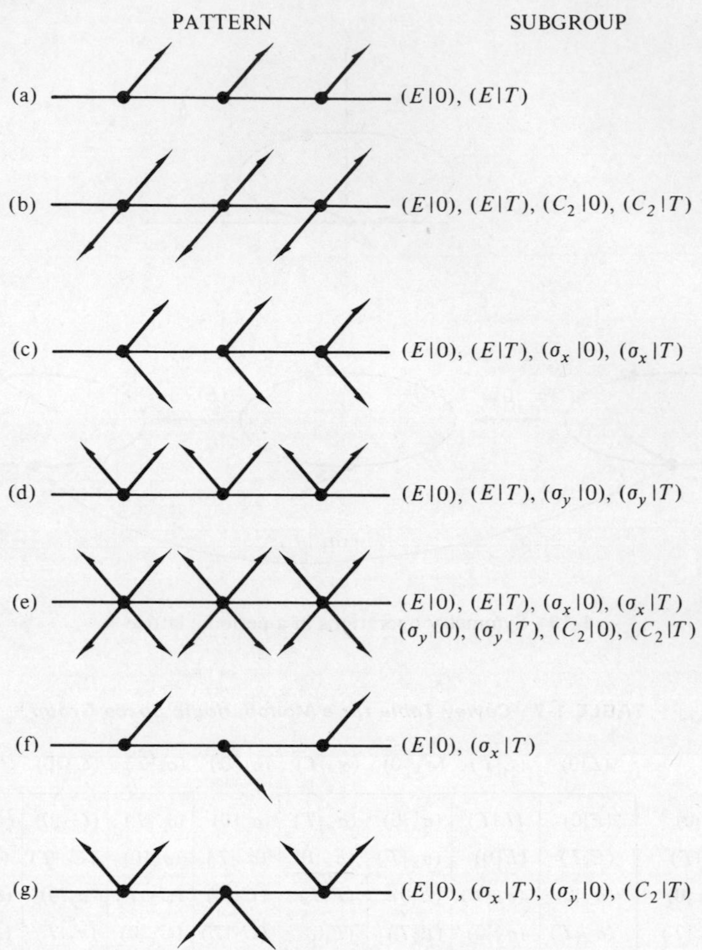

1-20 The seven frieze groups

of appreciating the answer is to try constructing a pattern from the operation consisting of a reflection in the y-axis and a translation. In Fig. 1-21, we show what happens. The operation $(\sigma_y|T)$ when performed twice in succession returns the arrow to its original position.

PROBLEM 1-23

Verify that this result is independent of the order in which the operations σ_y and T are performed.

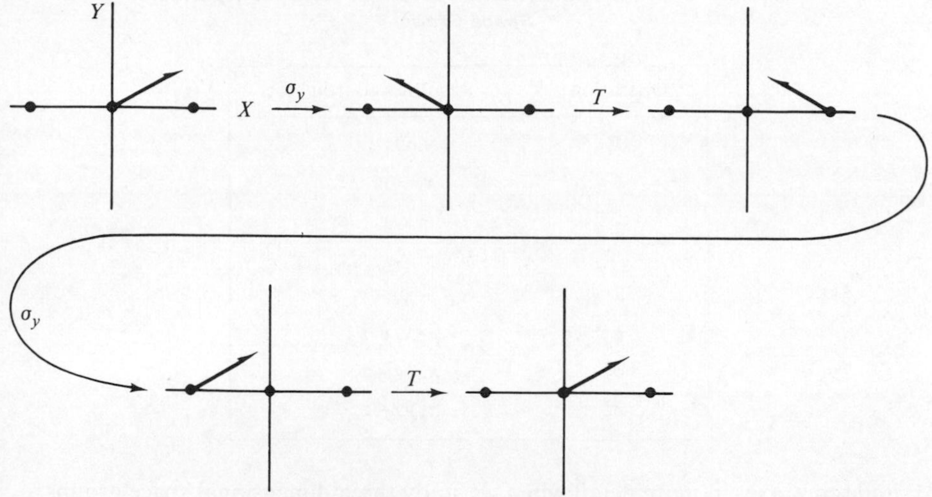

1-21 The operation $(\sigma_y | T)$ performed twice

This means that this subgroup cannot generate a linear pattern, and the only two groups containing glides are those of Fig. 1-20(f) and (g). These seven groups form the basis of strip decorations or *friezes*. A very common pattern is shown in Fig. 1-22(a); it corresponds to the group of Fig. 1-20(a). A more symmetrical one is that of Fig. 1-22(b).

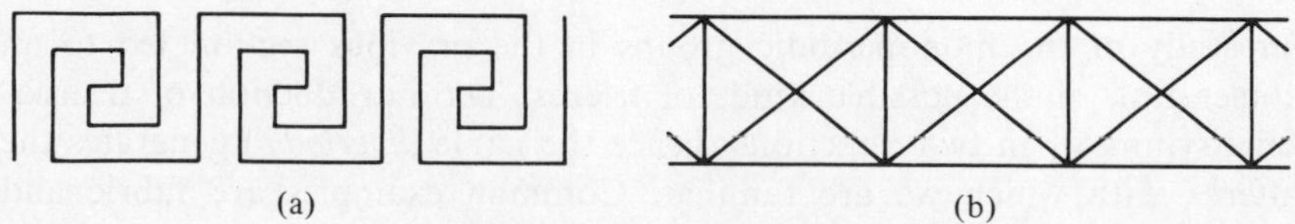

1-22 Some typical friezes

This discussion can be generalized by introducing the possibility of variation with respect to a third dimension. That is, we may permit the friezes to have elevations and depressions. To express this quantitatively, consider a point with coordinates (x, y, z). A symmetry operation with respect to the origin may produce the seven additional permutations of the coordinates listed in Table 1-8. To explain the symbols, we note, for example, that C_2^x is a 180° rotation about the x-axis. These eight operations form a group D_{2h} with 16 subgroups. Each subgroup leads to a three-dimensional linear pattern. In addition, there is also an operation analogous to the glide which consists of a translation combined with a 180° rotation about the translation line. This is called a *screw axis* and gives 15 more patterns for a total of 31 groups. Rather than enumerate them, we shall consider glides

TABLE 1-8 ***Three-Dimensional Operations for a Monoperiodic Space Group***

Operation:	Resulting coordinate:
E	x, y, z
C_2^x	$x, -y, -z$
C_2^y	$-x, y, -z$
C_2^z	$-x, -y, z$
J	$-x, -y, -z$
σ_v^x	$-x, y, z$
σ_v^y	$x, -y, z$
σ_v^z	$x, y, -z$

and screw axes in more detail when we study three-dimensional space groups. These linear groups that we have been discussing are also known as *monoperiodic groups*; the seven groups of Fig. 1-20 are two-dimensional monoperiodic groups and the 31 groups just mentioned are the three-dimensional monoperiodic groups.

1-6 Diperiodic space groups

Our study of the monoperiodic groups in the previous section led to an enumeration of the possible kinds of friezes. The introduction of translational symmetry in two directions (hence the name *diperiodic*) generates the patterns with which we are familiar. Common examples are fabric and

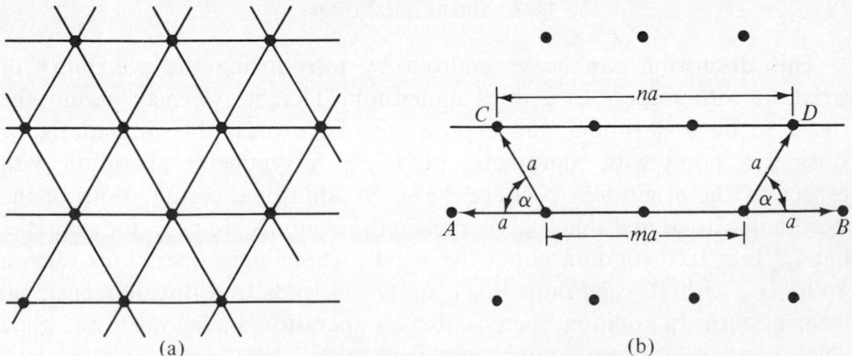

1-23 (a) A hexagonal lattice; (b) hexagonal lattice used to establish Table 1-9

wallpaper designs. Using two-dimensional arrangements of asymmetric arrows, we shall find that there are 17 distinct groups, generally known as *plane groups*, although the complete name should be diperiodic, two-dimensional space groups.

One of the interesting aspects of this set of groups is the fact that the total number is rather small; the reason for this is that there are only a few kinds of rotational symmetry which are consistent with translational symmetry. To see why, let us consider the two-dimensional hexagonal lattice of Fig. 1-23(a). This lattice has translational symmetry in two directions forming an angle of 60°, and it also has six-fold rotation symmetry about each lattice point.

PROBLEM 1-24

What other rotation symmetries does the lattice of Fig. 1-23(a) have?

We can express the rotational symmetry in terms of a vector of length a terminating at A in Fig. 1-23(b); after a clockwise rotation through an angle $\alpha = 60°$, it will terminate at C. Similarly, a counterclockwise rotation will move the vector terminating at B so that its endpoint is at D. The separation of the lattice points C and D must be some integral multiple na of a, and the separation of the centers of rotation is some other integral multiple ma of a. Hence

$$na = ma + 2a \cos \alpha$$

or

$$\cos \alpha = \frac{n - m}{2}$$

Since the quantity $(n - m)$ must be an integer, this equation can have only the solutions listed in Table 1-9. Although our argument is based on the hex-

TABLE 1-9 ***Rotational Symmetries Consistent with Translational Symmetry for Diperiodic Groups***

$(n - m)$	$\cos \alpha$	α	Symmetry
-2	-1	180°	Two-fold
-1	$-\frac{1}{2}$	120°	Three-fold
0	0	90°	Four-fold
1	$\frac{1}{2}$	60°	Six-fold
2	1	0°	One-fold

agonal lattice, it actually holds for any kind with rotational symmetry. We see then that translation symmetry permits only four kinds of rotational symmetries, the last entry in the table corresponding to a one-dimensional lattice.

PROBLEM 1-25

The above principle is the reason behind the fact that only three kinds of regular figures can be used to tile a floor.

(a) Show by a sketch that pentagons and octagons do not fit together.

(b) Derive a formula for the angle between adjacent sides of a regular figure in terms of the number n of sides.

(c) State mathematically the condition that a group of identical, regular figures fit around a point without an overlap or gap.

(d) Use the results of (b) and (c) to prove that integral values of n limit the tiles to equilateral triangles, squares, and hexagons.

An elegant demonstration of the rotational symmetries consistent with a space lattice has been given by Jaswon.[4] He starts with an n-fold axis through the origin and lets

$$z = x + iy$$

be the location of a lattice point. Then other lattice points lie at $ze^{2\pi i/n}$, $ze^{4\pi i/n}$, ... Any one of these points may be expressed as a linear combination of the first two. For example, $ze^{4\pi i/n}$ may be written as

$$ze^{4\pi i/n} = az + bze^{2\pi i/n}$$

where a and b are integers. Then

$$a = \left[\cos\left(\frac{4\pi}{n}\right) - b\cos\left(\frac{2\pi}{n}\right)\right] + i\left[\sin\left(\frac{4\pi}{n}\right) - b\sin\left(\frac{4\pi}{n}\right)\right]$$

The imaginary term gives

$$b = \frac{\sin(4\pi/n)}{\sin(2\pi/n)} = 2\cos\left(\frac{2\pi}{n}\right)$$

Since b is an integer, this is satisfied only by

[4] M. A. Jaswon, *An Introduction to Mathematical Crystallography*, American Elsevier, New York, 1965.

$$n = 1, 2, 3, 4, 6$$

where $n = 1$ means that z and $ze^{2\pi i/n}$ are identical. That is, a one-fold lattice has only translational symmetry.

Two-dimensional lattices can be constructed by putting one-dimensional lattices together in all possible ways. If we start with a row of points [Fig. 1-24(a)] and repeat it by translation, we obtain the *oblique lattice* of Fig. 1-24(b). The initial translations a_1 and a_2 are called the *lattice constants*; we notice that they will define an arbitrary parallelogram provided they are unequal in length and that the points in the second (and succeeding rows) do not fall on perpendiculars through the points or the midpoints of the first row. A more symmetrical arrangement is obtained when the displacement corresponding to a_2 is at right angles to that of a_1, producing the *rectangular lattice* of Fig. 1-24(c); if the two displacements are equal, we obtain the *square lattice* [Fig. 1-24(d)]. If the oblique lattice is formed by having the second row of points fall on the perpendicular-bisectors of the first row, the result [Fig. 1-24(e)] is called a *centered rectangular lattice*, and we then describe Fig. 1-24(c) as *simple rectangular* to distinguish the two lattices. Finally, if the two translations are equal and make an angle of 60° to one another, the lattice is *hexagonal* [Fig. 1-24(f)].

PROBLEM 1-26

Enumerate the rotational symmetries associated with the rectangular, square, and hexagonal lattices.

The properties of these plane lattices are summarized in Table 1-10. The four categories in the table are called *systems*; only one system contains

TABLE 1-10 ***The Diperiodic Lattices and Systems***

Name	Lattice angle α	Lattice constants
Oblique	$\alpha \neq 90°$	$a_1 \neq a_2$
Simple rectangular	$\alpha = 90°$	$a_1 \neq a_2$
Centered rectangular	$\alpha \neq 60°$	$a_1 = 2a_2 \cos \alpha$
Hexagonal	$\alpha = 60°$	$a_1 = a_2$
Square	$\alpha = 90°$	$a_1 = a_2$

1-24 (a) A one-dimensional lattice used to generate the plane lattices; (b) an oblique lattice; (c) the simple rectangular lattice; (d) the square lattice; (e) the centered rectangular lattice; (f) the hexagonal lattice

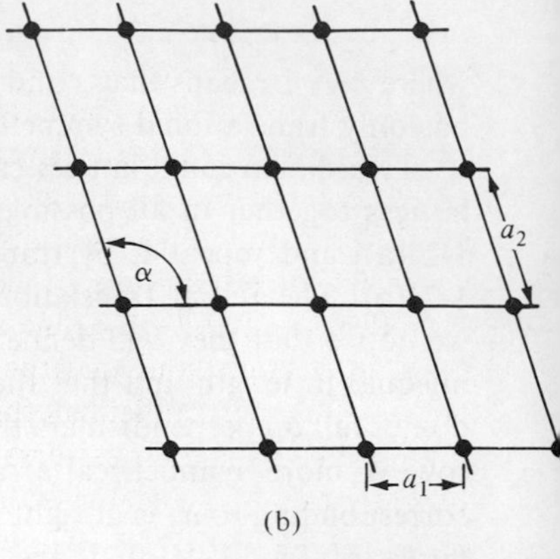

(a) (b)

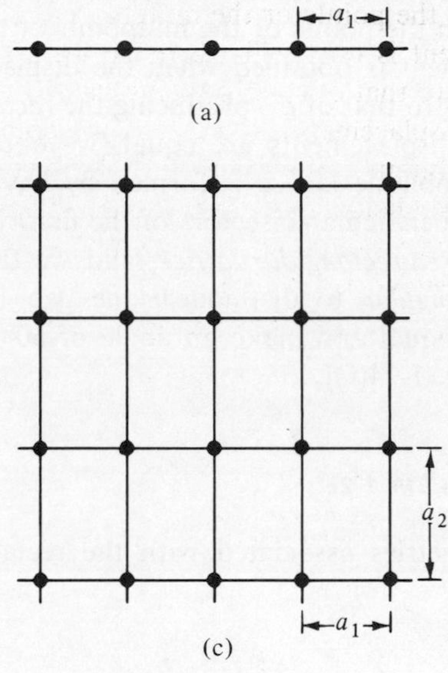

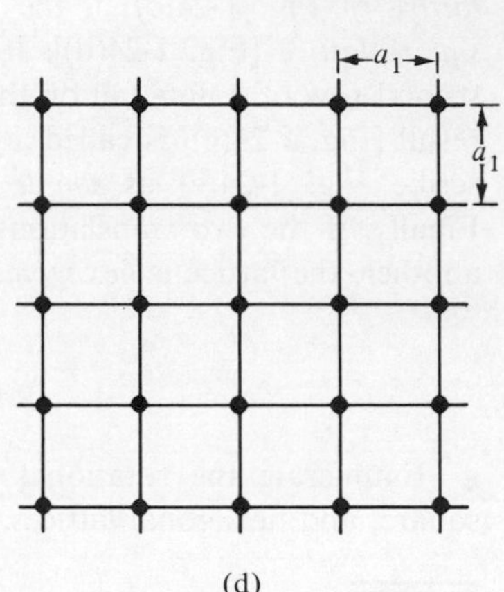

(c) (d)

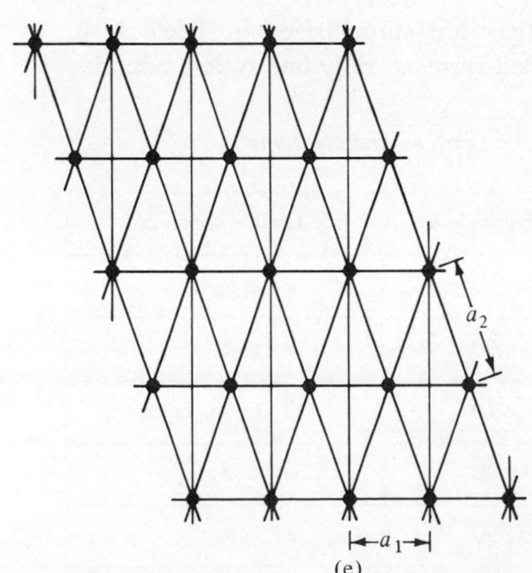

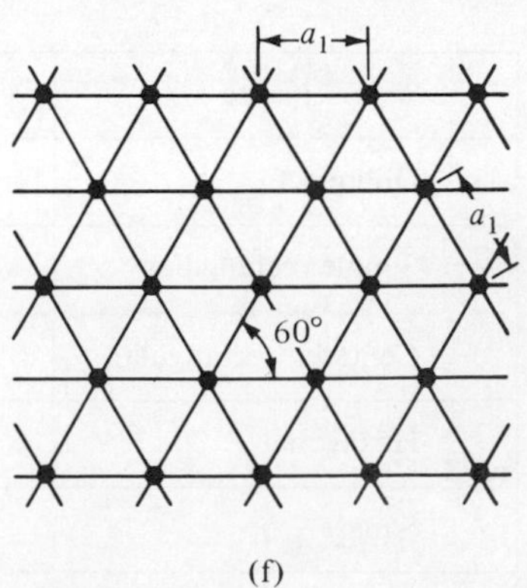

(e) (f)

more than one lattice, but we shall see that in three dimensions, most systems have several lattices.

To find the plane groups which can be generated by combining point operations with the five translation groups of Table 1-10, we proceed just as we did for the monoperiodic case. That is, we must find groups and subgroups of compound operations $(X|T)$ which will generate each lattice and a repeated pattern at every point in that lattice. To start with the oblique lattice as the simplest case, the subgroup $(E|0)$, $(E|T_1)$, $(E|T_2)$, where T_1 and T_2 are the translation operators in each of the two directions, will produce the pattern of Fig. 1-25(a). We notice, however, that the lattice has two-fold symmetry at each point. Hence, the full group $(E|0)$, $(E|T_1)$, $(E|T_2)$, $(C_2|0)$, $(C_2|T_1)$, $(C_2|T_2)$ will give Fig. 1-25(b), and these are the only two arrangements belonging to the oblique lattice. In a similar way, we generate the other plane lattices of Fig. 1-25. Some of these are obvious and others require some explanation. For example, the lattice of lowest symmetry consistent with a reflection, and hence a glide operation, is the simple rectangular [Fig. 1-25(e)]. Note that we show the lattice points as the positions of all the arrows like the original one, and the arrows due to a glide are halfway between these points. This is permissible because the lattice points are arbitrary and need not be shown at all, in fact. What the figure indicates is that a group containing the operations $(E|0)$, $(\sigma_x|T_1)$, $(E|T_2)$ will produce a simple rectangular pattern with arrows at distances T_1 and T_2 apart and reflected arrows at distances $T_1/2$ from each original one.

PROBLEM 1-27

(a) Which of the 17 groups of Fig. 1-25 contain glide operations?

(b) Why is there not a pattern which bears the same relation to Fig. 1-25(e) that 1-25(g) bears to 1-25(f)?

Another interesting situation concerns the difference between Figs. 1-25(n) and (o); both are examples of the point group D_3, but in one the two-fold axes are through the corners of the hexagon and in the other they pass through the centers of the sides.

As with the linear groups, we may also add symmetry operations in a third direction, obtaining 80 diperiodic three-dimensional groups. These have recently become important in connection with research on the surface properties of single crystals. The surface of a crystal may be regarded as a two-dimensional pattern with some of the atoms above and some below the plane of the pattern. Hence, it is equivalent to one of these 80 groups. More details may be found in a paper by Wood[5].

[5]E. A. Wood, *Bell Syst. Tech. J.* **43**, 541 (1964).

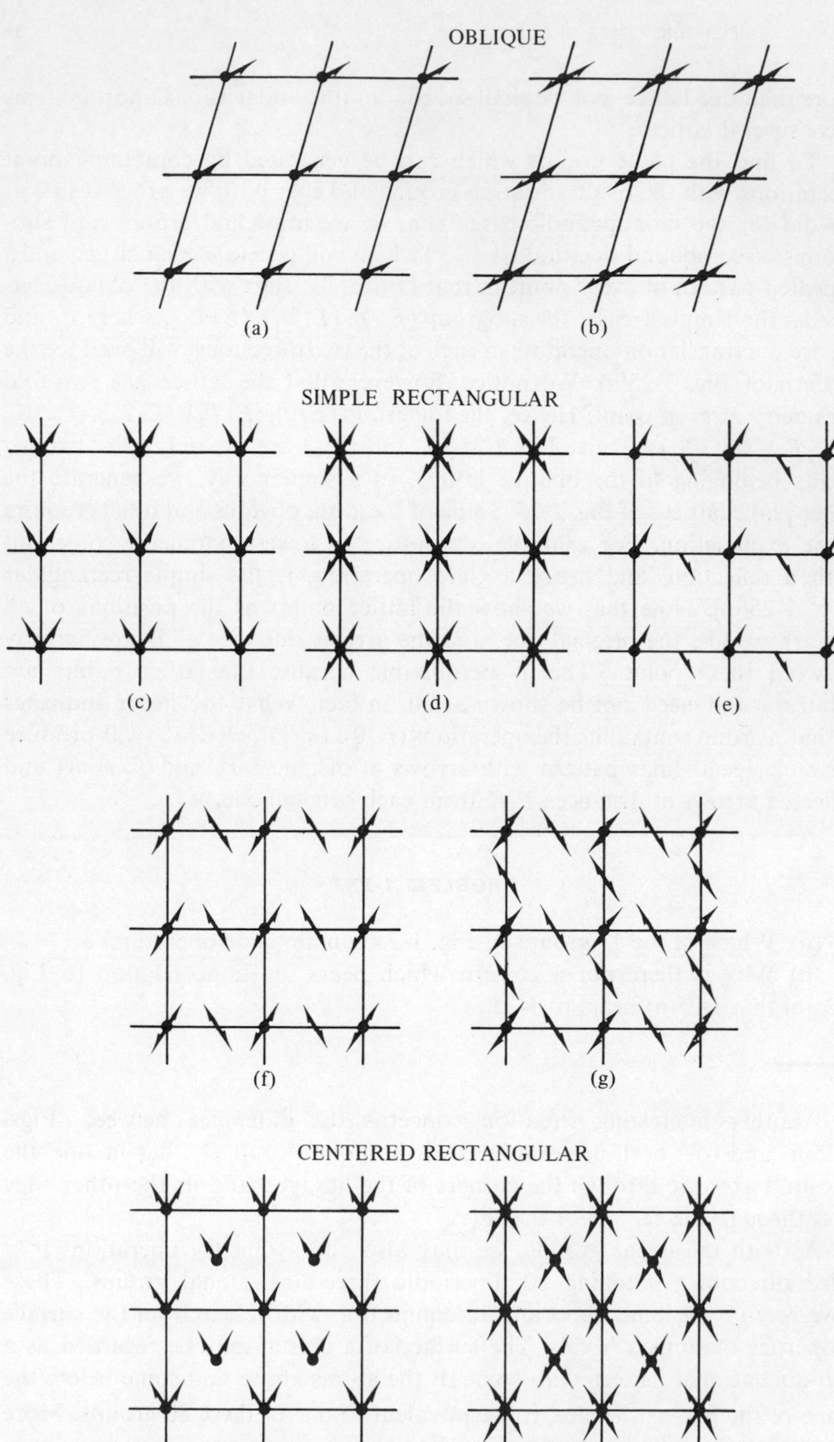

1-25 The seventeen plane groups (see also the frontispiece)

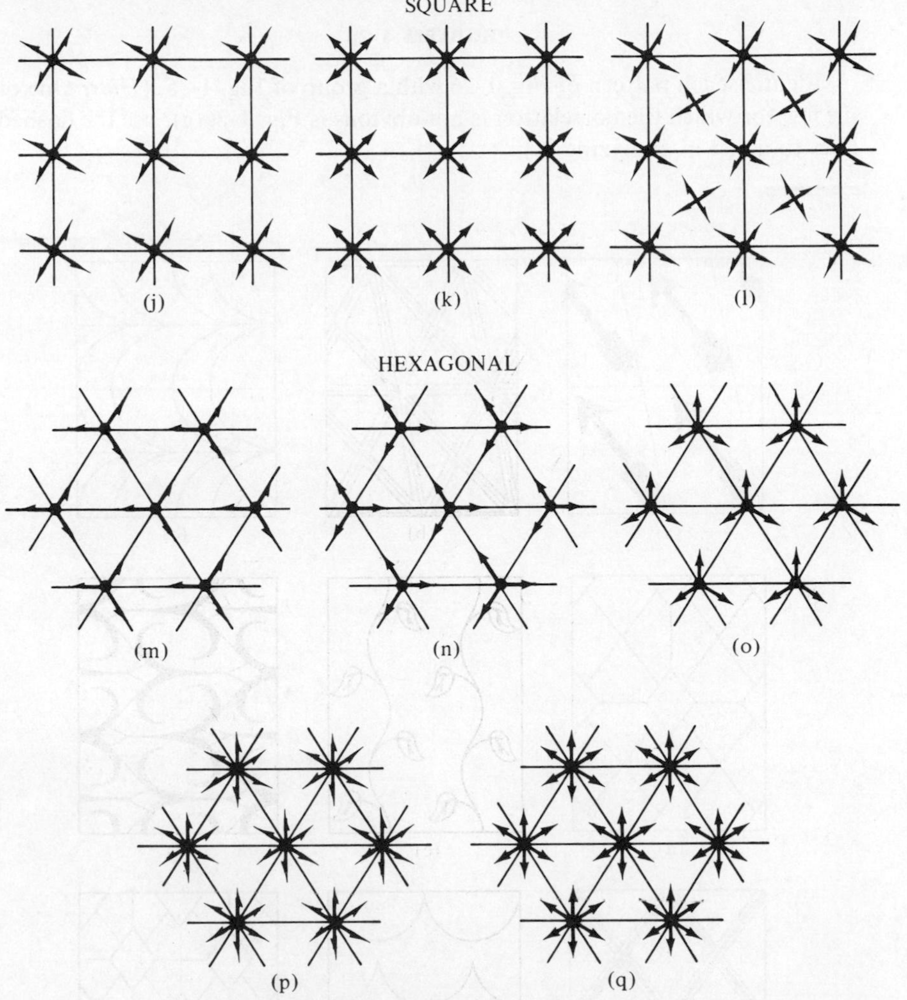

1-25 The seventeen plane groups (cont.)

As we have already mentioned, the plane groups form the basis of two-dimensional designs. A number of interesting examples have been devised to illustrate the aesthetic, and useful, properties of the plane groups. Figure 1-26 shows possible wallpaper or fabric patterns made by arranging specific design elements in 17 different ways. An alternate set of patterns is shown in the frontispiece. These are due to Arthur L. Loeb and Philippe LeCorbeiller and were produced at the Ledgemont Laboratory of the Kennecott Copper Corporation. Perhaps the most remarkable examples of groups are those due to the Dutch artist M. C. Escher. His designs, involving cleverly interlocked animals, flowers, and other figures, are used as the basis for a textbook on symmetry by the crystallographer, Caroline H. MacGillavry (*Symmetry Aspects of M. C. Escher's Drawings*, A. Oosthoek, Utrecht, 1965). A few of the Escher drawings have been reproduced in Coxeter.[1]

PROBLEM 1-28

Identify each pattern of Fig. 1-26 with a group of Fig. 1-25. (*Hint:* One of the few for which the correlation is not obvious is Fig. 1-26(g); use the dashed lines to assist in analyzing it.)

1-26 Wallpaper patterns corresponding to the possible plane groups

1-7 The crystallographic point groups

Although we developed a listing of the possible point groups through the use of the equivalent pole concept, an alternate approach, and one that is quite common in the literature, involves the use of *stereograms*. We have

1-26 Wallpaper patterns corresponding to the possible plane groups (cont.)

seen that translation symmetry permits the existence of two-fold, three-fold, four-fold, and six-fold rotation symmetry. By a *crystal*, we mean a regular arrangement of atoms. Hence the point groups limited to the four kinds of rotation symmetry just mentioned are known as the *crystallographic point groups* or *crystal classes*, and we shall see that there are 32 of them.

The stereogram, following Falicov[6], is a device for picturing on a plane the motion of a point on a sphere. Figure 1-27 shows a sphere with an equatorial plane and an arbitrary point P in the upper hemisphere. The *stereographic projection* R of the point P is the intersection of the line PS with the equatorial plane, where S is the south pole. The dashed circle shown in the *stereogram* of Fig. 1-28 represents the intersection of the sphere and the plane, while the point R is indicated by a cross, X. If the north-south axis is one with two-fold rotational symmetry, then the new position of point P will have a stereographic position indicated by the second cross in Fig. 1-29. The boat-shaped symbol ⬭ indicates a two-fold axis. These last two figures are

[6]L. M. Falicov, *Group Theory and Its Physical Applications*, University of Chicago Press, Chicago, 1966.

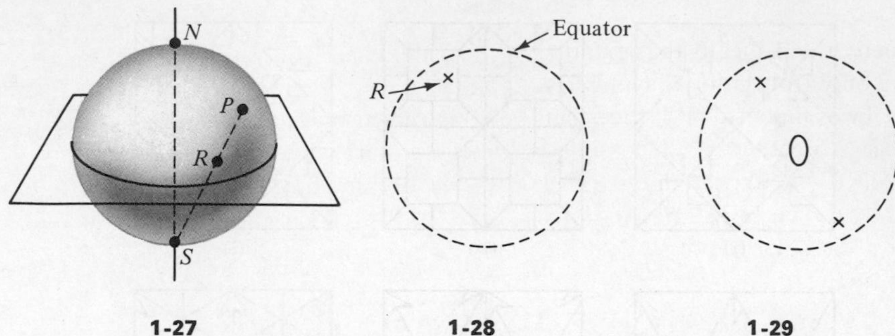

1-27 The stereographic projection R of a point P
1-28 Stereogram obtained from Fig. 1-27 by orienting the NS axis normal to plane of page
1-29 Stereogram obtained from Fig. 1-27 by adding a two-fold axis

repeated in Figs. 1-30(a) and (b), and Figs. 1-30(c), (d), and (e) give the corresponding stereograms for three-fold, four-fold, and six-fold symmetry (the meaning of the central symbols should be obvious). The first five are the cyclic groups C_1, C_2, C_3, C_4, and C_6. With each stereogram we also show the *Hermann-Mauguin* or *international* notation, which is gradually replacing the Schoenflies symbols. The correlation between the two for the cyclic groups is obvious; the international symbol for a six-fold cyclic group (or axis), for example, is simply 6.

If we now add a horizontal reflection plane to each cyclic group, the

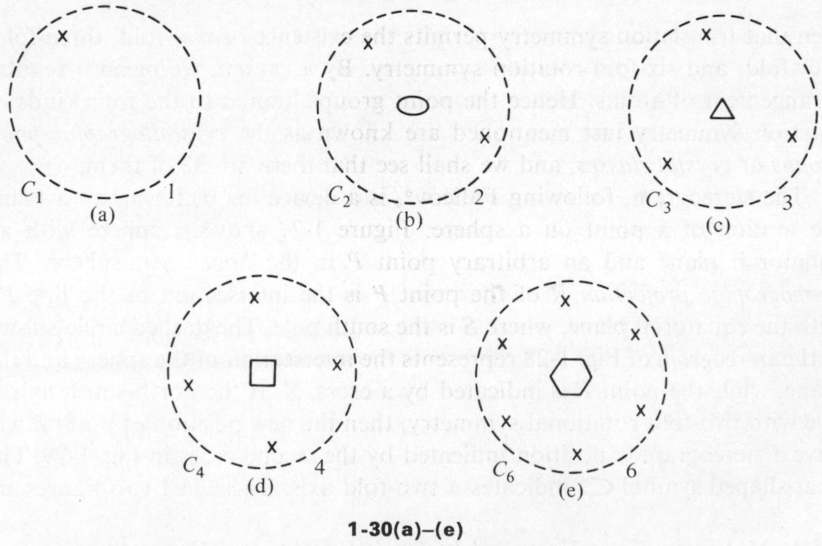

1-30 Stereograms of the 32 crystallographic points

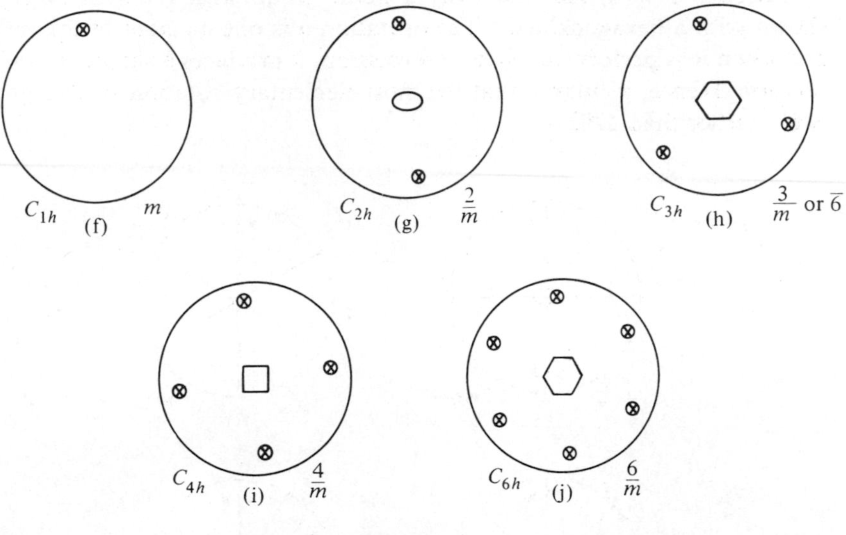

1-30(f)–(j)

resulting stereograms appear as shown in Fig. 1-30(f)–(j). The reflection of the point P in the equatorial plane of Fig. 1-27 produces a point P' whose stereographic projection is obtained by running a line to the north pole.

PROBLEM 1-29

Show that the stereographic projections of P and P' coincide.

The projection of P' is indicated by a small circle surrounding the cross, and the horizontal reflection plane is denoted by making the large circle solid rather than dashed. The international symbol for a reflection plane is m and, for a reflection plane normal to an n-fold axis, it is n/m. The group C_{3h} or either $\bar{6}$ or $3/m$ (Fig 1-30h), which we would expect to have a three-fold principal axis, is shown with the hexagonal symbol. The notation $\bar{6}$ is the international symbol for an improper operation with an inversion. The Schoenflies notation is then the compound symbol JC_6, and it should be distinguished from an operation like S_6, which is the combination of C_6 and σ_h. An operation like $\bar{6}$ is called an *alternating axis.*

PROBLEM 1-30

(a) Show that $3/m$ and $\bar{6}$ are indentical operations.
(b) Prove that

$$\bar{6}^2 = 3$$

Problem 1-30(b) indicates why a point group with the symbol C_{3h} is shown with a hexagonal axis. The operation $\bar{6}$ is one member of this group and when it is performed twice in succession, it produces a simple three-fold rotation. Hence, it follows that the most elementary rotation in this group is 60° rather than 120°.

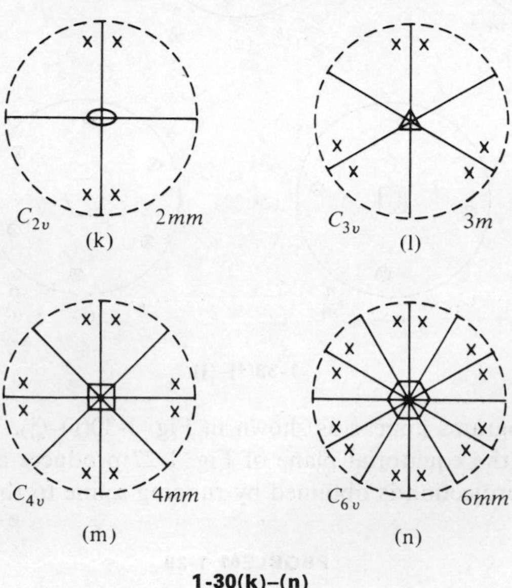

1-30(k)–(n)

Mirror planes may also be added vertically rather than horizontally to give the groups of Figs. 1-30(k) through (n).

PROBLEM 1-31

Why does this procedure give four rather than five groups?

An n-fold axis in the plane of a mirror is denoted nm; two nonequivalent reflection planes give nmm, as in Figs. 1-30(k) and (m).

PROBLEM 1-32

Why does Fig. 1-30(l) show the symbol $3m$ rather than $3mm$?

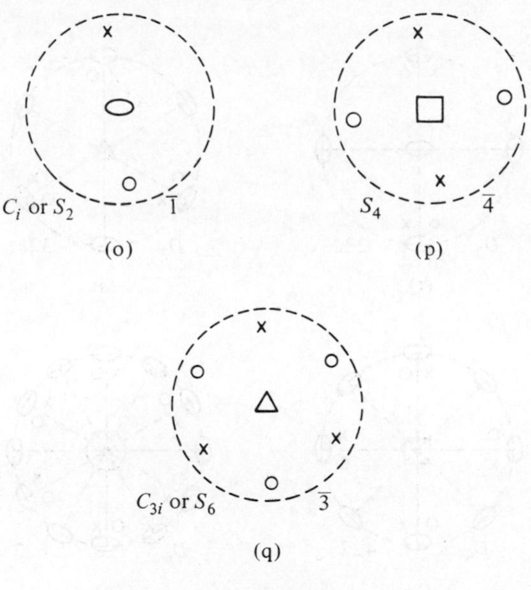

1-30(o)–(q)

Returning to the groups C_n, we may also add improper rotations. The three new groups are S_2 (also designated as C_i), S_4, and S_6 (or C_{3i}); the alternate notation indicates that the inversion is automatically generated.

PROBLEM 1-33

(a) To what previous groups do S_1 and S_3 correspond?
(b) Show that $\bar{1}$, $\bar{4}$, $\bar{3}$ are equivalent to S_2, S_4, S_6, respectively.

A two-fold axis can also be added to the five cyclic groups, producing four dihedral groups (C_1 plus a two-fold axis is simply C_2). The international notation for the groups D_n of Figs 1-30(r)–(u) specifies the multiplicity of nonequivalent axes; 422, for example, means dihedral axes through the corners of the square and also bisecting its sides (we recall that nonequivalent axes are those not related by a symmetry operation). Dihedral axes can also be incorporated into the groups of Figs. 1-30(g) through (j) giving Figs. 1-30(v)–(y).

PROBLEM 1-34

Show that when a dihedral axis is added to C_{1h}, the result is C_{2v}.

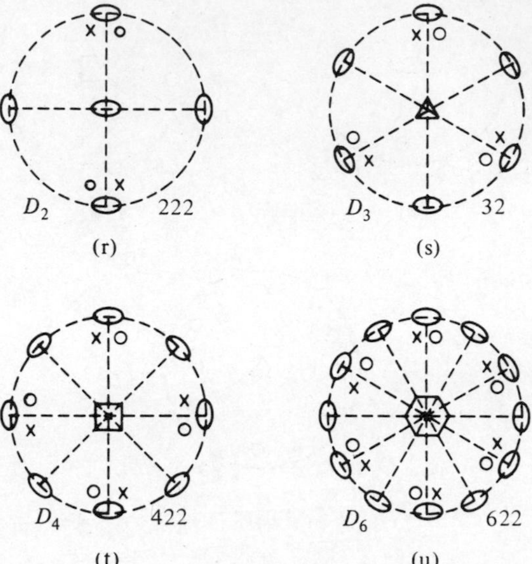
D_2
222
(r)
D_3
32
(s)
D_4
422
(t)
D_6
622
(u)

1-30(r)–(u)

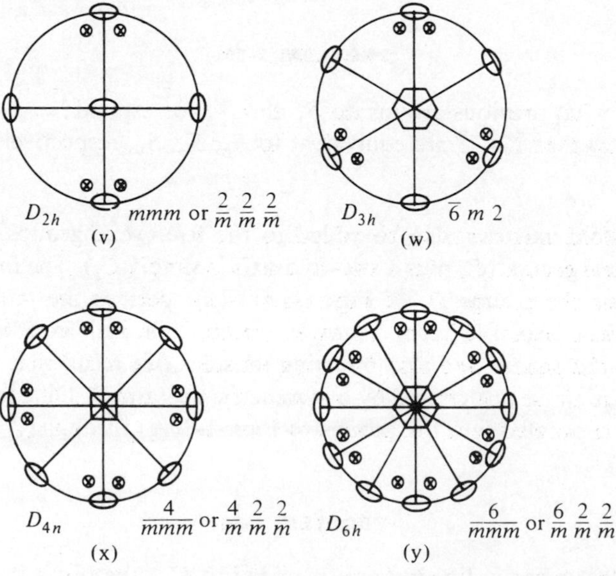
D_{2h}
mmm or $\frac{2}{m}\frac{2}{m}\frac{2}{m}$
(v)
D_{3h}
$\bar{6}\,m\,2$
(w)
D_{4n}
$\frac{4}{mmm}$ or $\frac{4}{m}\frac{2}{m}\frac{2}{m}$
(x)
D_{6h}
$\frac{6}{mmm}$ or $\frac{6}{m}\frac{2}{m}\frac{2}{m}$
(y)

1-30(v)–(y)

The international notation for the group D_{2h} is $2/m\ 2/m\ 2/m$, which denotes three sets of two-fold axes with perpendicular mirror planes. As a matter of fact, the existence of three mirrors in a group means that they must be mutually perpendicular, so that the two-fold axes come automatically. This is the basis of the *abbreviated international symbol mmm.*

PROBLEM 1-35

Verify that *mmm* fully specifies the group also known as D_{2h}.

Similarly, a group such as D_{6h} or $6/m\ 2/m\ 2/m$ can be abbreviated as $6/mmm$, which indicates a principal six-fold axis and three kinds of mirror planes.

PROBLEM 1-36

Show that $6/mmm$ fully describes the group of Fig. 1-30(y).

The last thing that can be done in a simple way is to add a two-fold axis to the groups S_n. For S_2, we obtain C_{2h}. For S_4 and S_6, we obtain a diagonal reflection plane called a *diagonal mirror* (rather than vertical) since it passes through the corners of the square in Fig. 1-30(z) and the triangle in Fig. 1-30(aa). This is the plane designated as σ_d in connection with Fig. 1-13.

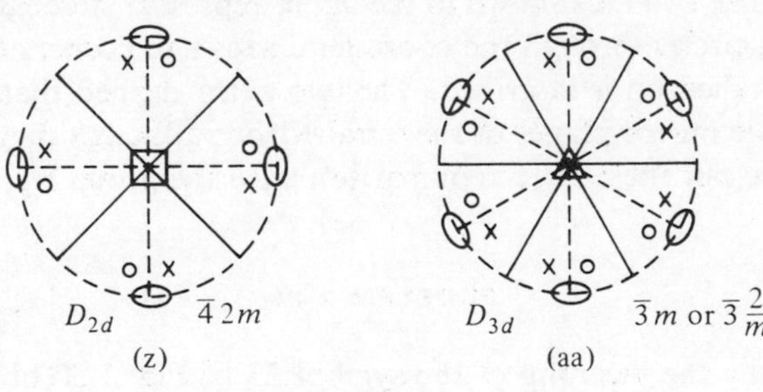

1-30(z)–(aa)

PROBLEM 1-37

Show that the notation $\bar{4}2m$ of Fig. 1-30(z) is self-explanatory.

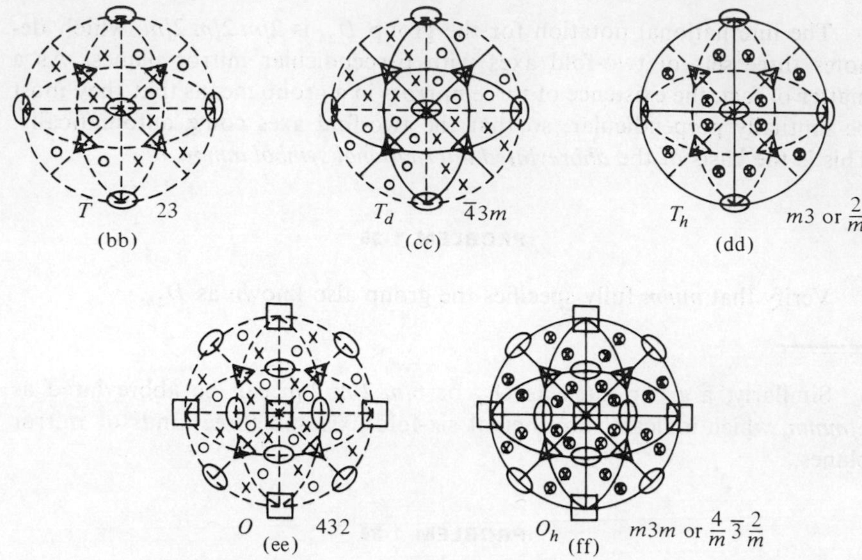

1-30(bb)–(ff)

The process above has enumerated 27 crystallographic groups; the remaining five are the cubic groups T, O, T_d, T_h, and O_h, as discussed at the end of Sec. 1-4. These groups do not have a principal axis; rather, they have four three-fold axes oriented like the diagonals of a cube. The stereogram of the twelve-element group T or 23 is shown in Fig. 1-30(bb). All the lines are dashed since there is no reflection symmetry; this is a group composed solely of rotations. The two arcs shown in the figure represent stereographic projections of great circles through the coordinate axes and corners of the cube in which the tetrahedron is inscribed. The two extra dashed diameters are the positions of the mirror planes of the tetrahedron; although they do not enter this group, we put them in for comparison with the group T_d.

PROBLEM 1-38

Explain why the meaning of the symbol 23 of Fig. 1-30(bb) differs from that of 32 in Fig. 1-30(s).

The holohedral, or full tetrahedral, group consists of T plus the reflection planes, as in Fig. 1-30(cc). Adding inversion symmetry, instead, gives T_h or $2/m\ \bar{3}$ of Fig. 1-30(dd). Note that this group does not specify a tetrahedron, which has no inversion symmetry. The octahedral group O may be fully described by 432, since three such noncollinear axes generate the 24 operations of the group. Finally, adding inversion symmetry to O or to T_d

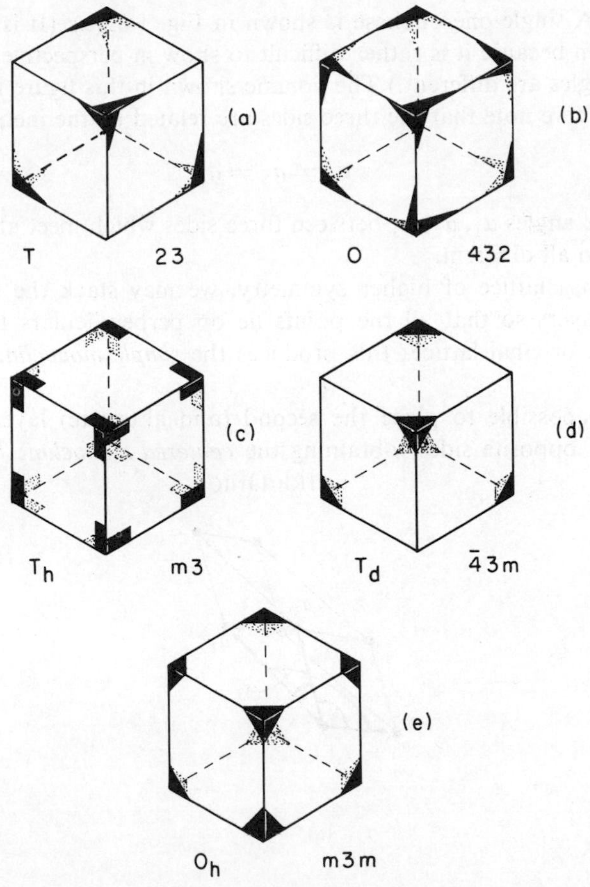

1-31 The five cubic groups

gives the full cubic group O_h The five cubic groups may be represented pictorially by cubes whose symmetry has been lowered in the manner of Fig. 1-31; the abbreviated international symbols are shown in this figure.

1-8 The Bravais lattices

Our next problem is the development of the three-dimensional translation groups. These are known as *Bravais lattices* and there are fifteen of them. Starting with the oblique plane lattice of Fig. 1-24(b), we can displace it parallel to itself at a repeated distance a_3 in an arbitrary direction and generate a three-dimensional net. This lattice, called the *triclinic Bravais lattice*, can be imagined as a stack of parallelepipeds with arbitrary sides and angles. A single one of these is shown in Fig. 1-32(a). (It is drawn in a tilted position because it is rather difficult to show in perspective that all the sides and angles are different.) The volume shown in this figure is called the *primitive cell*; we note that the three sides are related by the inequality

$$a_1 \neq a_2 \neq a_3$$

and the three angles α_1, α_2, α_3 between three sides which meet at a common point are also all different.

To obtain a lattice of higher symmetry, we may stack the second and succeeding layers so that all the points lie on perpendiculars through the points of the original lattice; this produces the *simple monoclinic lattice* of Fig. 1-32(b).

It is also possible to place the second (and alternate) layers over the midpoints of opposite sides, obtaining the *centered monoclinic lattice* [Fig.

TRICLINIC

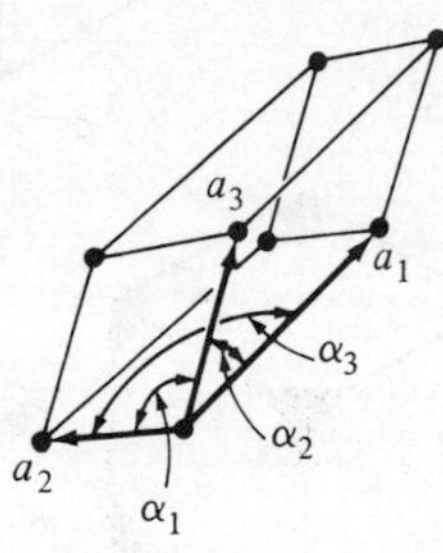

(a)

MONOCLINIC

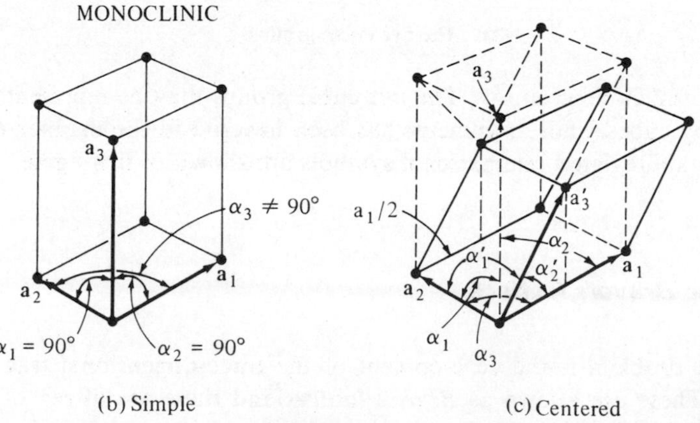

(b) Simple

(c) Centered

1-32(a)–(c)

1-32 The Bravais lattices

1-32(c)]. We could actually call this a *base-centered lattice* (interpreting "base" to mean the side towards the viewer and the one opposite). The figure shows two types of cells; the one determined by the translations $\mathbf{a}_1$, $\mathbf{a}_2$, and $\mathbf{a}_3$—which are identical to the translations of Fig. 1-32(b)—determines what we shall now call the *unit cell.* The unit cell, which contains lattice points at its corners and the two centered points, will generate the complete lattice when translated. However, a smaller cell which also will generate the lattice is the one determined by $\mathbf{a}_1$, $\mathbf{a}_2$, and $\mathbf{a}_3'$, and this one is called the primitive cell, as defined above. It is the smallest and simplest volume which generates the centered lattice. That is, the primitive cell is the smallest such cell; the unit cell is the smallest one which generates the lattice *and* which

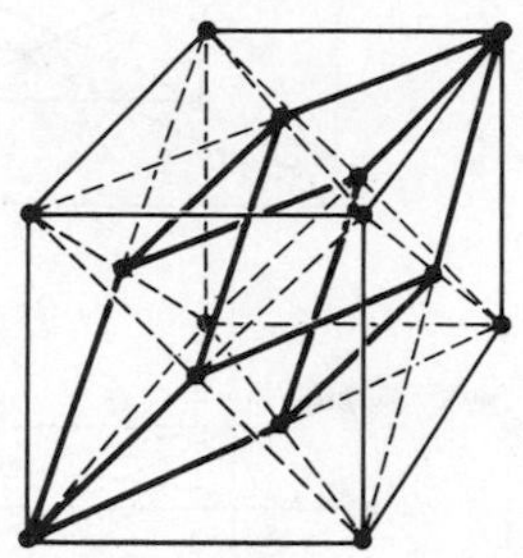

1-33 The primitive cell and the unit cell for the face-centered cubic lattice

also displays the symmetry of the lattice. Consider, for example, Fig. 1-33, showing the two types of cells for the face-centered cubic. The primitive cell, indicated by heavy lines, is a *rhombohedron*, which appears to bear no resemblance to a cube; it is rather surprising, in fact, that the repetition of this structure generates a cubic lattice. Regarding the lattice points as small spheres, we can see geometrically that the fractions of each sphere which lie within the rhombohedron add up to form a single complete sphere; that is, the primitive cell contains the equivalent of one lattice point. On the other hand, the unit cell of Fig. 1-33, the cube, contains the equivalent of four lattice points. For a simple cubic lattice, then, it is quite obvious that the unit cell and primitive cell are identical, and both contain the equivalent of a single lattice point.

An alternate way of obtaining the centered monoclinic lattice is to place the second layer directly over the center of the rhombic figures in the first layer; the resultant structure could be called *body-centered*, but there is no difference in symmetry for body centering or base centering for this Bravais lattice.

PROBLEM 1-39

Verify that the above procedure gives a monoclinic lattice.

The three translation groups created so far are all those which we may regard as being derived from the oblique plane lattice. Moving on to the simple rectangular lattice of Fig. 1-24(d), an arbitrary translation perpendicular to itself gives the *simple orthorhombic* lattice of Fig. 1-32(d). In addition, by stacking alternately over the centers of the sides or the centers of the rectangles, we obtain the *base-centered* or the *body-centered* structures of Figs. 1-32(e) and (f).

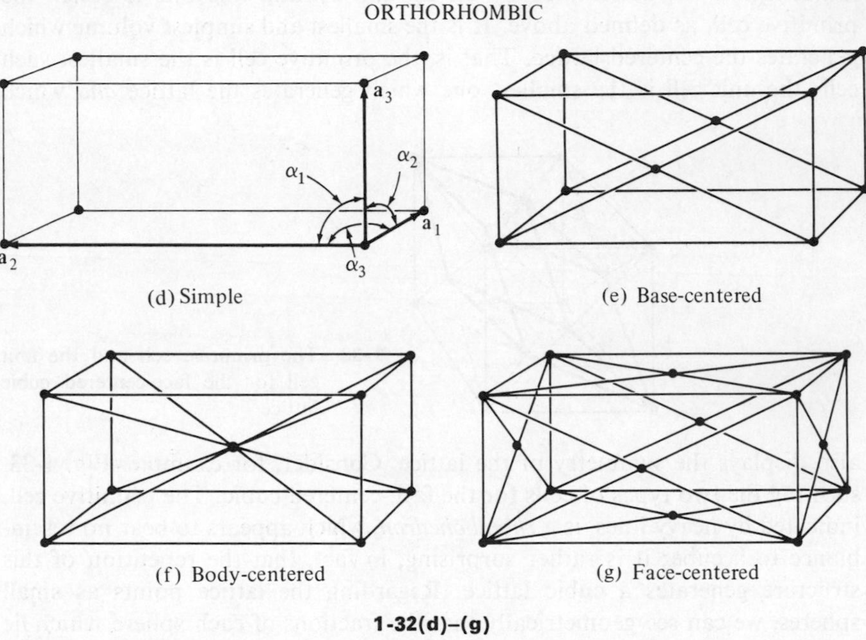

1-32(d)–(g)

PROBLEM 1-40

Figure 1-32(g) shows the *face-centered* orthorhombic lattice. How could such a structure be generated?

Square lattices may be stacked at perpendicular distances not equal to the side of the squares to obtain the two *tetragonal* structures of Figs. 1-32(h) and (i), or they may be placed at the distance which makes $a_1 = a_2 = a_3$, producing the three cubic lattices shown [Figs. 1-32(j), (k), (l)]. Finally, we come to hexagonal nets, which are a source of confusion and controversy. If hexagonal plane nets are arranged one above another, they produce either the *simple trigonal* or the *hexagonal* Bravais lattices of Figs. 1-32(m) and (o), respectively.

TETRAGONAL

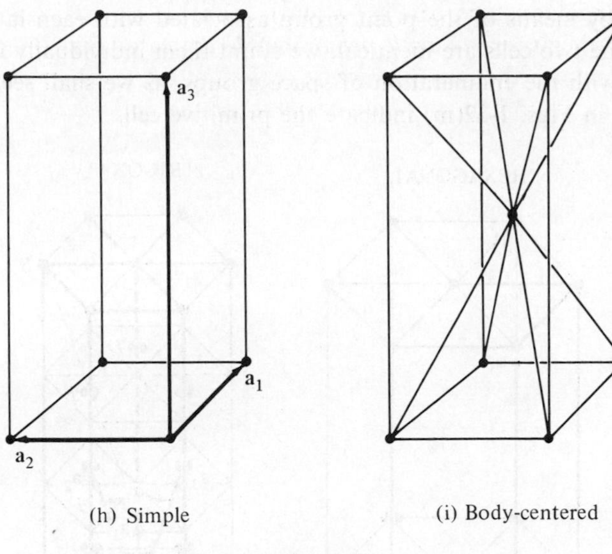

(h) Simple

(i) Body-centered

1-32(h)–(i)

CUBIC

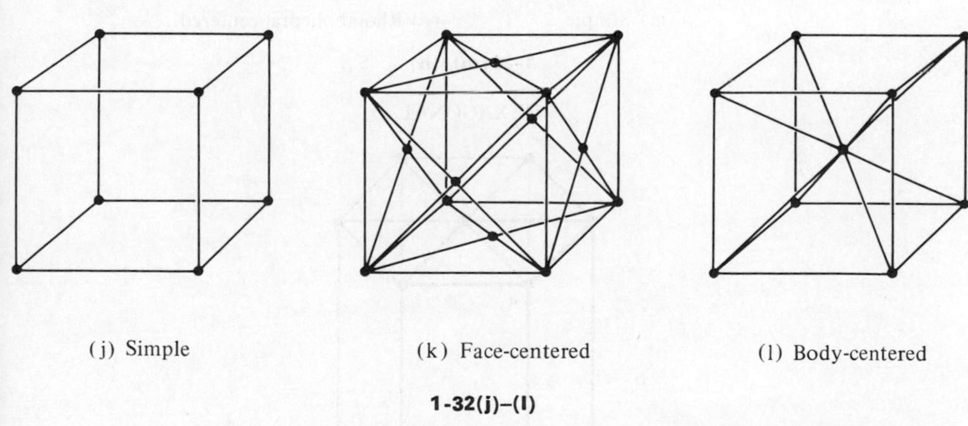

(j) Simple

(k) Face-centered

(l) Body-centered

1-32(j)–(l)

PROBLEM 1-41

Demonstrate by means of a sketch that it is not possible to have a two-dimensional translation group which has three-fold symmetry without also having six-fold symmetry.

As Problem 1-41 indicates, both the simple trigonal and hexagonal Bravais unit cells have six-fold symmetry. The way we differentiate between the two is by means of the point group associated with each lattice point. Although the two cells are identical, we count them individually for reasons connected with the enumeration of space groups, as we shall see later. The heavy lines in Figs. 1-32(m) indicate the primitive cell.

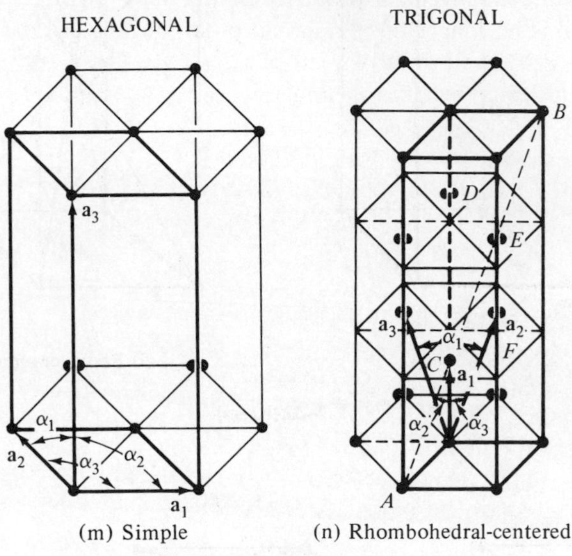

1-32(m)–(n)

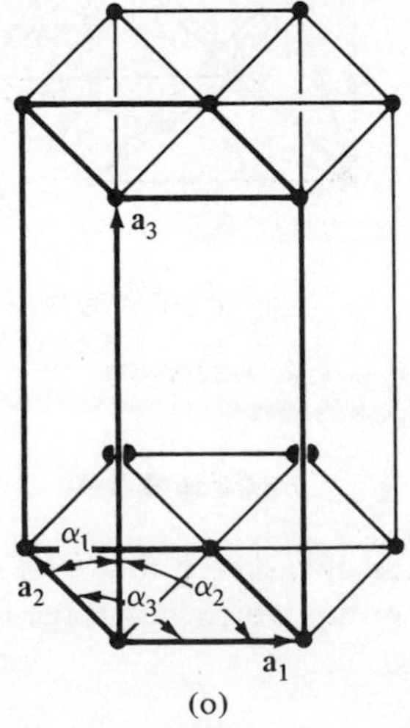

1-32(o)

Another way of stacking hexagonal nets is indicated in Fig. 1-32(n). Lattice points for the second layer go above the midpoints of the triangles in the first layer, and those in the third layer bear the same relation to the second as the second does to the first. That is, points such as *C* and *F* are in the centers of the equilateral triangles, and those like *D* and *E* are over the midpoints of triangles in the second layer which do not contain lattice points. This alternating arrangement of lattice points has only three-fold symmetry. The unit cell is no longer the hexagonal prism but instead is the rhombohedron specified by the primitive vectors $\mathbf{a}_1$, $\mathbf{a}_2$, $\mathbf{a}_3$. These vectors all have the same magnitude and make equal angles with one another.

PROBLEM 1-42

Explain the restriction on the angles α_1, α_2, α_3 given in Table 1-11, lattice number 14.

To avoid cluttering Fig. 1-32(n), the rhombohedron is shown separately (and with a new orientation for clarity) in Fig. 1-34. It is quite clear that this figure has three-fold symmetry, so that the rhombohedron is both the primitive and the unit cell. Returning to Fig. 1-32(n), the parallelepiped indicated with heavy lines has a structure unlike any we have previously encountered. The lattice points *C* and *E* are inside the cell and they divide the line *AB* into three equal parts. This arrangement is called *rhombohedral centering*, but this name is confusing, since the structure is not a rhombohedron, and the rhombohedron of Fig. 1-34 is not centered.

The question of whether there are fourteen or fifteen Bravais lattices appears to have been ignored in the literature on crystallography. Dickenson[7], for example, states that there are fourteen, and then enumerates fifteen.

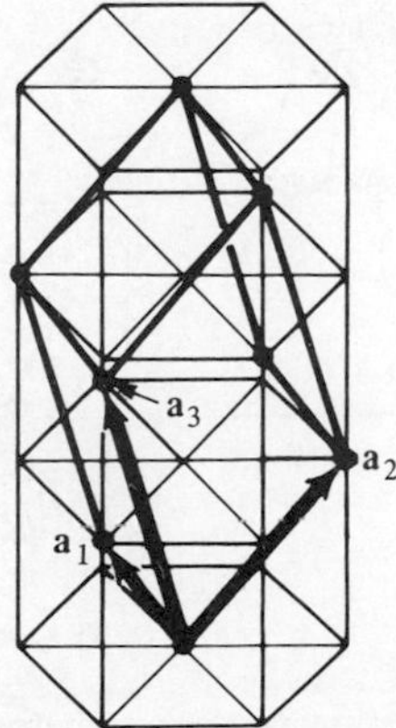

1-34 The rhombohedron corresponding to Fig. 1-32(n).

[7]S. Dickinson, Jr., *Guide to the Interpretation of Space Group Symbols*, Cambridge Research Laboratories, USAF, 1965 (Government document AD-616-246).

Cornwell[8] states that the trigonal and hexagonal systems have one lattice each. As we have seen, there are in fact fifteen, two of which are identical. Although this seems to be a minor point, it will be shown in the next section that a failure to understand the situation leads to further contradictions among the various textbooks.

We notice that Fig. 1-32 shows fifteen Bravais lattices or translational space groups divided into seven categories (triclinic, monoclinic, etc.) which are known as *crystal systems*. An interesting way of visualizing the relations

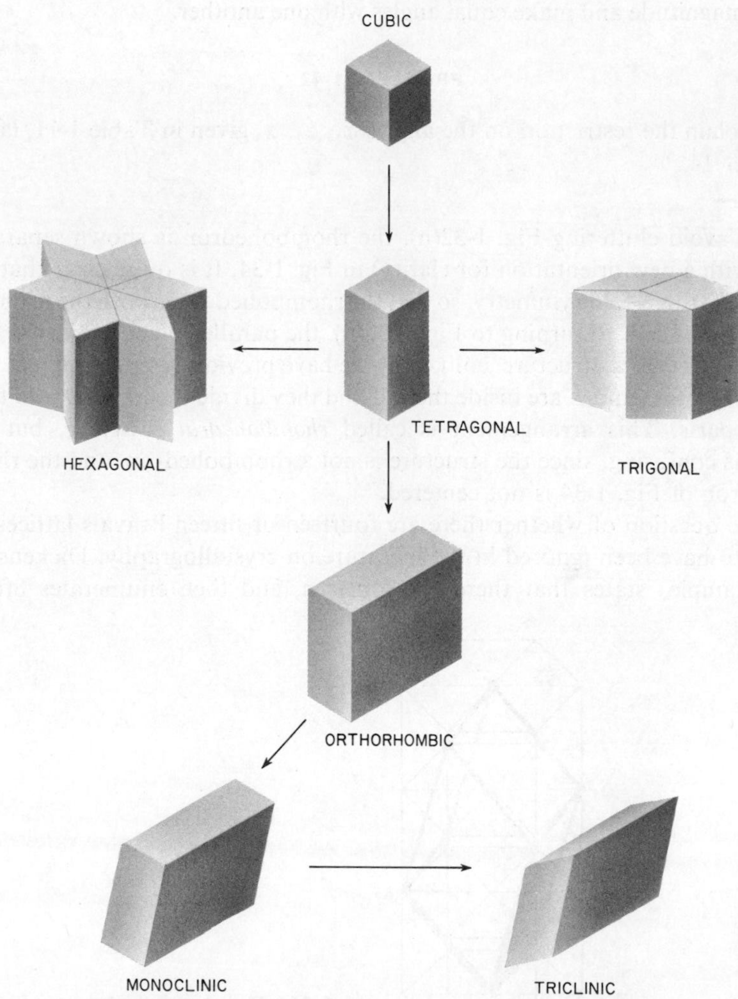

1-35 The seven crystal systems, showing the symmetry hierarchy

[8] J. F. Cornwell, *Group Theory and Electronic Energy Bands in Solids*, Wiley Interscience, New York, 1969.

TABLE 1-11 ***Bravais Lattices and Crystal Systems***

System	Bravais lattice	Cell geometry
1. Triclinic	1. Simple	$a_1 \neq a_2 \neq a_3$ $\alpha_1 \neq \alpha_2 \neq \alpha_3 \neq 90°$
2. Monoclinic	2. Simple 3. Centered	$a_1 \neq a_2 \neq a_3$ $\alpha_1 = \alpha_2 = 90° \neq \alpha_3$
3. Orthorhombic	4. Simple 5. Base-centered 6. Body-centered 7. Face-centered	$a_1 \neq a_2 \neq a_3$ $\alpha_1 = \alpha_2 = \alpha_3 = 90°$
4. Tetragonal	8. Simple 9. Body-centered	$a_1 = a_2 \neq a_3$ $\alpha_1 = \alpha_2 = \alpha_3 = 90°$
5. Cubic	10. Simple 11. Face-centered 12. Body-centered	$a_1 = a_2 = a_3$ $\alpha_1 = \alpha_2 = \alpha_3 = 90°$
6. Trigonal	13. Simple[a] 14. Rhombohedral-centered	$a_1 = a_2 \neq a_3$ $\alpha_1 = \alpha_2 = 90°$ $\alpha_3 = 120°$ $a_1 = a_2 = a_3$ $120° > \alpha_1 = \alpha_2$ $= \alpha_3 \neq 90°$
7. Hexagonal	15. Simple[a]	$a_1 = a_2 \neq a_3$ $\alpha_1 = \alpha_2 = 90°$ $\alpha_3 = 120°$

[a] These two lattices are identical.

TABLE 1-12 ***Definitions of the Seven Crystal Systems***

System	Minimum symmetry
Triclinic	None
Monoclinic	A two-fold axis
Orthorhombic	Two orthogonal two-fold axes or two orthogonal mirror planes
Tetragonal	A four-fold or a four-fold alternating axis
Cubic	Four three-fold axes related like the diagonals of a cube
Trigonal	A three-fold or three-fold alternating axis
Hexagonal	A six-fold or six-fold alternating axis

TABLE 1-13 ***The 32 Crystallographic Point Groups***

System	Schoenflies symbol	Hermann-Mauguin or international symbol	
		Full	Abbreviated
Triclinic	C_1	1	1
	C_i or S_2	$\bar{1}$	$\bar{1}$
	C_{1h}	m	m
Monoclinic	C_2	2	2
	C_{2h}	$2/m$	$2/m$
	C_{2v}	$2mm$	mm
Orthorhombic	D_2	222	222
	D_{2h}	$2/m\,2/m\,2/m$	mmm
	C_4	4	4
	S_4	$\bar{4}$	$\bar{4}$
	C_{4h}	$4/m$	$4/m$
Tetragonal	D_{2d}	$\bar{4}2m$	$\bar{4}2m$
	C_{4v}	$4mm$	$4mm$
	D_4	422	42
	D_{4h}	$4/m\,2/m\,2/m$	$4/mmm$
	T	23	23
	T_h	$2/m\,\bar{3}$	$m3$
Cubic	T_d	$\bar{4}3m$	$\bar{4}3m$
	O	432	432
	O_h	$4/m\,3\,2/m$	$m3m$
	C_3	3	3
	C_{3i} or S_6	$\bar{3}$	$\bar{3}$
Trigonal	C_{3v}	$3m$	$3m$
	D_3	32	32
	D_{3d}	$\bar{3}\,2/m$	$\bar{3}m$
	C_6	6	6
	C_{3h}	$\bar{6}$ or $3/m$	$\bar{6}$ or $3/m$
	C_{6h}	$6/m$	$6/m$
Hexagonal	D_{3h}	$\bar{6}m2$	$\bar{6}m2$
	C_{6v}	$6mm$	$6mm$
	D_6	622	62
	D_{6h}	$6/m\,2/m\,3/m$	$6/mmm$

of the systems to one another is shown in Fig. 1-35. If we start with the cube at the top and distort it by pulling apart two opposite faces, we obtain the tetragonal figure. Squeezing on two of the opposite long edges, we obtain a figure with a rhombic base and top; three of these may be placed together to give the trigonal structure on the right and six give the hexagonal one on the left. A stretching of the tetragonal unit cell in a new direction gives the orthorhombic lattice, and squeezing two opposite sides produces the monoclinic structure, while a second squeeze gives the triclinic shape.

We may summarize the geometric properties of the seven crystal systems in the fashion of Table 1-11, which gives specifications on the edges and angles of the unit cell. We should like to emphasize that this table, although it lists the geometrical properties of the crystal systems, does *not* define them. That is, a crystal in which all the lattice translations were equal and all the edges were 90° might not be cubic. The point group of the atomic arrangement at each lattice point might have lower symmetry. Hence, we explicitly define a system in terms of the minimum symmetry of the groups which generate it. These properties are listed in Table 1-12. The system produced by adding compound operations to each point group is indicated in Table 1-13, which also summarizes the three kinds of notation used.

1-9 Space groups

We have now seen that there are a total of fifteen three-dimensional translation groups or Bravais lattices, one of which appears twice in Table 1-11, and thirty-two ways of arranging a completely unsymmetrical element about each point in the Bravais lattice. We might naively say, then, that there are $15 \times 32 = 480$ space groups, but this ignores the fact that a given point group produces only a particular crystal system. Referring to Tables 1-11 and 1-13, we see that the point groups C_1 and C_i when repeated in three independent directions generate a triclinic lattice; the next three groups in the table generate a simple or centered monoclinic lattice, etc. Hence, we might expect that there are two space groups in the triclinic system, six in the monoclinic, and so on, for a total of 66. Many authors, such as McWeeny[9] for example, overlook the fact that there are two Bravais lattices associated with the trigonal system, and they obtain only 61 lattices by counting all the possibilities indicated in Table 1-13. Either figure (61 or 66) is too low and 480 is too high; it was discovered independently about 1890 by Schoenflies in Germany, Federov in Russia, and Barlow in England, that there are in fact 230 space groups. The existence of these space groups is of primary interest to crystallographers, for this result means that, ignoring differences in chemical constitution, there are precisely 230 ways of arranging

[9] R. McWeeny, *Symmetry*, Pergamon Press, Oxford, 1963.

atoms or groups of atoms to form single crystals. This appears to be astonishing, in view of the thousands of crystals that have been found in nature or synthesized. However, it seems more reasonable when we stop to realize that crystals such as the alkali halides (NaCl, KI, etc.) all have the same *rocksalt* structure, so that they form a single space group. Further, the elimination of five-fold, seven-fold, and higher symmetries because of their conflict with translational symmetry reduces the possible spatial arrangements to a comparatively small number. Nevertheless, the enumeration of 230 space groups is quite a tedious process, and what we shall do here is discuss how it may be done and consider a few examples. First, however, we should note that there are two kinds of space groups. If we use the procedure mentioned just above, of taking a point group and repeating it in three independent directions, we would expect to obtain 66 simple or *symmorphic* groups. Actually, there are 73 such groups, and the additional ones arise from the fact that it is possible to orient some point groups in more than one way with respect to a lattice. An example of Weinreich[10] should clarify this point. Figure 1-36 shows four identical atoms arranged with the symmetry of the point group D_{2d} or $\bar{4}\,2\,m$.

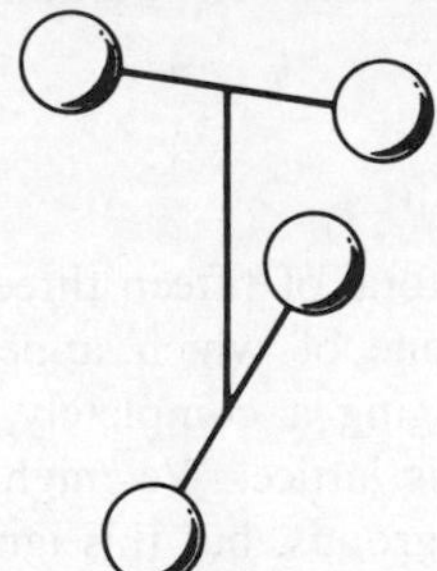

1-36 A basis for the group D_{2d} or $\bar{4}2m$

PROBLEM 1-43

Correlate this figure with the stereogram for D_{2d} or $\bar{4}\,2\,m$.

This atom grouping can generate a tetragonal lattice by repetition, as Table 1-13 indicates, and Fig. 1-37(a) shows the resulting crystal. The set of four atoms is found at each lattice point oriented in such a way that the minimum symmetry required by Table 1-12 is present; the figure shows the two two-fold axes. However, this set of four atoms, called the *basis* of the crystal, may also be oriented as shown in Fig. 1-37(b) and still meet the min-

[10] G. Weinreich, *Solids: Elementary Theory for Advanced Students*, John Wiley & Sons, Inc., New York, 1965.

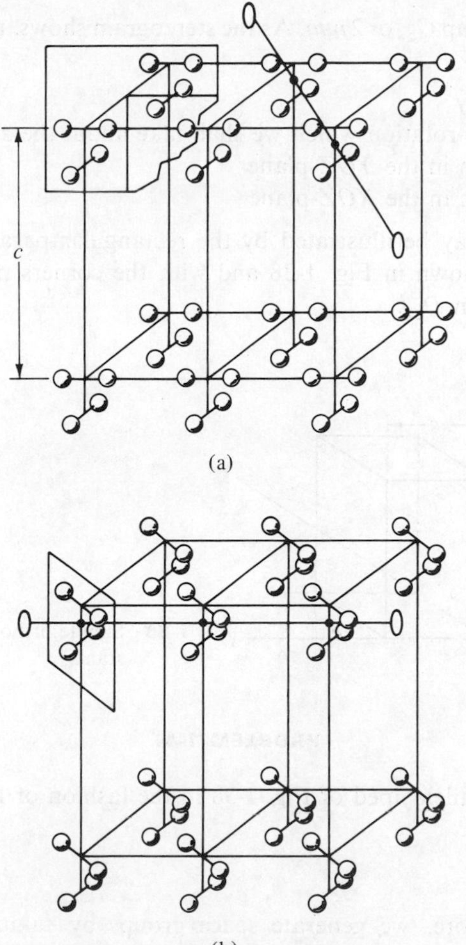

1-37 Two possible symmorphic space groups using the basis of Fig. 1-36

imum symmetry requirements. Each figure shows the required pair of two-fold axes and it is easy to see that no other orientation of the basis with respect to the directions of translation will produce this symmetry. This situation is like that of Figs. 1-25(n) and (o), where two orientations were possible.

In order to get some idea of the complexities involved in listing all the space groups, let us consider an example of Buerger[11] which shows the origin of the space groups based on the simple orthorhombic lattice and derivable

[11] M. J. Buerger, *Elementary Crystallography*, John Wiley & Sons, Inc. New York, 1956.

from the point group C_{2v} or $2mm$. As the stereogram shows, this group has the operations

E, the identity
C_2, a two-fold rotation which we shall take about the z-axis
m_x, a reflection in the YOZ-plane
m_y, a reflection in the XOZ-plane

This symmetry may be illustrated by the rectangular parallelepiped whose dimensions are shown in Fig. 1-38 and with the corners painted to reduce the symmetry from D_{2h}.

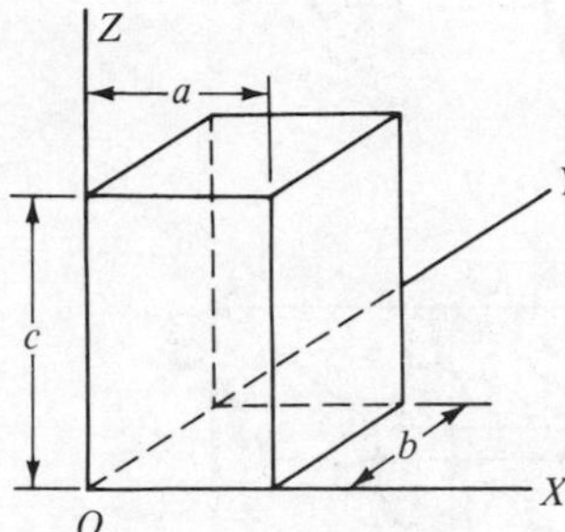

1-38 Simple orthorhombic lattice constants

PROBLEM 1-44

Mark the parallelepiped of Fig. 1-38 in the fashion of Fig. 1-7 to obtain the symmetry C_{2v}.

As stated before, we generate space groups by taking the symmetry operations of point groups and combining them with the translation operations of translation groups to form compound operations of the form $(X|T)$. Sometimes this will simply reproduce a cluster of atoms to form a regular spatial array, thus defining the lattice. At other times, the group of atoms is reproduced periodically but with the orientation changed, and we have seen that there are two possibilities here. One is the *glide plane*, consisting of a translation parallel to a given plane and a reflection in that plane; the other is the *screw axis*, which is a displacement in a given direction and a rotation (180°, 120°, 90° or 60°) about the displacement direction. A three-fold screw axis is shown in Fig. 1-39. The international notation for a right-handed axis [Fig. 1-39(a)] is 3_1, designating a 120° turn to the right combined with a translation of $a/3$, where a is the lattice spacing. In Fig. 1-39(b), a 120° rotation to the left (which is equivalent to 240° to the right) is designated as 3_2. It is the presence of these two elements—glide planes and screw axes—which produces the 157 *nonsymmorphic* space groups.

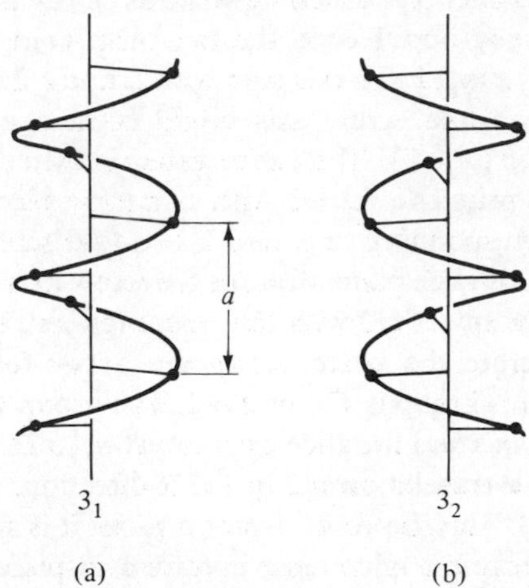

1-39 A right-handed (3_1) and a left-handed (3_2) screw axis

Considering now the groups produced by translation of C_{2v}, we anticipate that some will be simple or symmorphic and others will contain screw axes, glide planes, or both. Actually, the z-axis has been chosen as the two-fold axis, so that it may be simple or a two-fold screw axis, denoted 2_1. Similarly, the two vertical coordinate planes may be mirror or glide planes. The simplest group that we can generate from compound operations of the form $(X|T)$ consists simply of an atomic arrangement with symmetry C_{2v} at each point of the orthorhombic lattice.

PROBLEM 1-45

Make a sketch like Fig. 1-37, showing a simple orthorhombic lattice with a group of atoms having C_{2v} symmetry at each point.

The *Niggli* notation for this space group is C_{2v}^1; the notation system is to number all the possible space groups derived from the point group by adding a superscript. We shall find that there are ten such space groups, indicated by C_{2v}^1 through C_{2v}^{10}. The international notation is *Pmm*2, and is more informative. The *P* indicates a simple (or primitive) lattice, and the *mm*2 conveys the same information as for the point group. The second space group, C_{2v}^2, comes by adding a two-fold screw axis to the previous operations. Hence, one of the operations in the space group will be $(C_2\,|\,c/2)$, where c is the lattice constant in the z-direction. In order for the product of this operation with the other ones to be a member of the group, one of the reflections must also have a displacement of $c/2$ associated with it, so that the total displacement for the product is c, which we write as $(E\,|\,c)$, and this is a member of the translation subgroup. Hence, the two other compound operations must be either $(m_x\,|\,c/2)$, $(m_y\,|\,E)$ or this pair with E and $c/2$ reversed. That is, the group with the two-fold screw axis could contain a glide operation $(m_y|c/2)$ and a reflection $(m_x\,|\,E)$. The entire group is then denoted as $Pmc2_1$ or $Pcm2_1$, indicating a primitive lattice with a mirror plane, a glide plane along the direction corresponding to c, and a two-fold screw axis. Next, let us ask if we can have this glide plane without the screw axis. We can, provided we associate a displacement $c/2$ with the operation m_y, so that the three operations which generate the space group are a two-fold axis and two vertical glide planes. This group is C_{2v}^3 or $Pcc2$. The group C_{2v}^4 we obtain by considering a two-fold axis and the glide operation $(m_y\,|\,a/2)$. The third operation must also involve a translation $a/2$ in the x-direction, so that the compound form is $(m_x\,|\,a/2)$. This, however, is not a glide; it is simply a reflection and a displacement, because a glide must involve a displacement parallel to the glide plane. However, the symbol for this group in the international system is $Pma2$, where the a specifically indicates a glide along the x-direction. The appearance of this glide in the x-direction will be explained in Problem 1-46. Carrying on in this fashion, we obtain the ten space groups of Table 1-14. We note that in order to satisfy the requirements that the total translation produce a lattice point, a combination of $c/2$ for the screw axis and $a/2$ for a glide plane gives the third displacement as $(a/2) + (c/2)$. This displacement, when introduced into the space group, leads to the possibility of $(a/2) + (b/2) + (c/2)$, and we find two of these in the group C_{2v}^{10} or $Pnn2$, where the n denotes a *diagonal glide*. This term refers to a glide in a direction other than along a coordinate axis.

As Table 1-14 and the preceding discussion indicate, the enumeration of the 230 space groups is a rather complex process, and we have tried here to give only a rough description of the general principles. A complete listing and description of all these groups will be found in the *International X-Ray Tables*, Vol. I.[12] We shall show in a later chapter how to make use of the information in this reference and discuss in detail the properties of some specific space groups. Problem 1-46 provides a start in this direction.

PROBLEM 1-46

A more concrete idea of the nature of the space group C_{2v}^4 or $Pma2$ may be obtained in the following two ways:

(a) The space group operations have been shown above to be: (1) the identity, (2) a reflection in the YOZ-plane combined with a translation $a/2$ along

[12] N. F. M. Henry and K. Lonsdale, editors, *International Tables for X-Ray Crystallography*, Kynoch Press, Birmingham, 1952.

TABLE 1-14 ***The Space Groups Derived from the Point Group C_{2v}***

Schoenflies symbol	International symbol	Translation associated with		
		m_x	m_y	C_2
C_{2v}^1	*Pmm*2	0	0	0
C_{2v}^2	$Pmc2_1$	0	$\frac{c}{2}$	$\frac{c}{2}$
C_{2v}^3	*Pcc*2	$\frac{c}{2}$	$\frac{c}{2}$	0
C_{2v}^4	*Pma*2	$\frac{a}{2}$	$\frac{a}{2}$	0
C_{2v}^5	$Pbc2_1$	$\frac{b}{2}$	$\frac{b}{2}+\frac{c}{2}$	$\frac{c}{2}$
C_{2v}^6	*Pcn*2	$\frac{a}{2}+\frac{c}{2}$	$\frac{a}{2}+\frac{c}{2}$	0
C_{2v}^7	$Pmn2_1$	$\frac{a}{2}$	$\frac{a}{2}+\frac{c}{2}$	$\frac{c}{2}$
C_{2v}^8	*Pba*2	$\frac{a}{2}+\frac{b}{2}$	$\frac{a}{2}+\frac{b}{2}$	0
C_{2v}^9	$Pna2_1$	$\frac{a}{2}+\frac{b}{2}+\frac{c}{2}$	$\frac{a}{2}+\frac{b}{2}$	$\frac{c}{2}$
C_{2v}^{10}	*Pnn*2	$\frac{a}{2}+\frac{b}{2}+\frac{c}{2}$	$\frac{a}{2}+\frac{b}{2}+\frac{c}{2}$	0

the x-axis, (3) a reflection in the XOZ-plane combined with the translation $a/2$ along the x-axis, (4) a two-fold rotation about the z-axis, (5) translations of b along the y-axis and c along the z-axis, and (6) combinations of these operations with themselves and each other. Place an asymmetric arrow at the origin with its point at some arbitrary location (x, y, z) and sketch the spatial pattern that operations (1) through (5) generate. Show that it is an orthorhombic lattice with glide planes parallel to XOZ at distances $y = 0$, $b/2$, and b from this plane and with pure reflection planes parallel to YOZ at distances $x = a/4$ and $3a/4$.

(b) The International Tables state that *equivalent points* for this space group have coordinates x, y, z; $\bar{x}, \bar{y}, \bar{z}$; $\frac{1}{2} - x, y, z$; $\frac{1}{2} + x, \bar{y}, z$, where the notation $\bar{x}$ means $-x$. The meaning of this information is that one of the symmetry operations (1)–(4) above will move the point of the arrow from x, y, z to $-x, -y, -z$, and another will move the point to $-x, y, z$ and then displace the entire arrow parallel to itself a distance $a/2$ in the x-direction. Show that this information agrees with your diagram in part (a).

We should make it clear here that a space group is not a crystal; a real crystal is obtained by placing an atom or a group of atoms at every point in the lattice. As a reasonably complex example, consider lithium sulfate monohydrate, $Li_2SO_4 \cdot H_2O$. According to Wyckoff[13], this compound crystallizes in the monoclinic form, as in Fig. 1-40. The unit cell has two $Li_2SO_4 \cdot H_2O$

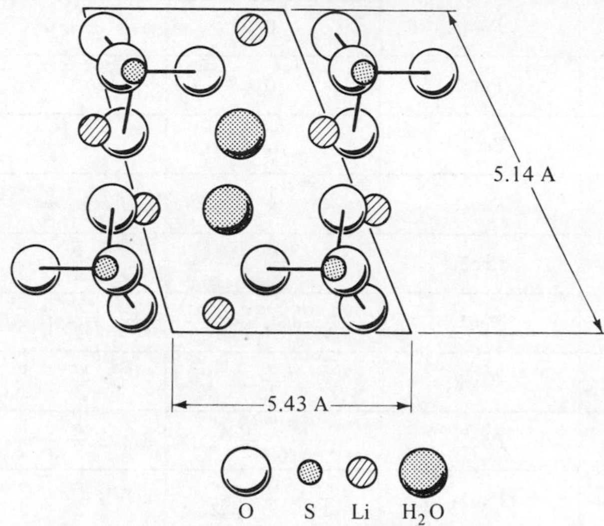

1-40 The unit cell of $Li_2SO_4 \cdot H_2O$

molecules, arranged to have two-fold symmetry. The Schoenflies notation for the space group is C_2^1; the abbreviated international symbol is $P2$. Note that the detailed structure of the water molecule is not shown; the determination of crystal structure is done by the use of X-ray methods, and it would take extremely refined methods to determine the exact orientation of the atoms in the H_2O molecule.

A concise and useful summary of the space groups is given in a report by Dickinson[7]; the chart included with this report shows the symmorphic and nonsymmorphic groups which go with each Bravais lattice.

PROBLEM 1-47

Show the position of the screw axis in the group C_{2v}^2 or $Pmc2_1$.

---

[13]R. W. G. Wyckoff, *Crystal Structures*, Interscience, New York, 1960.

1-10 Group theory and space groups

Our approach to the study of space groups has been primarily descriptive. To put some of these ideas on a firmer foundation, let us first show how to expand the usefulness of the Seitz notation. The symbol $(X|T)$ denotes the dual operation of a rotation, reflection, or inversion X and a translation T. The point symmetries X, when operating on a vector **r**, convert it into a new vector **r**′. If we follow this by a translation T, corresponding to a vector **t**, then **r**′ becomes **r**″, where

$$\mathbf{r}' = X\mathbf{r}$$

and

$$\mathbf{r}'' = \mathbf{r}' + \mathbf{t}$$

or

$$\mathbf{r}'' = X\mathbf{r} + \mathbf{t} \tag{1-15}$$

Going from **r** to **r**″ in the reverse order, we obtain

$$\mathbf{r}'' = \mathbf{r} + \mathbf{t}$$

and

$$\mathbf{r}'' = X(\mathbf{r} + \mathbf{t}) = X\mathbf{r} + X\mathbf{t} \tag{1-16}$$

The vectors **r**″ in (1-15) and in (1-16) are identical only if

$$X\mathbf{t} = \mathbf{t} \quad \text{or} \quad X = E$$

Hence, in three dimensions we define the symbol $(X|T)$ by (1-15); that is, we must stipulate that X is to be performed first.

Next, let there be two successive operations $(X_1|T_1)$ and $(X_2|T_2)$. Then

$$\begin{aligned}(X_2|T_2)(X_1|T_1)\mathbf{r} &= (X_2|T_2)(X_1\mathbf{r} + \mathbf{t}_1)\\ &= X_2(X_1\mathbf{r} + \mathbf{t}_1) + \mathbf{t}_2\\ &= X_2X_1\mathbf{r} + (X_2\mathbf{t}_1 + \mathbf{t}_2)\end{aligned} \tag{1-17}$$

The term in parentheses is a translation obtained by performing T_1, X_2, and T_2 in succession, so that (1-17) is equivalent to

$$(X_2|T_2)(X_1|T_1) = (X_2X_1|X_2T_1 + T_2) \tag{1-18}$$

We may use (1-18) to find an expression for the inverse $(X|T)^{-1}$ of an operation $(X|T)$, where

$$(X|T)^{-1}(X|T) = (X|T)(X|T)^{-1} = (E|0) \tag{1-19}$$

Let $(X|T)^{-1}$ have elements W and S, so that

$$(X|T)^{-1} = (W|S)$$

Then
$$(W|S)(X|T) = (E|0)$$

By (1-18), this becomes

$$(WX|WT + S) = (E|0)$$

or
$$WX = E, \qquad WT + S = 0$$

from which

$$W = X^{-1} \quad \text{and} \quad S = -W^{-1}T = -X^{-1}T$$

so that
$$(X|T)^{-1} = (X^{-1}|-X^{-1}T) \tag{1-20}$$

PROBLEM 1-48

Verify (1-19).

It is easy to see from (1-18) and (1-20) that Table 1-7, which we obtained in a very intuitive fashion, actually obeys the conditions required of a group. However, it is more instructive to consider the Cayley table for a simple diperiodic group. Figure 1-41 shows a rectangular lattice to which we apply

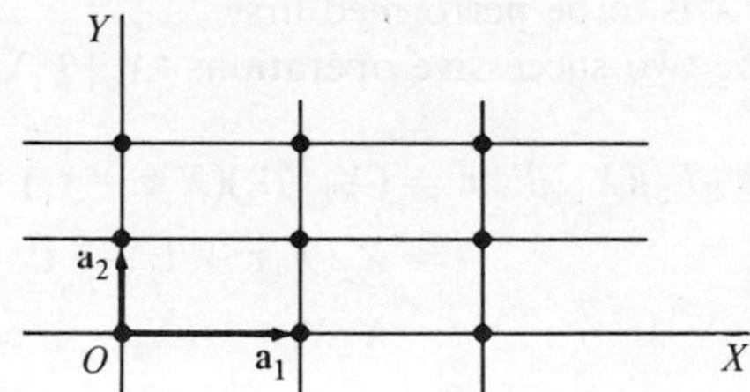

1-41 Primitive vectors of a rectangular lattice

the Born-von Karman cyclic boundary conditions in two directions. That is, we take

$$T_x^2 = E, \qquad T_y^2 = E$$

or in vector form

$$\mathbf{a}_1 + \mathbf{a}_1 = 0, \qquad \mathbf{a}_2 + \mathbf{a}_2 = 0$$

TABLE 1-15 ***Portion of the Cayley Table for a Diperiodic Space Group***

	$(E\|0)$	$(E\|T_x)$	$(E\|T_y)$	$(E\|T_x+T_y)$	$(\sigma_x\|0)$	$(\sigma_x\|T_x)$	$(\sigma_x\|T_y)$	$(\sigma_x\|T_x+T_y)$
$(E\|0)$	$(E\|0)$	$(E\|T_x)$	$(E\|T_y)$	$(E\|T_x+T_y)$	$(\sigma_x\|0)$	$(\sigma_x\|T_x)$	$(\sigma_x\|T_y)$	$(\sigma_x\|T_x+T_y)$
$(E\|T_x)$	$(E\|T_x)$	$(E\|0)$	$(E\|T_x+T_y)$	$(E\|T_y)$	$(\sigma_x\|T_x)$	$(\sigma_x\|0)$	$(\sigma_x\|T_x+T_y)$	$(\sigma_x\|T_y)$
$(E\|T_y)$	$(E\|T_y)$	$(E\|T_x+T_y)$	$(E\|0)$	$(E\|T_x)$	$(\sigma_x\|T_y)$	$(\sigma_x\|T_x+T_y)$	$(\sigma_x\|0)$	$(\sigma_x\|T_x)$
$(E\|T_x+T_y)$	$(E\|T_x+T_y)$	$(E\|T_y)$	$(E\|T_x)$	$(E\|0)$	$(\sigma_x\|T_x+T_y)$	$(\sigma_x\|T_y)$	$(\sigma_x\|T_x)$	$(\sigma_x\|0)$
$(\sigma_x\|0)$	$(\sigma_x\|0)$	$(\sigma_x\|T_x)$	$(\sigma_x\|T_y)$	$(\sigma_x\|T_x+T_y)$	$(E\|0)$	$(E\|T_x)$	$(E\|T_y)$	$(E\|T_x+T_y)$
$(\sigma_x\|T_x)$	$(\sigma_x\|T_x)$	$(\sigma_x\|0)$	$(\sigma_x\|T_x+T_y)$	$(\sigma_x\|T_y)$	$(E\|T_x)$	$(E\|0)$	$(E\|T_x+T_y)$	$(E\|T_y)$
$(\sigma_x\|T_y)$	$(\sigma_x\|T_y)$	$(\sigma_x\|T_x+T_y)$	$(\sigma_x\|0)$	$(\sigma_x\|T_y)$	$(E\|T_y)$	$(E\|T_x+T_y)$	$(E\|0)$	$(E\|T_x)$
$(\sigma_x\|T_x+T_y)$	$(\sigma_x\|T_x+T_y)$	$(\sigma_x\|T_y)$	$(\sigma_x\|T_y)$	$(\sigma_x\|0)$	$(E\|T_x+T_y)$	$(E\|T_y)$	$(E\|T_x)$	$(E\|0)$
$(\sigma_y\|0)$	$(\sigma_y\|0)$	$(\sigma_y\|T_x)$	$(\sigma_y\|T_y)$	$(\sigma_y\|T_x+T_y)$	$(C_2\|0)$	$(C_2\|T_x)$	$(C_2\|T_y)$	$(C_2\|T_x+T_y)$
$(\sigma_y\|T_x)$	$(\sigma_y\|T_x)$	$(\sigma_y\|0)$	$(\sigma_y\|T_x+T_y)$	$(\sigma_y\|T_y)$	$(C_2\|T_x)$	$(C_2\|0)$	$(C_2\|T_x+T_y)$	$(C_2\|T_y)$
$(\sigma_y\|T_y)$	$(\sigma_y\|T_y)$	$(\sigma_y\|T_x+T_y)$	$(\sigma_y\|0)$	$(\sigma_y\|T_x)$	$(C_2\|T_y)$	$(C_2\|T_x+T_y)$	$(C_2\|0)$	$(C_2\|T_x)$
$(\sigma_y\|T_x+T_y)$	$(\sigma_x\|T_x+T_y)$	$(\sigma_y\|T_y)$	$(\sigma_y\|T_x)$	$(\sigma_y\|0)$	$(C_2\|T_x+T_y)$	$(C_2\|T_y)$	$(C_2\|T_x)$	$(C_2\|0)$

Hence, two translations such as $(E|T_x)$, $(E|T_y)$ combine to form

$$(E|T_y)(E|T_x) = (E|ET_x + T_y) = (E|T_x + T_y)$$

and $(E|T_x)$ followed by $(E|T_x + T_y)$ gives

$$(E|ET_x + T_x + T_y) = (E|T_y)$$

In a similar way, we generate the upper left 4×4 block of Table 1-15. We show only 96 of the 256 possible entries, since writing out this table is tedious and not very instructive. As we see, the rotational part of each Seitz symbol is identical for every entry in each 4×4 block and the translational part is the upper left-hand corner, repeated 15 times. In fact, Table 1-15 may be regarded as the *direct product* of the point group C_{2v} of Table 1-6 and of the two-dimensional rectangular translation group T_2 whose multiplication table is Table 1-16. Denoting the resulting space group by S, we symbolize this direct product as

$$S = C_{2v} \times T_2$$

That is, the 16×16 complete Table 1-15 is obtained by combining each operation of the 4×4 Table 1-6 with *each* of the 16 operations in Table 1-16 to obtain the table for the space group.

TABLE 1-16 ***Cayley Table for a Diperiodic Translation Group***

	E	T_x	T_y	$T_x + T_y$
E	E	T_x	T_y	$T_x + T_y$
T_x	T_x	E	$T_x + T_y$	T_y
T_y	T_y	$T_x + T_y$	E	T_x
$T_x + T_y$	$T_x + T_y$	T_y	T_x	E

Some of the entries in this table require that we be able to handle all possible products of the form XT. For example,

$$\sigma_x T_x = T_x, \qquad \sigma_x T_y = -T_y \qquad \textbf{(1-21)}$$

Since the lattice of Fig. 1-41 contains only two unit cells in each direction, then

$$-T_y = T_y$$

so that

$$\sigma_x T_y = T_y \qquad \textbf{(1-22)}$$

If we had a lattice with, say, three cells in each direction, then the situation would be more complicated. For example, we would write

$$\sigma_x T_y = -T_y = +2T_y$$
$$\sigma_x(2T_y) = -2T_y = +T_y$$
$$\sigma_x(3T_y) = -3T_y = 3T_y$$

and so on. Thus, the principles are the same, but the Cayley table would be unmanageably large.

Returning to Table 1-15, we combine (1-18) with relations like (1-20) and (1-21) to show that

$$(\sigma_y | T_y)(\sigma_x | T_x + T_y) = (C_2 | T_x + T_y + T_y)$$
$$= (C_2 | T_x)$$

as an example. All the entries follow this general pattern and hence we have only shown a small number explicitly. As a matter of fact, we realize the structure of this table is like that of Table 1-7; each 2×2 block in Table 1-7 has a corresponding 4×4 block in the present table.

PROBLEM 1-49

Prove that every member of the translation subgroup in Table 1-15 is in a class by itself.

The group in this table is a combination of the point group C_{2v} and the simple or primitive rectangular translation group, so that it is a symmorphic group, as defined in Sec. 1-9. Falicov[6] has pointed out that we can recognize this fact from the Cayley table by realizing that all point group operations have the form $(X|0)$. If we can choose an origin so that all members of the point group appear also in the space group, the group is symmorphic. As Table 1-15 shows, this has already been accomplished for this group.

To see what happens in a nonsymmorphic group, Fig. 1-42 shows a plane group with a glide line. The unit cell is chosen as indicated in the lower left-hand corner; note that $\mathbf{a}_3$ cannot be a primitive translation since $2\mathbf{a}_3$ is not a lattice vector. The origin of the unit cell is placed symmetrically, and this is often done when the basis consists of two atoms. The point group for this lattice consists of the four operations

$$(E|0), \quad (\sigma_x|0), \quad (\sigma_y|0), \quad (C_2|0).$$

The space group will then contain $(E|0)$ and $(C_2|0)$; also, the glide line has

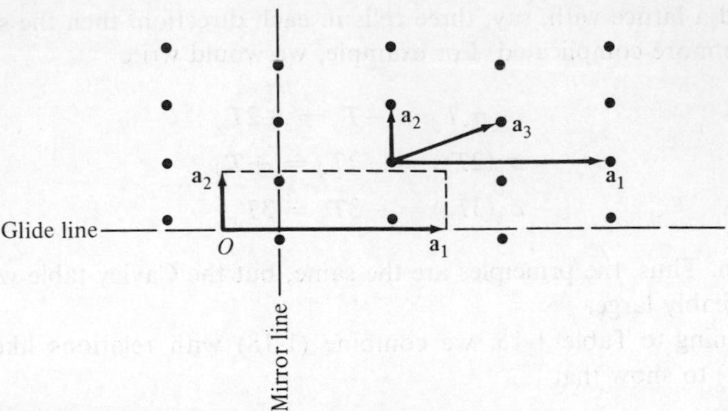

1-42 A nonsymmorphic plane lattice

to be expressed as $(\sigma_x|\mathbf{a}_1/2)$. For the mirror, the operation $(\sigma_y|0)$ is not expressed in terms of the origin we have chosen; one of the mirror planes is located as shown. However, a line through the origin used as a mirror followed by a translation $\mathbf{a}_1/2$ is also a symmetry operation; that is $(\sigma_y|-\mathbf{a}_1/2)$ is the space group operation we need. We see this analytically because the mirror indicated may be transformed to a new reference position by (1-5). That is,

$$\begin{aligned}(E|\mathbf{a}_1/4)^{-1}(\sigma_y|0)(E|\mathbf{a}_1/4) &= (E|-\mathbf{a}_1/4)(\sigma_y|-\mathbf{a}_1/4)\\ &= (\sigma_y|-\mathbf{a}_1/2)\end{aligned} \tag{1-23}$$

Hence, this is the correct operation for the space group.

Now let us see if it is possible to find an origin such that these four symmetry operations have the form $(X|0)$. If this new origin has a position $\mathbf{r}$ with respect to the one in Fig. 1-41. then the new operations are obtainable from the original ones by the use of (1-5). This follows from the fact that a change in coordinates does not alter the symmetry properties of the members of any class. Hence

$$\begin{aligned}(E|-\mathbf{r})(E|0)(E|\mathbf{r}) &= (E|0)\\ (E|-\mathbf{r})(C_2|0)(E|\mathbf{r}) &= (C_2|-2\mathbf{r})\\ (E|-\mathbf{r})(\sigma_x|\mathbf{a}_1/2)(E|\mathbf{r}) &= (\sigma_x|(\mathbf{a}_1/2)-2y\mathbf{j})\\ (E|-\mathbf{r})(\sigma_y|-\mathbf{a}_1/2)(E|\mathbf{r}) &= (\sigma_y|(-\mathbf{a}_1/2)-2x\mathbf{i})\end{aligned} \tag{1-24}$$

where $\mathbf{r} = x\mathbf{i} + y\mathbf{j}$.

Since no value of x and y will cause all of the last three translations to vanish, this group must be nonsymmorphic. The source of this criterion is that if we have a group containing operations such as $(\sigma_x|\mathbf{a}_1/2)$, which represents a glide along the x-axis, and if a simple shift of origin reduces all these compound operations to pure rotations or reflections, then we have incorrectly determined the inherent symmetries.

We may use the symmetry elements of Eq. (1-24) to establish the Cayley table, assuming as before that there are two cells in each of the two directions. The translation subgroup contains $(E|0)$, $(E|\mathbf{a}_1)$, $(E|\mathbf{a}_2)$, and $(E|\mathbf{a}_1 + \mathbf{a}_2)$. Multiplying each of these in turn by $(C_2|0)$, $(\sigma_x|\mathbf{a}_1/2)$, and $(\sigma_y|-\mathbf{a}_1/2)$ gives 12 additional elements, and we obtain a 16 element group divided into four classes. We could, of course, have a structure with M unit cells in one direction and N in the other direction. This would just increase the size of the Cayley table without adding anything to our understanding of the fundamentals. A portion of the 16×16 table is shown as Table 1-17. We notice that this is *not* a direct product of the translation group and the point group, as was the case for the symmorphic group of Table 1-15. Otherwise, the structures are similar; that is, each row and each column contains a given element once only and each 4×4 block corresponds to products involving a single point symmetry operation E, σ_x, σ_y, or C_2 with the translation subgroup.

To introduce another important concept from formal group theory, we recall that the group D_{3h} of Table 1-2 has a number of subgroups, such as C_3 and C_{3v} (which are explicitly outlined) or the group E, σ_v^a. The subgroups C_3 and C_{3v} are seen to consist of complete classes—K_1, K_2 and K_1, K_2, K_3, respectively—and are called *invariant subgroups*. Hence, the two-element subgroup E, σ_v^a is not invariant but the subgroup E, σ_h is.

We also recall the coset decomposition of this group, as expressed in Eq. (1-8). The table shows that Eq. (1-8), filling in the unknowns A, B, and C, is equivalent to

$$(D_{3h}) = (C_3) + (C_3)(\sigma_v) + (C_3)(\sigma_h) + (C_3)(C_2) \tag{1-25}$$

where the notation $(C_3)(\sigma_v)$, for example, means the product of any member of class K_3 by any member of the group C_3. Since there are three elements in C_3 and three elements in class K_3, this product has nine possibilities, of which three are σ_v^a, three are σ_v^b, and three are σ_v^c. Using this same notation, Eq. (1-25) is also equivalent to

$$\begin{aligned}(D_{3h}) &= (C_3)(E) + (C_3)(\sigma_v) + (C_3)(\sigma_h) + (C_3)(C_2) \\ &= (C_3)[(E) + (\sigma_v) + (\sigma_h) + (C_2)]\end{aligned} \tag{1-26}$$

where (E) is any member of the subgroup C_3, since the coset of C_3 with any one of its own members is C_3, itself. Note that (C_3) and (E) both represent the group C_3 (or the elements E, C_3, C_3^2) but there is a good reason for this dual notation. Denoting each set of three elements in Table 1-2 as indicated

TABLE 1-17 ***Portion of the Cayley Table for a Non-symmorphic Group***

	$(\sigma_x\|\mathbf{a}_1/2)$	$(\sigma_x\|3\mathbf{a}_1/2)$	$(\sigma_x\|\mathbf{a}_1/2+\mathbf{a}_2)$	$(\sigma_x\|3\mathbf{a}_1/2+\mathbf{a}_2)$	$(\sigma_y\|-\mathbf{a}_1/2)$	$(\sigma_y\|\mathbf{a}_1/2)$	$(\sigma_y\|-\mathbf{a}_1/2+\mathbf{a}_2)$	$(\sigma_y\|\mathbf{a}_1/2+\mathbf{a}_2)$
$(\sigma_x\|\mathbf{a}_1/2)$	$(E\|\mathbf{a}_1)$	$(E\|0)$	$(E\|\mathbf{a}_1+\mathbf{a}_2)$	$(E\|\mathbf{a}_2)$	$C_2\|0)$	$(C_2\|\mathbf{a}_1)$	$(C_2\|\mathbf{a}_2)$	$(C_2\|\mathbf{a}_1+\mathbf{a}_2)$
$(\sigma_x\|3\mathbf{a}_1/2)$	$(E\|0)$	$(E\|\mathbf{a}_1)$	$(E\|\mathbf{a}_2)$	$(E\|\mathbf{a}_1+\mathbf{a}_2)$	$(C_2\|\mathbf{a}_1)$	$(C_2\|0)$	$(C_2\|\mathbf{a}_1+\mathbf{a}_2)$	$(C_2\|\mathbf{a}_2)$
$(\sigma_x\|\mathbf{a}_1/2+\mathbf{a}_2)$	$(E\|\mathbf{a}_1+\mathbf{a}_2)$	$(E\|\mathbf{a}_2)$	$(E\|\mathbf{a}_1)$	$(E\|0)$	$(C_2\|\mathbf{a}_2)$	$(C_2\|\mathbf{a}_1+\mathbf{a}_2)$	$(C_2\|0)$	$(C_2\|\mathbf{a}_1)$
$(\sigma_x\|3\mathbf{a}_1/2+\mathbf{a}_2)$	$(E\|\mathbf{a}_2)$	$(E\|\mathbf{a}_1+\mathbf{a}_2)$	$(E\|0)$	$(E\|\mathbf{a}_1)$	$(C_2\|\mathbf{a}_1+\mathbf{a}_2)$	$(C_2\|\mathbf{a}_2)$	$(C_2\|\mathbf{a}_1)$	$(C_2\|0)$

TABLE 1-18 ***Cayley Table in Terms of Cosets***

	(E)	(σ_v)	(σ_h)	(C_2)
(E)	(E)	(σ_v)	(σ_h)	(C_2)
(σ_v)	(σ_v)	(C_3)	(C_2)	(σ_h)
(σ_h)	(σ_h)	(C_2)	(C_3)	(σ_v)
(C_2)	(C_2)	(σ_h)	(σ_v)	(C_3)

in Eq. (1-26), we may condense it to the 4×4 form of Table 1-18, This table has the structure of the Cayley table for a four-element group, each member of which is either the subgroup C_3 or one of its cosets. In this group, the invariant subgroup C_3 plays the role of the unit element, which is why it is denoted by (E). A group like D_{3h} containing an invariant subgroup is called a *factor group*; thus, a factor group is any group which may be expressed in the *supergroup* form of Table 1-18. A notation commonly used is

$$[(E), (\sigma_v), (\sigma_h), (C_2)] = (D_{3h}/C_3)$$

which is why an alternate name is *quotient group*. However, the "factor" part of factor group seems more logical in view of Eq. (1-26), where C_3 is actually the common factor in the coset decomposition.

Problem 1-49 indicates that the translation subgroup of Table 1-15 is an invariant subgroup, so that this space group is also a factor group. The unit element is the group $(E|T)$ and the cosets are the collections $(X_1|T)$, $(X_2|T), \ldots$, where $E, X_1, X_2, \ldots$ is the underlying point group. We can give a simple example of how space groups are generated from invariant subgroups by considering Fig. 1-20(f). The infinite group corresponding to this figure is

$$(E|0)(\sigma_x|T)(\sigma_x|T)^2(\sigma_x|T)^3, \ldots$$

or

$$(E|0)(\sigma_x|T)(E|T^2)(\sigma_x|T^3), \ldots$$

Since T is equivalent to $a/2$, where a is the lattice spacing, the group becomes

$$(E|0), (\sigma_x|a/2), (E|a), (\sigma_x|3a/2), (E|2a), \ldots$$

which may be written in two parts as

$$(E|0), (E|a), (E|2a), \ldots$$

and

$$(\sigma_x|a/2), (\sigma_x|3a/2), \ldots$$

The first part is the monoperiodic translation group. The second part is the coset of this group and the glide operation $(\sigma_x \mid a/2)$, since

$$(E \mid a)(\sigma_x \mid a/2) = (\sigma_x \mid E(a/2) + a), \quad \text{etc.}$$

Hence, we may generate the nonsymmorphic group of Fig. 1-20(f) by forming the cosets of the translation group with the operator $(\sigma_x \mid a/2)$.

In order to see how this procedure works for a real crystal space group, let us apply it to the structure shown in Fig. 1-43. This is the lattice for either tellurium or one crystalline form of selenium, which are semiconducting elements lying in column VIa of the periodic table. The other members of this column are the insulators oxygen and sulfur, and polonium, a metal. Figure 1-44 shows a perspective view of the tellurium crystal. The atoms are arranged in spiral chains, with three atoms to each turn, so that the fourth atom lies directly over the first, the fifth is directly over the second, and so on. The chains are arranged to form hexagons, but the symmetry of the crys-

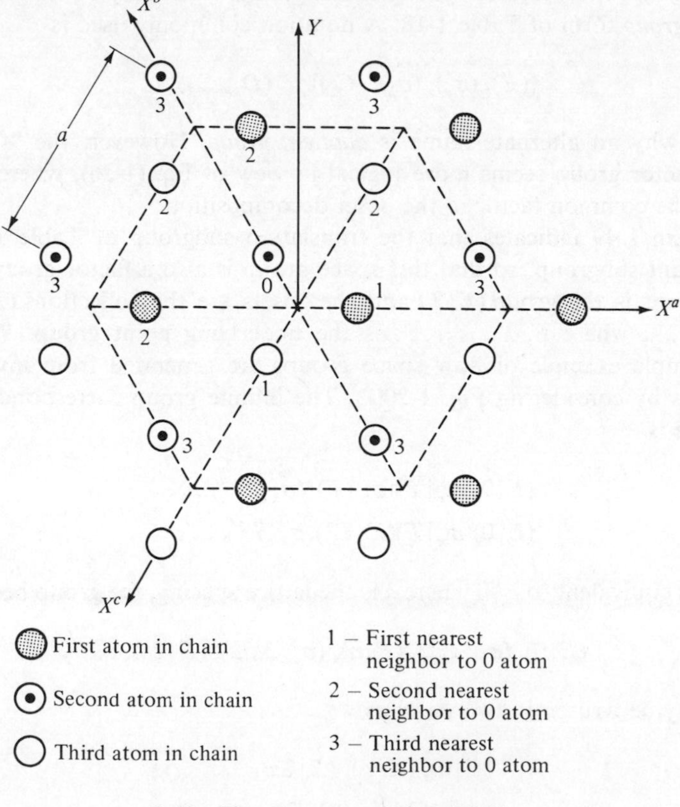

1-43 The space group D_4^3 or $P3_121$

tal is trigonal because of the threefold nature of the chains. The space group is designated as D_3^4 or $P3_121$ for the right-handed screw axis and D_3^6 or $P3_221$ for the left-handed form (both *enantiomorphic* crystal forms of tellurium have been grown). The symbol $P3_121$ denotes a primitive or simple trigonal lattice, containing a 3_1 screw axis and a two-fold dihedral axis. Normally, a hexagon has two classes of dihedral axis: a set of six through the centers of opposite sides and another six through pairs of opposite corners, explaining the notation 622 on the next-to-last line of Table 1-13. As the figure shows, the arrangement of atoms is symmetrical only with respect to the latter type of dihedral axis, so that we write $P3_121$ rather than $P3_122$.

Further, and this is more difficult to see, the structure appears to have the translational group of the simple hexagonal Bravais lattice rather than the simple trigonal, since there is a black dot, for example, just to the right of each corner and of the center of the hexagon. To explain this, we realize that Fig. 1-43 is essentially an X-ray pattern of a Te or Se single crystal. The seven atoms shown by dots in circles lie in the plane of the film, the seven shown as solid dots in a circle lie at a distance $c/3$ below the plane, where c is the lattice constant in the z-direction, and the seven open circles are at a distance $c/3$ above the base plane. As we go from dot to dot in the order indicated by the legend, we generate a screw axis 3_1, with the fourth atom over the first one and so on. Although all seven screw axes are identical in structure and orientation, their symmetry with respect to the hexagon divides itself into two sets. The three chains lying along the positive X^a, X^b, and X^c axes appear to us to be triangles pointing outward, whereas the other three point inward. Hence, the screw axis symmetry cannot be combined with the hexagonal translation group; instead, we have a trigonal crystal.

The symmetry operations which seem obvious from the figure are then $(E\,|\,0)$, $(C_3\,|\,c/3)$, and $(C_3^2\,|\,2c/3)$. Since the space group notation is based on the point group D_3 or 32, we also expect to find three dihedral axes.

PROBLEM 1-50

Show that three of the symmetry operations of the tellurium lattice consist of 180° rotations about the axes X^a, X^b, or X^c followed by translations $nc/3$, where c is the lattice constant along the trigonal axis and $n = 0, 1,$ or 2.

As the problem shows, the dihedral operations, although not screw axes, also involve translations. We write them as $(C_2\,|\,c/2)$, $(C_2\,|\,0)$, and $C_2(2c/3)$ about X^a, X^b, and X^c, respectively. Hence, we would expect the space group to contain six operations, all derived from the corresponding point group D_3. These are:

1. E, the identity.
2. $(C_3 | c/3)$, a 120° rotation about the z-axis, followed by a translation $c/3$ along the axis.
3. $(C_3 | c/3)^2$.
4. $(C_2^a | c/3)$, a 180° rotation about the axis X^a of Fig. 1-43, followed by a translation $c/3$.
5. $(C_2^b | 0)$, a 180° rotation about X^b, followed by a translation $0c/3$.
6. $(C_2^c | 2c/3)$, a 180° rotation about X^c, followed by a translation $2c/3$.

These six operations do not form a group, however, for if we multiply $(C_3 | c/3)$ and $(C_3^2 | 2c/3)$ in either order, we obtain

$$(C_3 | c/3)(C_3^2 | 2c/3) = (E | c)$$

which is a pure translation of one lattice distance c along the z-axis, as we would expect. This translation $(E | c)$ has a peculiar property: it is its own inverse. To show why, consider Fig. 1-45 which represents three consecutive atoms in a chain, with the three levels in the z-direction labeled 0, 1, and 2. This denotes their height above the xy-plane as 0, $c/3$ and $2c/3$, respectively. The corresponding atoms are labelled A, B, and C. The result of $(E | c)$ is shown in Fig. 1-45(b), where the new levels are denoted 3, 4, and 5 to indicate

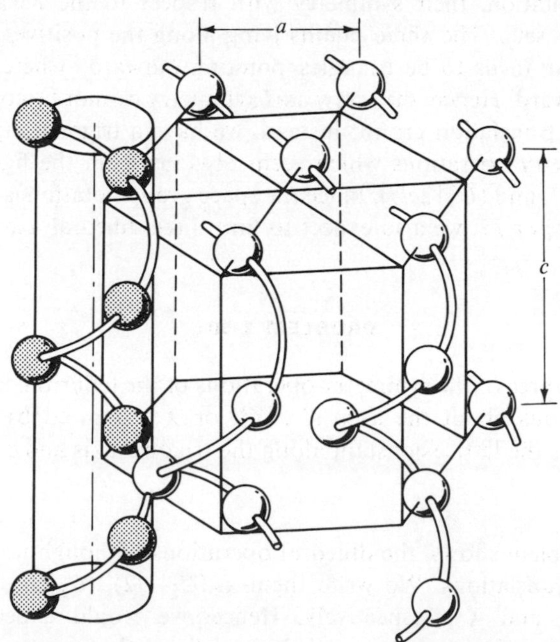

1-44 Perspective view of the tellurium crystal

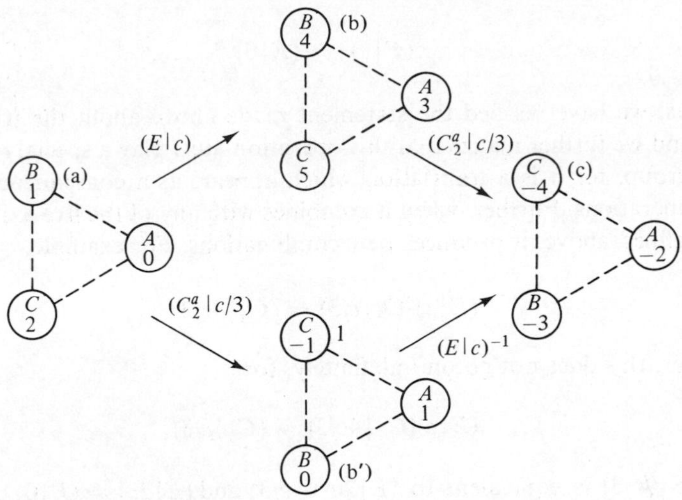

1-45 Demonstration of the periodicity of tellurium along the z direction

that each atom has moved a distance $3c/3$ along the axis. Applying $(C_2^a | c/3)$ gives the result shown in Fig. 1-45(c). Now proceeding in the reverse order, $(C_2^a | c/3)$ produces Fig. 1-45(b′) and the inverse $(E | c)^{-1} = (E | -c)$ once again gives Fig. 1-45(c), so that

$$(C_2^a | c/3)(E | c) = (E | -c)(C_2^a | c/3) \tag{1-27}$$

Multiplying this from the left by $(E | c)$ gives

$$(E | c)(C_2^a | c/3)(E | c) = (C_2^a | c/3) \tag{1-28}$$

and multiplying from the right by $(E | -c)$ gives

$$(C_2^a | c/3) = (E | -c)(C_2^a | c/3)(E | -c) \tag{1-29}$$

so that

$$(E | c)(C_2^a | c/3)(E | c) = (E | -c)(C_2^a | c/3)(E | -c) \tag{1-30}$$

PROBLEM 1-51

Show that (1-30) is also a consequence of the fact that

$$(E | c) = (E | c)^{-1} \tag{1-31}$$

From (1-31), we see that

$$(E\,|\,c)^2 = (E\,|\,0) \tag{1-32}$$

Thus, we have verified the statement made above about the translation $(E\,|\,c)$ and we further realize that this operation must play a special role in the space group, for it is a translation which appears as a consequence of two screw operations. Further, when it combines with any of the five axial operations defined above, it produces new combinations. For example,

$$(E\,|\,c)(C_3\,|\,c/3) = (C_3\,|\,4c/3)$$

However, this does not go on indefinitely, for

$$(E\,|\,c)(C_3\,|\,4c/3) = (C_3\,|\,c/3)$$

since $(E\,|\,7c/3)$ is equivalent to $(E\,|\,2c + c/3)$ and $(E\,|\,2c) = (E\,|\,0)$ by (1-32). Proceeding in this fashion, we find that a total of 12 elements are generated and these form a group of six classes, as shown in Table 1-19. To save space, the translations are indicated in units of $c/3$, as was done with Fig. 1-45. Although this is *not* the complete tellurium space group, it would be if we were to add translations in the xy-plane. We may conveniently choose them as the lattice distance a along the X^a and X^b directions. Denoting these translations by T^a and T^b, respectively, we find, for example, that

$$(C_3\,|\,c/3)(E\,|\,T^a) = (C_3\,|\,T^b + c/3)$$

and

$$(C_3\,|\,c/3)(E\,|\,T^b) = (C_3\,|\,-T^a-T^b + c/3)$$

Then, using two cells in each direction as before, we would generate a multiplication table similar to, but larger than, Table 1-16. That is, the translation subgroup is an invariant subgroup and the space group has the form of a factor group. The coset decomposition is

$$\begin{aligned} D_3^4 &= P3_121 \\ &= \{(E\,|\,\mathbf{T})\}[\{(E\,|\,\mathbf{T})\} + \{(C_3\,|\,c/3)\} + \{(C_3^2\,|\,2c/3)\} \\ &\quad + \{(C_2^a\,|\,c/3)\} + \{(C_2^b\,|\,0)\} + \{(C_2^c\,|\,2c/3)\}] \end{aligned} \tag{1-33}$$

where $\{(E\,|\,\mathbf{T})\}$ is used to denote the translation subgroup, containing translations $\mathbf{a}_1, \mathbf{a}_2, \mathbf{c}, \mathbf{a}_1 + \mathbf{a}_2, \mathbf{a}_2 + \mathbf{c}, \mathbf{c} + \mathbf{a}_1, \mathbf{a}_1 + \mathbf{a}_2 + \mathbf{c}$ and the other symbols are the cosets. Thus we see that all space groups, symmorphic or nonsymmorphic, are also factor groups and this is of extreme importance not only for applications to come later but because it enables us to generate all

TABLE 1-19 *Partial Cayley Table for Tellurium*

	$(E\|0)$	$(E\|3)$	$(C_3\|1)$	$(C_3^2\|5)$	$(C_3^2\|2)$	$(C_3\|4)$	$(C_2^a\|1)$	$(C_2^b\|3)$	$(C_2^c\|5)$	$(C_2^a\|4)$	$(C_2^b\|0)$	$(C_2^c\|2)$
$(E\|0)$	$(E\|0)$	$(E\|3)$	$(C_3\|1)$	$(C_3^2\|5)$	$(C_3^2\|2)$	$(C_3\|4)$	$(C_2^a\|1)$	$(C_2^b\|3)$	$(C_2^c\|5)$	$(C_2^a\|4)$	$(C_2^b\|0)$	$(C_2^c\|2)$
$(E\|3)$	$(E\|3)$	$(E\|0)$	$(C_3\|4)$	$(C_3^2\|2)$	$(C_3^2\|5)$	$(C_3\|1)$	$(C_2^a\|4)$	$(C_2^b\|0)$	$(C_2^c\|2)$	$(C_2^a\|1)$	$(C_2^b\|3)$	$(C_2^c\|5)$
$(C_3\|1)$	$(C_3\|1)$	$(C_3\|4)$	$(C_3^2\|2)$	$(E\|0)$	$(E\|3)$	$(C_3^2\|5)$	$(C_2^c\|2)$	$(C_2^a\|4)$	$(C_2^b\|0)$	$(C_2^c\|5)$	$(C_2^a\|1)$	$(C_2^b\|3)$
$(C_3^2\|5)$	$(C_3^2\|5)$	$(C_3^2\|2)$	$(E\|0)$	$(C_3\|4)$	$(C_3\|1)$	$(E\|3)$	$(C_2^b\|0)$	$(C_2^c\|2)$	$(C_2^a\|4)$	$(C_2^b\|3)$	$(C_2^c\|5)$	$(C_2^a\|1)$
$(C_3^2\|2)$	$(C_3^2\|2)$	$(C_3^2\|5)$	$(E\|3)$	$(C_3\|1)$	$(C_3\|4)$	$(E\|0)$	$(C_2^b\|3)$	$(C_2^c\|5)$	$(C_2^a\|1)$	$(C_2^b\|0)$	$(C_2^c\|2)$	$(C_2^a\|4)$
$(C_3\|4)$	$(C_3\|4)$	$(C_3\|1)$	$(C_3^2\|5)$	$(E\|3)$	$(E\|0)$	$(C_3^2\|2)$	$(C_2^c\|5)$	$(C_2^a\|1)$	$(C_2^b\|3)$	$(C_2^c\|2)$	$(C_2^a\|4)$	$(C_2^b\|0)$
$(C_2^a\|1)$	$(C_2^a\|1)$	$(C_2^a\|4)$	$(C_2^b\|0)$	$(C_2^c\|2)$	$(C_2^c\|5)$	$(C_2^b\|3)$	$(E\|0)$	$(C_3\|4)$	$(C_3^2\|2)$	$(E\|3)$	$(C_3\|1)$	$(C_3^2\|5)$
$(C_2^b\|3)$	$(C_2^b\|3)$	$(C_2^b\|0)$	$(C_2^c\|2)$	$(C_2^a\|4)$	$(C_2^a\|1)$	$(C_2^c\|5)$	$(C_3^2\|2)$	$(E\|0)$	$(C_3\|4)$	$(C_3^2\|5)$	$(E\|3)$	$(C_3\|1)$
$(C_2^c\|5)$	$(C_2^c\|5)$	$(C_2^c\|2)$	$(C_2^a\|4)$	$(C_2^b\|0)$	$(C_2^b\|3)$	$(C_2^a\|1)$	$(C_3\|4)$	$(C_3^2\|2)$	$(E\|0)$	$(C_3\|1)$	$(C_3^2\|5)$	$(E\|3)$
$(C_2^a\|4)$	$(C_2^a\|4)$	$(C_2^a\|1)$	$(C_2^b\|3)$	$(C_2^c\|5)$	$(C_2^c\|2)$	$(C_2^b\|0)$	$(E\|3)$	$(C_3\|1)$	$(C_3^2\|5)$	$(E\|0)$	$(C_3\|4)$	$(C_3^2\|2)$
$(C_2^b\|0)$	$(C_2^b\|0)$	$(C_2^b\|3)$	$(C_2^c\|5)$	$(C_2^a\|1)$	$(C_2^a\|4)$	$(C_2^c\|2)$	$(C_3^2\|5)$	$(E\|3)$	$(C_3\|1)$	$(C_3^2\|2)$	$(E\|0)$	$(C_3\|4)$
$(C_2^c\|2)$	$(C_2^c\|2)$	$(C_2^c\|5)$	$(C_2^a\|1)$	$(C_2^b\|3)$	$(C_2^b\|0)$	$(C_2^a\|4)$	$(C_3\|1)$	$(C_3^2\|5)$	$(E\|3)$	$(C_3\|4)$	$(C_3^2\|2)$	$(E\|0)$
	K_1	K_2	K_3		K_4		K_5			K_6		

possible space groups from compatible combinations of point groups, translation groups, and the screw or glide operations. The construction of the 230 groups in this fashion, presented in a condensed and logical way, will be found in Jaswon.[4]

We should also remark here why we bother with Table 1-19, which is only a subgroup of the complete space group for tellurium, and not a factor group. We have included this table for two reasons: it shows how to combine compound symmetry operations in an actual crystal and it will be seen later, when we study the application of group theory to the electronic properties of solids, to specify the properties of the material along the principal symmetry axis in reciprocal space, where the term "reciprocal space" will be explained at the beginning of Chapter 5.

1-11 Final remarks

Let us conclude this chapter by considering the explanation that Weinreich[10] has given of the connection between the symmetry of a crystal lattice and the symmetry of the atomic arrangement at each lattice point. It is customary to call the atomic grouping at each lattice point the basis of the crystal, because it is the repeated translation of this basis which generates the actual crystal. We note that the basis may in fact be only a single atom, and while an isolated atom is spherically symmetrical, one which is part of a crystal generally will not be. That is, the orbits of the valence electrons tend to distort along the directions to neighboring atoms. Sodium in NaCl, for example, goes from spherical to cubic symmetry, and we shall see the theoretical reason for this in a later chapter. The same is true for the tellurium crystal of Fig. 1-44. In fact, we have shown in Fig. 1-43 that a given atom in this lattice has two nearest neighbors and four second-nearest neighbors. Since column VI atoms have six valence electrons, we would expect bonding by the sharing of electron pairs and these bonds have directional properties.

There is, of course, no reason why spheres cannot be arranged in a triclinic lattice or identical asymmetric objects in a cubic lattice and, as we have already pointed out, it is the overall point symmetry which determines the crystal system, so that both of these crystals are triclinic. However, we would expect the symmetry of the interatomic forces, that is, the basis symmetry, to determine the lattice symmetry. For example, a monoclinic lattice must have a two-fold axis at the very least (Table 1-12) and it could have a mirror plane normal to the C_2-axis (Table 1-13). The lattice does not distort to triclinic because of the symmetry of these forces. Hence, we can say that the basis can have *no* symmetry element not possessed by the lattice and it must have *at least one* not possessed by a less symmetrical lattice. By "less symmetrical," we mean lattices related as indicated in Fig. 1-35. In the language of group theory, this statement is equivalent to saying that *the point group of the basis must be a subgroup of the point group of its lattice* (*where a subgroup may be the group itself*) *but not a subgroup of any less symmetrical lattice.*

Another aspect of the application of group theory to crystallography which often puzzles students is the existence of a criterion (other than intuition) for recognizing if a collection of elements actually forms a group. On looking over the many examples of linear, plane, and three-dimensional space groups covered in this chapter, we appreciate that the underlying common feature is that *all repeated elements in each pattern find themselves in an identical environment.* The X in the dotted rectangle of Fig. 1-26(g), for instance, is surrounded by a blank rhombus in the $\pm x$-, $\pm y$-directions and by an identical X along its four legs. The same principle holds for every asymmetric arrow of the 17 patterns in Fig. 1-25. But in Fig. 1-46, this is not

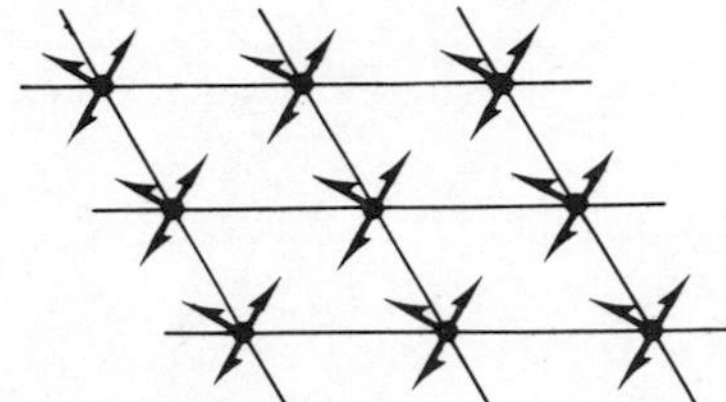

1-46 A lattice with a basis which may be inconsistent

the case if we regard the single arrow as the repeated pattern. If we discard one arrow, we then have an oblique lattice with two-fold symmetry; if we treat all three arrows as a single element, the oblique lattice has one-fold symmetry. This principle also explains why the orientation of the basis in Fig. 1-25(j), for example, is arbitrary, but in Fig. 1-25(k), the repeated pattern must be oriented at an angle of 45° to the lattice. The relation of these patterns to the lattice is also seen to obey the condition of Weinreich as quoted above.

As a last example of the problems associated with an understanding of point, translation, and space groups, let us return to Tables 1-11 and 1-12. We obtained the Bravais lattices and the systems enumerated in these tables by constructing three-dimensional combinations of the plane lattices. No mention was made, however, of possibilities not listed in the middle column of Table 1-11. There is no base-centered cubic or face-centered tetragonal entry, to name just two. The reason for this apparent oversight is that these are not new structures; they are identical to one of the fourteen Bravais lattices already given. Weinreich shows with some excellent drawings that the simple, base-centered, body-centered, and face-centered triclinic lattices can be converted into one another by proper choice of the unit cell. The problem which follows extends this idea to other systems.

PROBLEM 1-52

(a) Show that the face-centered and body-centered tetragonal lattices are identical.

(b) To what lattice is the base-centered tetragonal equivalent, where the square faces are the ones which are centered?

(c) Is it possible to have a hexagonal lattice with $a = c$ and is it possible to have a tetragonal lattice with the rectangular faces centered? (*Hint:* The criteria of Table 1-12 apply to both parts of this question.)

CHAPTER TWO REPRESENTATIONS AND VIBRATIONS

2-1 A review of matrix algebra

Although we shall assume that the reader is familiar with the concept of a matrix and the way in which matrices are multiplied, let us briefly review the essential ideas of matrix algebra. Consider the following set of simultaneous equations

$$\begin{aligned} z_1 &= a_{11}y_1 + a_{12}y_2 + a_{13}y_3 \\ z_2 &= a_{21}y_1 + a_{22}y_2 + a_{23}y_3 \\ z_3 &= a_{31}y_1 + a_{32}y_2 + a_{33}y_3 \end{aligned} \tag{2-1}$$

connecting three variables y_1, y_2, y_3 with three other variables z_1, z_2, z_3 through the use of nine constants $a_{11}, a_{12}, \ldots, a_{33}$. This set of equations may be written in the briefer form

$$\begin{pmatrix} z_1 \\ z_2 \\ z_3 \end{pmatrix} = \begin{pmatrix} a_{11} & a_{12} & a_{13} \\ a_{21} & a_{22} & a_{23} \\ a_{31} & a_{32} & a_{33} \end{pmatrix} \begin{pmatrix} y_1 \\ y_2 \\ y_3 \end{pmatrix} \tag{2-2}$$

where the symbol

$$\mathbf{A} = \begin{pmatrix} a_{11} & a_{12} & a_{13} \\ a_{21} & a_{22} & a_{23} \\ a_{31} & a_{32} & a_{33} \end{pmatrix} \tag{2-3}$$

is called a 3×3 *matrix* and the symbols

$$\mathbf{z} = \begin{pmatrix} z_1 \\ z_2 \\ z_3 \end{pmatrix}, \qquad \mathbf{y} = \begin{pmatrix} y_1 \\ y_2 \\ y_3 \end{pmatrix} \tag{2-4}$$

are 3×1 *matrices*. Equation (2-2) will be equivalent to the three equations (2-1) if we define matrix multiplication by the following rule: to find the element in row i and column j of the product matrix, multiply the members of column i in the first matrix in the product by the corresponding members of row j in the second matrix being multiplied and sum these products. For example, the term z_2 in (2-2) is in row 2, column 1. Since row 2 of the first matrix is

$$a_{21} \quad a_{22} \quad a_{23}$$

and column 1 (the only column, in fact) of the second matrix is

$$\begin{matrix} y_1 \\ y_2 \\ y_3 \end{matrix}$$

the rule tells us that the z_2 is the sum of $a_{21}y_1$, $a_{22}y_2$, and $a_{23}y_3$, so that

$$z_2 = a_{21}y_1 + a_{22}y_2 + a_{23}y_3 \tag{2-5}$$

This result agrees with the second equation of the set (2-1).

Using the definitions (2-3) and (2-4), Eqs. (2-1) can be written very compactly as

$$\mathbf{z} = \mathbf{Ay} \tag{2-6}$$

and this is one advantage of matrix notation. Another comes when we consider a second set of relations like (2-1) connecting y_1, y_2, y_3 with a new group of variables x_1, x_2, x_3; that is, let

$$\begin{aligned} y_1 &= b_{11}x_1 + b_{12}x_2 + b_{13}x_3 \\ y_2 &= b_{21}x_1 + b_{22}x_2 + b_{23}x_3 \\ y_3 &= b_{31}x_1 + b_{32}x_2 + b_{33}x_3 \end{aligned} \tag{2-7}$$

In matrix notation, this may be written as

$$\mathbf{y} = \mathbf{Bx} \tag{2-8}$$

Then we may relate the variables x_1, x_2, x_3 to the variables z_1, z_2, z_3 by substituting (2-8) into (2-6), obtaining

$$\mathbf{ABx} = \mathbf{Cx} \tag{2-9}$$

where

$$\mathbf{C} = \mathbf{AB}$$

is the product of two 3×3 matrices. This product may be easily obtained by extending the rule given above for matrix multiplication. That is, the element c_{ij} in the product is obtained from row i of A

$$a_{i1} \quad a_{i2} \quad a_{i3}$$

and column j of B

$$\begin{matrix} b_{1j} \\ b_{2j} \\ b_{3j} \end{matrix}$$

to give

$$c_{ij} = a_{i1}b_{1j} + a_{i2}b_{2j} + a_{i3}b_{3j} \tag{2-10}$$

Hence, (2-9) expresses in a very simple way the complicated expressions which would result if Eqs. (2-7) were substituted into each of the Eqs. (2-1).

PROBLEM 2-1

(a) Show that

$$\begin{pmatrix} 1 & 4 & 7 \\ 2 & 0 & -3 \\ 4 & 1 & 1 \end{pmatrix}\begin{pmatrix} 2 & 3 & 8 \\ 1 & 2 & 0 \\ -1 & 0 & 1 \end{pmatrix} = \begin{pmatrix} -1 & 11 & 15 \\ 7 & 6 & 13 \\ 8 & 14 & 33 \end{pmatrix}$$

(b) Verify that the rule for matrix multiplication may be expressed as

$$c_{ij} = \sum_{k=1}^{3} a_{ik}b_{kj}, \quad (i, j = 1, 2, 3) \tag{2-11}$$

2-2 Eigenvectors and the diagonalization of matrices

If the product of a 2×2 matrix **M** and a 2×1 matrix **V** is equal to **V** times a constant λ, then **V** is called an *eigenvector* of **M** and λ is called an *eigenvalue* of **V** with respect to **M** (where the term *eigen* is German for *characteristic*). This is symbolized as

$$\mathbf{MV} = \lambda\mathbf{V}$$

or

$$\begin{pmatrix} m_{11} & m_{12} \\ m_{21} & m_{22} \end{pmatrix}\begin{pmatrix} v_1 \\ v_2 \end{pmatrix} = \lambda\begin{pmatrix} v_1 \\ v_2 \end{pmatrix} \tag{2-12}$$

The matrix **V** is called an eigenvector since it is specified by two numbers v_1 and v_2 which may be regarded as the components of a two-dimensional vector. If we write (2-12) as

$$\mathbf{MV} = \lambda\mathbf{V} \tag{2-13}$$

where **V** is taken as a vector, then (2-13) states that the effect of **M** on **V** is to stretch or contract **V** without changing its direction. This is a physical meaning of the concept of the eigenvector of a matrix, and the eigenvalue λ measures the change in length of **V**.

Equation (2-12) may be rewritten in the form

$$\begin{pmatrix} m_{11} - \lambda & m_{12} \\ m_{21} & m_{22} - \lambda \end{pmatrix} \begin{pmatrix} v_1 \\ v_2 \end{pmatrix} = 0 \tag{2-14}$$

or

$$\begin{aligned} (m_{11} - \lambda)v_1 + m_{12}v_2 &= 0 \\ m_{21}v_1 + (m_{22} - \lambda)v_2 &= 0 \end{aligned} \tag{2-15}$$

This pair of homogeneous equations has a solution only when the determinant of the coefficients vanishes, or

$$\begin{vmatrix} m_{11} - \lambda & m_{12} \\ m_{21} & m_{22} - \lambda \end{vmatrix} = 0 \tag{2-16}$$

The relation (2-16) is called a *secular* equation; the term comes from astronomy, and denotes long-term, or secular, variations in orbits. Equation (2-16) may also be written as

$$|\mathbf{M} - \lambda \mathbf{I}| = 0 \tag{2-17}$$

where **I** is the *unit matrix*

$$\mathbf{I} = \begin{pmatrix} 1 & 0 \\ 0 & 1 \end{pmatrix} \tag{2-18}$$

As an example, consider the matrix

$$\mathbf{M} = \begin{pmatrix} 0 & 2 \\ 2 & 3 \end{pmatrix} \tag{2-19}$$

Then the secular equation is

$$\begin{vmatrix} 0 - \lambda & 2 \\ 2 & 3 - \lambda \end{vmatrix} = 0$$

or

$$\lambda^2 - 3\lambda - 4 = 0 \tag{2-20}$$

with solutions

$$\lambda_1 = 4, \qquad \lambda_2 = -1 \tag{2-21}$$

Using λ_1 in (2-15) gives

$$\begin{aligned} -4v_1 + 2v_2 &= 0 \\ 2v_1 - v_2 &= 0 \end{aligned}$$

Either of these equations shows that

$$\frac{v_1}{v_2} = \frac{1}{2}$$

and this is all that we can determine about the eigenvector. That is, the ratio of components gives the direction but not the magnitude of the vector. Hence, we arbitrarily require that the eigenvectors have unit length. This *normalization* condition is then

$$v_1^2 + v_2^2 = 1 \tag{2-22}$$

from which $$v_1^2 + 4v_1^2 = 1 \quad \text{or} \quad v_1 = \frac{1}{\sqrt{5}}$$

Then $$\mathbf{V}_1 = \frac{1}{\sqrt{5}}\begin{pmatrix} 1 \\ 2 \end{pmatrix} \tag{2-23}$$

Similarly, $$\mathbf{V}_2 = \frac{1}{\sqrt{5}}\begin{pmatrix} -2 \\ 1 \end{pmatrix} \tag{2-24}$$

We note that $\mathbf{V}_1$ and $\mathbf{V}_2$ are *orthogonal*, or at right angles, for

$$\mathbf{V}_1 \cdot \mathbf{V}_2 = \frac{(1)(-2) + (2)(1)}{5} = 0$$

This is characteristic of the eigenvectors of a real, symmetric matrix.

Now let us form a matrix $\mathbf{S}$ from the two vectors $\mathbf{V}_1$ and $\mathbf{V}_2$ by using the components of these vectors as the columns. That is,

$$\mathbf{S} = \frac{1}{\sqrt{5}}\begin{pmatrix} 1 & -2 \\ 2 & 1 \end{pmatrix} \tag{2-25}$$

We note that the *inverse* $\mathbf{S}^{-1}$ of $\mathbf{S}$ is obtained by using $\mathbf{V}_1$ and $\mathbf{V}_2$ as the rows of a matrix; that is,

$$\mathbf{S}^{-1} = \frac{1}{\sqrt{5}}\begin{pmatrix} 1 & 2 \\ -2 & 1 \end{pmatrix} \tag{2-26}$$

for $$\mathbf{S}^{-1}\mathbf{S} = \mathbf{S}\mathbf{S}^{-1} = \begin{pmatrix} 1 & 0 \\ 0 & 1 \end{pmatrix} = \mathbf{I} \tag{2-27}$$

The matrix $\mathbf{M}'$ obtained from $\mathbf{M}$ through the *similarity transformation*

$$\mathbf{M}' = \mathbf{S}^{-1}\mathbf{M}\mathbf{S} \tag{2-28}$$

is $$\mathbf{M}' = \frac{1}{\sqrt{5}}\begin{pmatrix} 1 & 2 \\ -2 & 1 \end{pmatrix}\begin{pmatrix} 0 & 2 \\ 2 & 3 \end{pmatrix}\frac{1}{\sqrt{5}}\begin{pmatrix} 1 & -2 \\ 2 & 1 \end{pmatrix} = \begin{pmatrix} 4 & 0 \\ 0 & -1 \end{pmatrix} \tag{2-29}$$

The matrix **M** is said to be *diagonalized* by the similarity transformation (2-28), and the diagonal elements are the eigenvalues λ_1 and λ_2 of (2-21). Hence, the diagonalization of a symmetric matrix is equivalent to solving the eigenvalue equation.

PROBLEM 2-2

Given the matrix

$$\mathbf{T} = \tfrac{1}{6}\begin{pmatrix} 11 & -1 & -4 \\ -1 & 11 & -4 \\ -4 & -4 & 14 \end{pmatrix} \tag{2-30}$$

(a) Find the eigenvalues. (*Hint:* They are integers.)
(b) Find a right-handed set of orthogonal eigenvectors.

2-3 The representation of a group

The point group C_{3v} was generated in the previous chapter by starting with the three operations E, C_3 and σ_v^a; forming all possible combinations, such as C_3^2 or $\sigma_v^a C_3$, we obtain the remainder of the group. Another way of generating this group involves expressing each symmetry operation as a matrix.

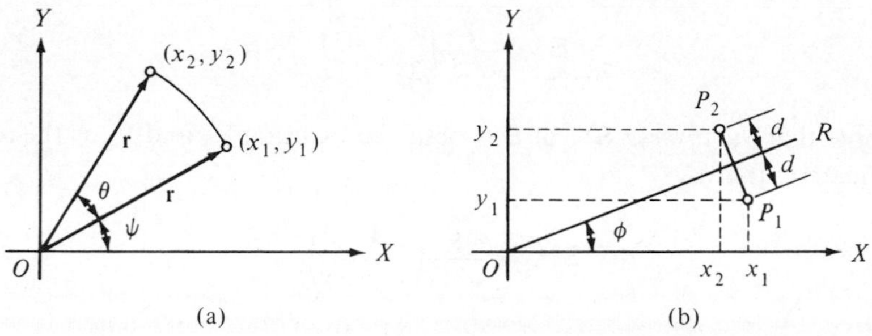

2-1 (a) Calculation of a rotation matrix; (b) calculation of a reflection matrix

To see how this is done, consider the vector **r** [Fig. 2-1(a)] whose initial point is the origin, whose terminal point is (x_1, y_1), and which makes an angle ψ with respect to the x-axis. If **r** is pivoted about O through an angle θ so that its endpoint moves to (x_2, y_2), the coordinates of the old and new positions of this terminal point are related by the equations

$$\begin{aligned} x_2 &= x_1 \cos\theta - y_1 \sin\theta \\ y_2 &= x_1 \sin\theta + y_1 \cos\theta \end{aligned} \tag{2-31}$$

These relations are established by multiplying the trigonometric identities

$$\begin{aligned} \cos(\theta+\psi) &= \cos\theta\cos\psi - \sin\theta\sin\psi \\ \sin(\theta+\psi) &= \sin\theta\cos\psi + \cos\theta\sin\psi \end{aligned} \tag{2-32}$$

by the magnitude r of $\mathbf{r}$; Eqs. (2-32) then become identical to Eqs. (2-31).

PROBLEM 2-3

Prove the statement just above.

We simplify Eqs. (2-31) by writing them in matrix form as

$$\begin{pmatrix} x_2 \\ y_2 \end{pmatrix} = \begin{pmatrix} \cos\theta & -\sin\theta \\ \sin\theta & \cos\theta \end{pmatrix} \begin{pmatrix} x_1 \\ y_2 \end{pmatrix} \tag{2-33}$$

Next, consider a point P_1 with coordinates (x_1, y_1) at a distance d from a line OR, as shown in Fig. 2-1(b). Let $P_2(x_2, y_2)$ be the reflection of P_1 in the line. The coordinates of P_1 and P_2 are related by

$$\begin{aligned} x_2 &= x_1 + 2d\sin\phi \\ y_2 &= y_1 - 2d\cos\phi \end{aligned}$$

from which

$$\begin{pmatrix} x_2 \\ y_2 \end{pmatrix} = \begin{pmatrix} \cos 2\phi & \sin 2\phi \\ \sin 2\phi & -\cos 2\phi \end{pmatrix} \begin{pmatrix} x_1 \\ y_1 \end{pmatrix} \tag{2-34}$$

PROBLEM 2-4

Prove (2-34).

We may find the explicit form of the matrices corresponding to the group C_{3v} from Eqs. (2-33) and (2-34). The identity operation E is equivalent to a rotation $\theta = 0$, so that

$$\cos\theta = 1, \qquad \sin\theta = 0$$

and
$$E = \begin{pmatrix} 1 & 0 \\ 0 & 1 \end{pmatrix} \tag{2-35}$$

That is, the unit matrix of (2-18) specifies E, as we would expect. Similarly, for $\theta = 120°$, we have

$$\sin\theta = \frac{\sqrt{3}}{2}, \qquad \cos\theta = -\frac{1}{2}$$

giving
$$C_3 = \begin{pmatrix} -\frac{1}{2} & -\frac{\sqrt{3}}{2} \\ \frac{\sqrt{3}}{2} & -\frac{1}{2} \end{pmatrix} \tag{2-36}$$

To identify the angle ϕ associated with each of the operations σ_v^a, σ_v^b, σ_v^c, we must superimpose axes on the prism of Fig. 1-4. Locating them as shown in Fig. 2-2, we see that the angles and operations correspond as indicated by Table 2-1.

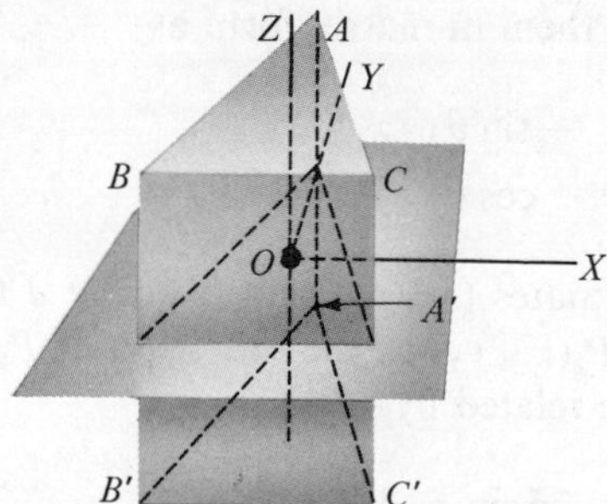

2-2 The prism belonging to the group C_{3v}

TABLE 2-1 ***Angle Corresponding to Reflection Operations in the Group*** C_{3v}

Operation	ϕ	2ϕ	$\cos 2\phi$	$\sin 2\phi$
σ_v^a	90°	180°	-1	0
σ_v^b	210°	420°	$\frac{1}{2}$	$\frac{\sqrt{3}}{2}$
σ_v^c	330°	660°	$\frac{1}{2}$	$-\frac{\sqrt{3}}{2}$

The matrices are then

$$\sigma_v^a = \begin{pmatrix} -1 & 0 \\ 0 & 1 \end{pmatrix} \tag{2-37}$$

$$\sigma_v^b = \begin{pmatrix} \frac{1}{2} & \frac{\sqrt{3}}{2} \\ \frac{\sqrt{3}}{2} & -\frac{1}{2} \end{pmatrix} \tag{2-38}$$

$$\sigma_v^c = \begin{pmatrix} \frac{1}{2} & -\frac{\sqrt{3}}{2} \\ -\frac{\sqrt{3}}{2} & -\frac{1}{2} \end{pmatrix} \tag{2-39}$$

Let us emphasize two points mentioned in Chapter 1 which are often the source of confusion: one is that the symbols σ_v^a, σ_v^b, and σ_v^c denote *both* the reflection planes and the associated operations, and the other is that the planes are *fixed* in space, and do not move with the prism.

PROBLEM 2-5

(a) Show in two ways that

$$C_3^2 = \begin{pmatrix} -\frac{1}{2} & \frac{\sqrt{3}}{2} \\ -\frac{\sqrt{3}}{2} & -\frac{1}{2} \end{pmatrix} \tag{2-40}$$

(b) Verify that the matrices of Eqs. (2-36), (2-37), and (2-38), obey the relation

$$\sigma_v^a = C_3\sigma_v^b \tag{2-41}$$

Equation (2-41) agrees with the entry in Table 2-1, and this result is typical of the product of any two of the matrices given above. In other words, the matrices multiply in the same way as the operations themselves. Any collection of quantities which obey the Cayley table of a group is called a *representation* of the group. The six 2×2 matrices of Eqs. (2-35) through (2-40) thus form a two-dimensional representation. A one-dimensional representation which appears to be trivial is

$$E = 1, \quad C_3 = 1, \quad C_3^2 = 1, \quad \sigma_v^a = 1, \quad \sigma_v^b = 1, \quad \sigma_v^c = 1 \tag{2-42}$$

where these numbers may be regarded as 1×1 matrices. This is called an *unfaithful representation* and is actually quite important, as we shall see. Another one, slightly more faithful, is

$$E = 1, \quad C_3 = 1, \quad C_3^2 = 1, \quad \sigma_v^a = -1, \quad \sigma_v^b = -1, \quad \sigma_v^c = -1 \tag{2-43}$$

The representation of Eqs. (2-42) is customarily labelled $\mathbf{\Gamma}_1$ (for a reason to be given in a later chapter), that of (2-43) is $\mathbf{\Gamma}_2$, and the matrices of (2-35) through (2-40) are denoted $\mathbf{\Gamma}_3$. These three are listed in Table 2-2.

It is also possible to find representations of higher dimensions. Suppose we specify the position of the corners of an equilateral triangle by the set of six coordinates $\mathbf{r}_1, \mathbf{r}_2, \mathbf{r}_3, \alpha_1, \alpha_2, \alpha_3$ of Fig. 2-3. The symmetry operations

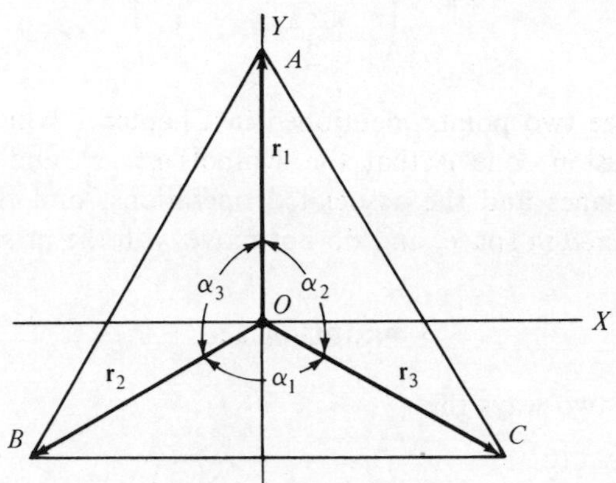

2-3 A six-dimensional basis

above can be then represented as 6×6 matrices, and the coordinates r_i and α_i ($i = 1, 2, 3$) are said to form a *basis*[1] for this representation. For example, the effect of σ_v^b (a reflection in the line OB) can be denoted by

$$\begin{pmatrix} 0 & 0 & 1 & 0 & 0 & 0 \\ 0 & 1 & 0 & 0 & 0 & 0 \\ 1 & 0 & 0 & 0 & 0 & 0 \\ 0 & 0 & 0 & 0 & 0 & 1 \\ 0 & 0 & 0 & 0 & 1 & 0 \\ 0 & 0 & 0 & 1 & 0 & 0 \end{pmatrix} \begin{pmatrix} r_1 \\ r_2 \\ r_3 \\ \alpha_1 \\ \alpha_2 \\ \alpha_3 \end{pmatrix} = \begin{pmatrix} r_3 \\ r_2 \\ r_1 \\ \alpha_3 \\ \alpha_2 \\ \alpha_1 \end{pmatrix} \qquad \textbf{(2-44)}$$

A similar 6×6 matrix can be obtained for the other five operations of the group, and it will be found that these six matrices obey the group multiplication table and hence form a representation $\mathbf{\Gamma}_4$. We shall denote the matrices in $\mathbf{\Gamma}_4$ by $\mathbf{\Gamma}_4(\sigma_v^b)$, etc., and these six matrices are listed in Table 2-3.

Another 6×6 representation which may be easily obtained is the *regular representation.* To produce this, we rewrite the multiplication table for the group C_{3v} so that the first operations (along the top of Table 1-1) are in the

[1]This use differs from the concept of the basis of a crystal, used in connection with Figs. 1-36 and 1-37.

TABLE 2-2 *The Irreducible Representations of the Group* C_{3v}

	E	C_3	C_3^2	σ_v^a	σ_v^b	σ_v^c
Γ_1	1	1	1	1	1	1
Γ_2	1	1	1	−1	−1	−1
Γ_3	$\begin{pmatrix}1 & 0\\ 0 & 1\end{pmatrix}$	$\begin{pmatrix}-\frac{1}{2} & -\frac{\sqrt{3}}{2}\\ \frac{\sqrt{3}}{2} & -\frac{1}{2}\end{pmatrix}$	$\begin{pmatrix}-\frac{1}{2} & \frac{\sqrt{3}}{2}\\ -\frac{\sqrt{3}}{2} & -\frac{1}{2}\end{pmatrix}$	$\begin{pmatrix}-1 & 0\\ 0 & 1\end{pmatrix}$	$\begin{pmatrix}\frac{1}{2} & \frac{\sqrt{3}}{2}\\ \frac{\sqrt{3}}{2} & -\frac{1}{2}\end{pmatrix}$	$\begin{pmatrix}\frac{1}{2} & -\frac{\sqrt{3}}{2}\\ -\frac{\sqrt{3}}{2} & -\frac{1}{2}\end{pmatrix}$

TABLE 2-3 *A Reducible Representation for* C_{3v}

	E	C_3	C_3^2	σ_v^a	σ_v^b	σ_v^c
Γ_4	$\begin{pmatrix}1&0&0&0&0&0\\0&1&0&0&0&0\\0&0&1&0&0&0\\0&0&0&1&0&0\\0&0&0&0&1&0\\0&0&0&0&0&1\end{pmatrix}$	$\begin{pmatrix}0&1&0&0&0&0\\0&0&1&0&0&0\\1&0&0&0&0&0\\0&0&0&0&1&0\\0&0&0&0&0&1\\0&0&0&1&0&0\end{pmatrix}$	$\begin{pmatrix}0&0&1&0&0&0\\1&0&0&0&0&0\\0&1&0&0&0&0\\0&0&0&0&0&1\\0&0&0&1&0&0\\0&0&0&0&1&0\end{pmatrix}$	$\begin{pmatrix}1&0&0&0&0&0\\0&0&1&0&0&0\\0&1&0&0&0&0\\0&0&0&1&0&0\\0&0&0&0&0&1\\0&0&0&0&1&0\end{pmatrix}$	$\begin{pmatrix}0&0&1&0&0&0\\0&1&0&0&0&0\\1&0&0&0&0&0\\0&0&0&0&0&1\\0&0&0&0&1&0\\0&0&0&1&0&0\end{pmatrix}$	$\begin{pmatrix}0&1&0&0&0&0\\1&0&0&0&0&0\\0&0&1&0&0&0\\0&0&0&0&1&0\\0&0&0&1&0&0\\0&0&0&0&0&1\end{pmatrix}$

TABLE 2-4 ***Cayley Table Used to Find the Regular Representation of*** C_{3v}

	E	C_3	C_3^2	σ_v^a	σ_v^b	σ_v^c
E	E	C_3	C_3^2	σ_v^a	σ_v^b	σ_v^c
C_3^2	C_3^2	E	C_3	σ_v^b	σ_v^c	σ_v^a
C_3	C_3	C_3^2	E	σ_v^c	σ_v^a	σ_v^b
σ_v^a	σ_v^a	σ_v^b	σ_v^c	E	C_3	C_3^2
σ_v^b	σ_v^b	σ_v^c	σ_v^a	C_3^2	E	C_3
σ_v^c	σ_v^c	σ_v^a	σ_v^b	C_3	C_3^2	E

order given, but the second operations (along the left-hand edge) are rearranged to correspond to the inverses of the first operations. Since the inverses of E, C_3, C_3^2, σ_v^a, σ_v^b, and σ_v^c are E, C_3^2, C_3, σ_v^a, σ_v^b, and σ_v^c, respectively, we obtain Table 2-4 from this rearrangement.

The regular representation of each operation is then obtained by writing a matrix with unity corresponding to the six positions of the operation E in Table 2-4 and zero everywhere else. For example,

$$E = \begin{pmatrix} 1 & 0 & 0 & 0 & 0 & 0 \\ 0 & 1 & 0 & 0 & 0 & 0 \\ 0 & 0 & 1 & 0 & 0 & 0 \\ 0 & 0 & 0 & 1 & 0 & 0 \\ 0 & 0 & 0 & 0 & 1 & 0 \\ 0 & 0 & 0 & 0 & 0 & 1 \end{pmatrix} \tag{2-45}$$

$$C_3 = \begin{pmatrix} 0 & 1 & 0 & 0 & 0 & 0 \\ 0 & 0 & 1 & 0 & 0 & 0 \\ 1 & 0 & 0 & 0 & 0 & 0 \\ 0 & 0 & 0 & 0 & 1 & 0 \\ 0 & 0 & 0 & 0 & 0 & 1 \\ 0 & 0 & 0 & 1 & 0 & 0 \end{pmatrix} \tag{2-46}$$

$$\sigma_v^a = \begin{pmatrix} 0 & 0 & 0 & 1 & 0 & 0 \\ 0 & 0 & 0 & 0 & 0 & 1 \\ 0 & 0 & 0 & 0 & 1 & 0 \\ 1 & 0 & 0 & 0 & 0 & 0 \\ 0 & 0 & 1 & 0 & 0 & 0 \\ 0 & 1 & 0 & 0 & 0 & 0 \end{pmatrix} \tag{2-47}$$

Note that the regular representation is similar to, but not identical with, the representation $\mathbf{\Gamma}_4$ of Table 2-3.

PROBLEM 2-6

Verify that the regular representation conforms to the Cayley table.

Let us use the symbol R_i ($i = 1, 2, \ldots, 6$) to denote the six elements of C_{3v}; that is, $R_1 = E$, $R_2 = C_3$, etc. Further, the matrices of the form (2-45), (2-46), and (2-47) will be denoted by $(R_i)_{jk}$ ($j, k = 1, 2, \ldots, 6$). For example, $(R_2)_{11} = 0$, $(R_2)_{12} = 1$, and so on. The definition of the regular representation can then be expressed as

$$(R_i)_{jk} = \begin{cases} 1 & \text{if } R_j^{-1}R_k = R_i \\ 0 & \text{otherwise} \end{cases} \tag{2-48}$$

For example, for $C_3 = R_2$, the elements in row 1 are $(R_2)_{11} = 0$, $(R_2)_{12} = 1$, $(R_2)_{13} = 0$, etc., where $R^{-1} = E^{-1} = E$ and $R_1^{-1}R_1 \neq R_2$, $R_1^{-1}R_2 = R_2$, $R_1^{-1}R_3 \neq R_3$, etc. From (2-11), the definition of the product R_n of two matrices R_i and R_m is

$$(R_n)_{jk} = \sum_l (R_i)_{jl}(R_m)_{lk} \tag{2-49}$$

PROBLEM 2-7

Use Eqs. (2-48) and (2-49) to prove that the regular representation obeys the Cayley table.

We shall shortly use the regular representation to determine some general properties of representations.

2-4 Reducible and irreducible representations

Returning to Fig. 2-1, let us consider the effect of an arbitrary operation A on the vector $\mathbf{r}$, which we shall now denote $\mathbf{r}_1$ for convenience. Let the matrix $\mathbf{A}$ convert $\mathbf{r}_1$ into $\mathbf{r}_2$ by a reflection, rotation, or inversion, so that

$$\mathbf{r}_2 = \mathbf{A}\mathbf{r}_1 \tag{2-50}$$

Instead of performing an operation on $\mathbf{r}_1$, we may rotate the coordinate

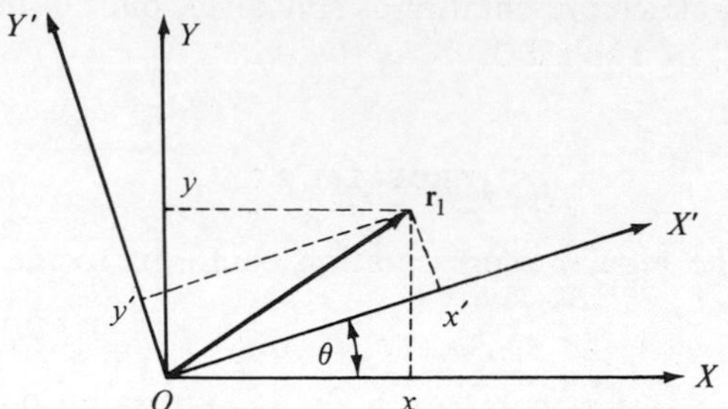

2-4 A coordinate transformation

system XOY about O through an angle θ by an operation denoted Θ (Fig. 2-4). Although $\mathbf{r}_1$ is unchanged by this operation, its coordinates in the two systems are (x, y) and (x', y') respectively, and we symbolize the connection between the two sets of coordinates by

$$\mathbf{r}_1 = \Theta\mathbf{r}'_1 \tag{2-51}$$

PROBLEM 2-8

By a shrewd choice of construction lines on Fig. 2-4, show that (2-51) is equivalent to

$$\begin{pmatrix} x \\ y \end{pmatrix} = \begin{pmatrix} \cos\theta & -\sin\theta \\ \sin\theta & \cos\theta \end{pmatrix} \begin{pmatrix} x' \\ y' \end{pmatrix} \tag{2-52}$$

Equation (2-52) indicates that the matrix Θ corresponding to a rotation of coordinates is identical to the matrix in (2-33) which represents a rotation of the vector by an equal amount.

The rotation of XOY affects the vector $\mathbf{r}_2$ of (2-50) in the same way as $\mathbf{r}_1$, and we have

$$\mathbf{r}_2 = \Theta\mathbf{r}'_2 \tag{2-53}$$

like (2-51). Inserting (2-51) and (2-53) into (2-50) gives

$$\mathbf{r}'_2 = \Theta^{-1}\mathbf{A}\Theta\mathbf{r}'_1 \tag{2-54}$$

However, $\mathbf{r}'_1$ and $\mathbf{r}'_2$ must be connected by the same symmetry operation as $\mathbf{r}_1$ and $\mathbf{r}_2$, since $\mathbf{r}'_1$ is merely $\mathbf{r}_1$ expressed in a different coordinate system. Hence,

$$\mathbf{r}'_2 = \mathbf{A}'\mathbf{r}'_1 \tag{2-55}$$

where $\mathbf{A}'$ is the matrix corresponding to $\mathbf{A}$ in $X'OY'$, Comparing (2-54) and (2-55) gives

$$\mathbf{A}' = \Theta^{-1}\mathbf{A}\Theta \tag{2-56}$$

Equation (2-56) defines a similarity transformation like that of (2-28); it tells us how to reexpress the matrix for an operation in a given coordinate system when that system is rotated. Although the operator Θ represents a rotation through an angle θ about the z-axis, the result (2-56) applies to an arbitrary orientation.

PROBLEM 2-9

Are $\mathbf{A}$ and $\mathbf{A}'$ in (2-56) members of the same class?

Let us consider a similarity transformation on one of the matrices of Table 2-3; take σ_v^c, for example. The matrix Θ (for a reason to be explained later) is chosen to have the form

$$\Theta = \begin{pmatrix} 1/\sqrt{3} & 1/\sqrt{2} & 1/\sqrt{6} & 0 & 0 & 0 \\ 1/\sqrt{3} & -1/\sqrt{2} & 1/\sqrt{6} & 0 & 0 & 0 \\ 1/\sqrt{3} & 0 & -\sqrt{2}/\sqrt{3} & 0 & 0 & 0 \\ 0 & 0 & 0 & 1/\sqrt{3} & 1/\sqrt{2} & 1/\sqrt{6} \\ 0 & 0 & 0 & 1/\sqrt{3} & -1/\sqrt{2} & 1/\sqrt{6} \\ 0 & 0 & 0 & 1/\sqrt{3} & 0 & -\sqrt{2}/\sqrt{3} \end{pmatrix} \tag{2-57}$$

To find Θ^{-1}, we may invoke the theorem from matrix theory that the elements b_{ij} of the inverse $\mathbf{A}^{-1} = \mathbf{B}$ of a matrix $\mathbf{A}$ are given by

$$b_{ij} = \frac{A_{ji}}{\det \mathbf{A}} \tag{2-58}$$

where A_{ji} is the *cofactor* of a_{ij}; i.e., the determinant formed by striking out row j and column i from the determinant det $\mathbf{A}$ of $\mathbf{A}$, and with a negative sign if $(i + j)$ is odd.

PROBLEM 2-10

Verify Eq. (2-58).

However, we might also guess that the argument which gave us (2-26) would work here; that is, that

$$\Theta^{-1} = \tilde{\Theta} \tag{2-59}$$

where $\tilde{\Theta}$ is called the *transpose* of Θ and is obtained by interchanging rows and columns of Θ. A matrix obeying (2-59) is said to be *orthogonal*, since the rows (or columns) act like orthogonal vectors. For example, row 1 of Θ, when used in a scalar product with itself, gives

$$\left(\frac{1}{\sqrt{3}} + 0 - \frac{\sqrt{2}}{\sqrt{3}} + 0 + 0 + 0\right)\cdot\left(\frac{1}{\sqrt{3}} + 0 - \frac{\sqrt{2}}{\sqrt{3}} + 0 + 0 + 0\right) = 1$$

while row 1 with row 2 (or any other row) vanishes. The fact that row 1 into itself was unity (rather than an arbitrary nonzero quantity) is due to the fact that all the rows and columns of Θ have been normalized. This leads us to suspect (and we may easily verify) that Θ is a matrix formed from a set of six unit vectors which are mutually orthogonal eigenvectors in a six-dimensional space.

Performing a similarity transformation with Θ on $\Gamma_4(\sigma_v^c)$ gives

$$\Gamma'_4(\sigma_v^c) = \Theta^{-1}\Gamma_4(\sigma_v^c)\Theta \tag{2-60}$$

$$\text{or} \qquad \Gamma'_4(\sigma_v^c) = \begin{pmatrix} 1 & 0 & 0 & 0 & 0 & 0 \\ 0 & 1/2 & -\sqrt{3}/2 & 0 & 0 & 0 \\ 0 & -\sqrt{3}/2 & -1/2 & 0 & 0 & 0 \\ 0 & 0 & 0 & 1 & 0 & 0 \\ 0 & 0 & 0 & 0 & 1/2 & -\sqrt{3}/2 \\ 0 & 0 & 0 & 0 & -\sqrt{3}/2 & -1/2 \end{pmatrix}$$

This matrix will be recognized from Table 2-2 as having the form

$$\begin{pmatrix} \boxed{\Gamma_1(\sigma_v^c)} & 0 & 0 & 0 & 0 & 0 \\ 0 & \multicolumn{2}{c}{\Gamma_3(\sigma_v^c)} & 0 & 0 & 0 \\ 0 & & & 0 & 0 & 0 \\ 0 & 0 & 0 & \boxed{\Gamma_1(\sigma_v^c)} & 0 & 0 \\ 0 & 0 & 0 & 0 & \multicolumn{2}{c}{\Gamma_3(\sigma_v^c)} \\ 0 & 0 & 0 & 0 & & \end{pmatrix} \tag{2-61}$$

where $\Gamma_1(\sigma_v^c)$ is a 1×1 block appearing twice on the main diagonal, and $\Gamma_3(\sigma_v^c)$ is a 2×2 block with the same property.

The matrix $\Gamma_4(\sigma_v^c)$ is said to be *reduced* (in group theory treatises, the term *completely reduced* is used). We then say that the representation Γ_4 is *reducible*. It is easy to verify that the same matrix Θ and its inverse Θ^{-1} will convert all the 6×6 matrices of Table 2-3 into the corresponding reduced form. On the other hand, it is not possible to find a similarity transformation which will simultaneously convert *all* the matrices of Γ_3 to diagonal form, so that Γ_3 is an *irreducible representation*. Since any one-dimensional representation is inherently irreducible, Table 2-2 lists only irreducible ones. As a consequence of Eq. (2-61), we say that the reducible representation Γ_4 *contains* the irreducible representations Γ_1 and Γ_3 twice each.

PROBLEM 2-11

Give a method of determining Θ. (*Hint:* Work with 3×3 matrices to simplify the algebra. Also note that the method of Sec. 2-2 does not apply here, for we are reducing $\Gamma_4(\sigma_v^c)$ rather than diagonalizing it.)

PROBLEM 2-12

Show that the set of matrices

$$\begin{pmatrix} 1 & 1 \\ 0 & 1 \end{pmatrix}, \quad \begin{pmatrix} 1 & 0 \\ 1 & 1 \end{pmatrix}$$

is irreducible.

2-5 Character systems

An important feature of a representation composed of a set of matrices is the *trace or character* of each matrix (the trace of a matrix is the sum of the elements on the main diagonal). The set of characters for the representations of a given group is called the *character system* and Table 2-5 shows this

TABLE 2-5 *Traces of the Matrices for C_{3v}*

	E	C_3	C_3^2	σ_v^a	σ_v^b	σ_v^c
Γ_1	1	1	1	1	1	1
Γ_2	1	1	1	−1	−1	−1
Γ_3	2	−1	−1	0	0	0
Γ_4	6	0	0	2	2	2

system for the group C_{3v}, as obtained from Tables 2-2 and 2-3. We note that the characters for any element in a given class are the same. Denoting characters by the symbol χ, it is seen, for example, that the characters for the representation Γ_3 of Table 2-2 are

$$\chi_3(E) = 1 + 1 = 2$$
$$\chi_3(C_3) = \chi_3(C_3^2) = -\tfrac{1}{2} + (-\tfrac{1}{2}) = -1$$
$$\chi_3(\sigma_v^a) = -1 + 1 = 0$$
$$\chi_3(\sigma_v^b) = \chi_3(\sigma_v^c) = \tfrac{1}{2} + (-\tfrac{1}{2}) = 0$$

Hence, Table 2-5 may be condensed to give Table 2-6.

TABLE 2-6 ***Character System for C_{3v}***

	E	C_3, C_3^2	$\sigma_v^a, \sigma_v^b, \sigma_v^c$
Γ_1	1	1	1
Γ_2	1	1	−1
Γ_3	2	−1	0
Γ_4	6	0	2

Table 2-6 illustrates a very important feature of Eq. (2-61). When Γ_4 is reduced by the method indicated in Eq. (2-60), the resulting matrices are seen to contain Γ_1 and Γ_3 twice each while Γ_2 does not appear at all. Letting these multiplicities be n_1, n_2, and n_3, they have the values

$$n_1 = 2, \qquad n_2 = 0, \qquad n_3 = 2 \tag{2-62}$$

Their values can be predicted by the theorem

$$n_k = \frac{\sum_R \chi(R)\chi_k(R)}{g} \tag{2-63}$$

where the n_k $(k = 1, 2, 3)$ are the numbers in (2-62), $\chi(R)$ is the character for the matrix R in the reducible representation (Γ_4 in our present example), the $\chi_k(R)$ are the characters in the irreducible representations, and g is the number of elements in the group. Since $g = 6$, the application of (2-63) to Γ_1 gives

$$n_1 = [\chi(E)\chi_1(E) + \chi(C_3)\chi_1(C_3) + \chi(C_3^2)\chi_1(C_3^2)$$
$$+ \chi(\sigma_v^a)\chi_1(\sigma_v^a) + \chi(\sigma_v^b)\chi_1(\sigma_v^b) + \chi(\sigma_v^c)\chi_1(\sigma_v^c)]/6$$

or $$n_1 = [6(1) + 0(1) + 0(1) + 2(1) + 2(1) + 2(1)]/6 = 2 \qquad \text{(2-64)}$$

where

$$\chi(E) = 6, \quad \chi(C_3) = \chi(C_3^2) = 0, \quad \chi(\sigma_v^a) = \chi(\sigma_v^b) = \chi(\sigma_v^c) = 2$$

as given by the row Γ_4 in Table 2-6. Similarly,

$$n_2 = 0, \qquad n_3 = 2 \qquad \text{(2-65)}$$

so that (2-64) and (2-65) verify (2-63). As stated in Chapter 1, we shall not prove theorems of this type (unless the arguments are simple) but rely on demonstration through examples.

Another important result is

$$\sum_k d_k^2 = g \qquad \text{(2-66)}$$

where d_k is the dimension of the kth irreducible representation, and g the order of the group. For the present example, we see that the relation

$$1^2 + 1^2 + 2^2 = 6 \qquad \text{(2-67)}$$

corresponds to Table 2-2. There are three irreducible representations for C_{3v}, with dimensions

$$d_1 = 1, \qquad d_2 = 1, \qquad d_3 = 2$$

thus verifying (2-66).

PROBLEM 2-13

Is there any other set of numbers whose squares satisfy (2-66) for a six-element group? Can these numbers correspond to a possible set of irreducible representations?

2-6 The calculation of character tables

The characters indicated in Table 2-6 were calculated by adding the elements on the main diagonals of the matrices in Tables 2-2 and 2-3. It is a rather tedious process to determine these matrices for all possible point groups, and we shall now show how to calculate a character system directly from the Cayley table. Let us write the three classes

$$E$$
$$C_3, \quad C_3^2$$
$$\sigma_v^a, \quad \sigma_v^b, \quad \sigma_v^c$$

as *class sums* $\mathscr{C}_i$, defined by

$$\begin{aligned}\mathscr{C}_1 &= E \\ \mathscr{C}_2 &= C_3 + C_3^2 \\ \mathscr{C}_3 &= \sigma_v^a + \sigma_v^b + \sigma_v^c\end{aligned} \tag{2-68}$$

We then see that

$$\begin{aligned}\mathscr{C}_2\mathscr{C}_3 &= (C_3 + C_3^2)(\sigma_v^a + \sigma_v^b + \sigma_v^c) \\ &= \sigma_v^b + \sigma_v^c + \sigma_v^a + \sigma_v^c + \sigma_v^a + \sigma_v^b \\ &= 2\mathscr{C}_3\end{aligned} \tag{2-69}$$

That is, the product of any two class sums will be of the form

$$\mathscr{C}_A\mathscr{C}_B = \sum_M c_{AB,M}\mathscr{C}_M \tag{2-70}$$

where, for this example

$$A = 2, \qquad B = 3, \qquad M = 1, \qquad c_{12,1} = 1$$

PROBLEM 2-14

Prove that

$$c_{33,1} = 3, \qquad c_{33,2} = 3, \qquad c_{33,3} = 0$$

Let $\mathscr{D}_3$ be the representation of class sum $\mathscr{C}_3$ derived from Γ_3. That is

$$\mathscr{D}_3 = \begin{pmatrix} -1 & 0 \\ 0 & 1 \end{pmatrix} + \begin{pmatrix} \frac{1}{2} & -\frac{\sqrt{3}}{2} \\ -\frac{\sqrt{3}}{2} & -\frac{1}{2} \end{pmatrix} + \begin{pmatrix} \frac{1}{2} & \frac{\sqrt{3}}{2} \\ \frac{\sqrt{3}}{2} & -\frac{1}{2} \end{pmatrix} = \begin{pmatrix} 0 & 0 \\ 0 & 0 \end{pmatrix}$$

Similarly,

$$\mathscr{D}_2 = \begin{pmatrix} -\frac{1}{2} & \frac{\sqrt{3}}{2} \\ -\frac{\sqrt{3}}{2} & -\frac{1}{2} \end{pmatrix} + \begin{pmatrix} -\frac{1}{2} & -\frac{\sqrt{3}}{2} \\ \frac{\sqrt{3}}{2} & -\frac{1}{2} \end{pmatrix} = \begin{pmatrix} -1 & 0 \\ 0 & -1 \end{pmatrix}$$

$$\mathscr{D}_1 = \begin{pmatrix} 1 & 0 \\ 0 & 1 \end{pmatrix}$$

These matrices are all *scalar*; they are matrices having the same element at every location on the main diagonal, and zeroes everywhere else. Thus, they have the form

$$\mathscr{D}_A = \begin{pmatrix} m_A & 0 & 0 & \dots & 0 \\ 0 & m_A & 0 & \dots & 0 \\ \dots & \dots & \dots & \dots & \dots \\ 0 & 0 & 0 & \dots & m_A \end{pmatrix} \tag{2-71}$$

where m_A is a number to be determined.

To find m_A, we first use the fact that $\mathscr{D}_A$ must have the dimensions of the representation, so that its character is given by

$$\chi(\mathscr{D}_A) = \chi_E m_A \tag{2-72}$$

where χ_E is the character of the matrix for E. For example,

$$\chi(\mathscr{D}_1) = 2(1) = 2, \qquad \chi(\mathscr{D}_2) = 2(-1) = -2, \qquad \chi(\mathscr{D}_3) = 2(0) = 0$$

as given above.

Next, we realize that if the class has r_A members and χ_A is their common character, then

$$\chi(\mathscr{D}_A) = r_A \chi_A \tag{2-73}$$

Again

$$\chi(\mathscr{D}_1) = 1(2) = 2, \qquad \chi(\mathscr{D}_2) = 2(-1) = -2, \qquad \chi(\mathscr{D}_3) = 3(0) = 0$$

Equating the right-hand sides of (2-72) and (2-73) gives

$$m_A = \frac{r_A \chi_A}{\chi_E} \tag{2-74}$$

Now the matrices $\mathscr{D}_A$ should satisfy (2-70), since they multiply according to the Cayley table. Hence,

$$\mathscr{D}_A \mathscr{D}_B = \sum_M c_{AB,M} \mathscr{D}_M \tag{2-75}$$

But the matrices $\mathscr{D}_A$ are scalar; they multiply like pure numbers and we can replace them with these numbers m_A, giving

$$m_A m_B = \sum_M c_{AB,M} m_M \tag{2-76}$$

or by (2-74)

$$\left(\frac{r_A \chi_A}{\chi_E}\right)\left(\frac{r_B \chi_B}{\chi_E}\right) = \sum_M c_{AB,M}\left(\frac{r_M \chi_M}{\chi_E}\right) \tag{2-77}$$

This relation is known as *Burnside's theorem*, and may be simplified to

$$r_A r_B \chi_A \chi_B = \chi_E \sum_M c_{AB,M} r_M \chi_M \tag{2-78}$$

Returning to C_{3v}, Eq. (2-70) leads to nine relations among the class sums. However, the following three

$$\mathscr{C}_2^2 = 2\mathscr{C}_1 + \mathscr{C}_2, \qquad \mathscr{C}_3^2 = 3\mathscr{C}_1 + 3\mathscr{C}_2, \qquad \mathscr{C}_2\mathscr{C}_3 = 2\mathscr{C}_3$$

are sufficient. From these, we obtain

$$c_{22,1} = 2, \qquad c_{22,2} = 1$$
$$c_{33,1} = 3, \qquad c_{33,2} = 3$$
$$c_{23,3} = 2$$

Then (2-78), for $A = 2$, $B = 2$, gives

$$2(2)\chi_2^2 = \chi_1[2(1)\chi_1 + 1(2)\chi_2]$$

Similarly,

$$3(3)\chi_3^2 = \chi_1[3(1)\chi_1 + 3(2)\chi_2]$$

and

$$3(2)\chi_2\chi_3 = \chi_1[2(3)\chi_3]$$

PROBLEM 2-15

(a) Show from (2-67) that

$$\chi_1 = 1, 1, \text{ or } 2$$

(b) Verify that (2-78) generates $\boldsymbol{\Gamma}_1$, $\boldsymbol{\Gamma}_2$, and $\boldsymbol{\Gamma}_3$ in Table 2-6.

The character systems for the irreducible representations of the other point groups discussed in Chapter 1 can be obtained from Burnside's theorem in the same way. Rather than enumerate them all here, we shall introduce them as we need them.

2-7 An introduction to normal coordinates

Consider a particle of mass m moving in the Earth's gravitational field. Its potential energy V is given by

$$V = -\int F\,dz = mg\int dz = mgz \tag{2-79}$$

and its kinetic energy is

$$T = m\int a\,dz = m\int \frac{dv}{dt}dz = m\int v\,dv = \frac{1}{2}mv^2 \tag{2-80}$$

Differentiating (2-79) with respect to z gives

$$\frac{dV}{dz} = mg \tag{2-81}$$

and differentiating (2-80) with respect to v and t gives

$$\frac{d}{dt}\frac{dT}{dv} = m\frac{dv}{dt} = ma \tag{2-82}$$

Since

$$a = -g$$

we see that (2-81) and (28-2) may be combined to give

$$\frac{d}{dt}\frac{dT}{dv} + \frac{dV}{dz} = 0 \tag{2-83}$$

This is known as *Lagrange's equation of motion*; for the simple applications we shall make, we may regard (2-83) as equivalent to Newton's second law $F = ma$.

PROBLEM 2-16

A *harmonic force* is defined by the law

$$F = -kx \tag{2-84}$$

where k is a constant. Show that Lagrange's equation leads to simple harmonic motion.

Problem 2-16 leads to a description of the motion of a mass m vibrating at the end of a spring with Hooke's law constant k. Let us consider three

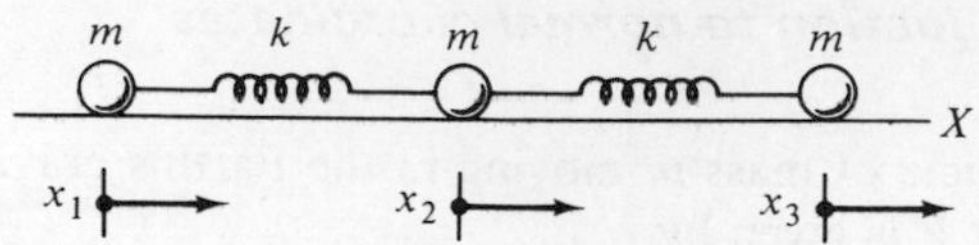

2-5 Model of a linear molecule

equal masses connected by two identical springs and constrained to vibrate along the x-axis (Fig. 2-5). The change in potential energy due to a displacement x, by (2-79) and (2-84), is

$$V = -\int(-kx)\,dx = \tfrac{1}{2}kx^2 \tag{2-85}$$

Letting x_1, x_2, x_3 denote the displacements from equilibrium of each mass, then the total change in the potential energy is

$$V = \frac{k}{2}[(x_2 - x_1)^2 + (x_3 - x_2)^2] \tag{2-86}$$

Using dots to denote time derivatives as

$$\dot{x} \equiv \frac{dx}{dt}, \qquad \ddot{x} \equiv \frac{d^2x}{dt^2} \tag{2-87}$$

the total kinetic energy becomes

$$T = \frac{m}{2}[\dot{x}_1^2 + \dot{x}_2^2 + \dot{x}_3^2] \tag{2-88}$$

There will be a Lagrange equation for each of the coordinates and these may be written in the compact form

$$\frac{d}{dt}\frac{\partial T}{\partial \dot{x}_s} + \frac{\partial V}{\partial x_s} = 0, \quad (s = 1, 2, 3) \tag{2-89}$$

where the use of partial derivatives is necessary since T depends on the three velocities $\dot{x}_s$ and V on the three displacements x_s. Substituting T and V into each of the three equations (2-89) gives

$$\begin{aligned} m\ddot{x}_1 - k(x_2 - x_1) &= 0 \\ m\ddot{x}_2 + k(x_2 - x_1) - k(x_3 - x_2) &= 0 \\ m\ddot{x}_3 + k(x_3 - x_2) &= 0 \end{aligned} \tag{2-90}$$

Differential equations with constant coefficients have solutions which may be expressed as exponential functions. Hence we try

$$x_r = a_r e^{i\omega t}, \quad (r = 1, 2, 3) \tag{2-91}$$

in (2-90), where the a_r are constants to be determined, obtaining

$$\begin{aligned} ma_1(-\omega^2) + ka_1 - ka_2 &= 0 \\ ma_2(-\omega^2) - ka_1 + 2ka_2 - ka_3 &= 0 \\ ma_3(-\omega^2) - ka_2 + ka_3 &= 0 \end{aligned} \tag{2-92}$$

Letting

$$\lambda = \frac{\omega^2 m}{k}$$

converts (2-92) into

$$\begin{aligned} -a_1\lambda + a_1 - a_2 &= 0 \\ a_2\lambda - a_1 + 2a_2 - a_3 &= 0 \\ -a_3\lambda - a_2 + a_3 &= 0 \end{aligned} \tag{2-93}$$

These simultaneous equations lead to a secular equation of the form of (2-16), so that

$$\begin{vmatrix} 1-\lambda & -1 & 0 \\ -1 & 2-\lambda & -1 \\ 0 & -1 & 1-\lambda \end{vmatrix} = 0 \tag{2-94}$$

The roots are

$$\lambda_1 = 0, \qquad \lambda_2 = 1, \qquad \lambda_3 = 3 \tag{2-95}$$

and

$$\omega_1 = 0, \qquad \omega_2 = \sqrt{\frac{k}{m}}, \qquad \omega_3 = \sqrt{\frac{3k}{m}} \tag{2-96}$$

Substituting the λ_r in turn into (2-93), we obtain

$$\begin{aligned} &a_1 = a_2 = a_3, \quad (\lambda_1 = 0) \\ &a_1 = -a_3, \qquad a_2 = 0, \quad (\lambda_2 = 1) \\ &a_1 = a_3, \qquad a_2 = -2a_1, \quad (\lambda_3 = 3) \end{aligned} \tag{2-97}$$

To interpret our results we note that the values of the a_r for $\omega = 0$ correspond to a motion of all three particles by the same amount in the same direction, as indicated in Fig. 2-6(a). This is a pure displacement. The second set of values of the a_r is illustrated in Fig. 2-6(b); the center mass is stationary and the two outer ones oscillate in opposite directions by equal amounts with an angular frequency $\omega_2 = \sqrt{k/m}$. This is what we would expect, since this situation corresponds to two independent harmonic oscillators of the

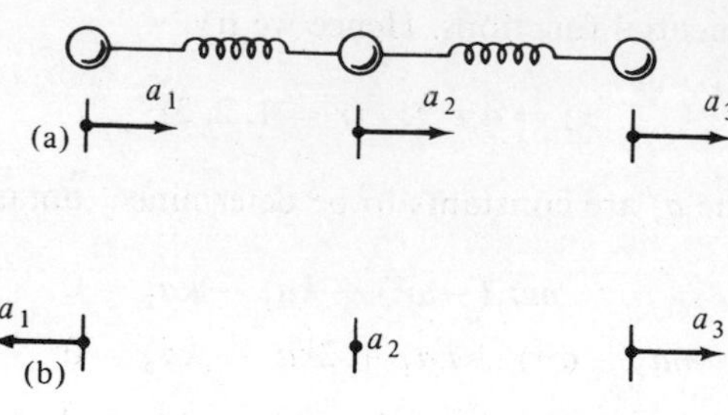

2-6 Normal modes of the linear molecule

type discussed in Problem 2-16. Figure 2-6(c) shows the center mass moving in a direction opposite to that of the end masses, with twice the amplitude.

These solutions to Lagrange's equations are called *modes.* A mode is the specification of the three values of a_r corresponding to each value of ω. Since the solutions produce only ratios of the a_r, we may impose the additional condition that

$$a_1^2 + a_2^2 + a_3^2 = 1 \tag{2-98}$$

for each value of ω, as we did in (2-22). The amplitudes are then normalized, and the normalization condition explicitly determines the a_r. Using (2-97) in (2-98) gives

$$\begin{aligned} a_1 = a_2 = a_3 &= \frac{1}{\sqrt{3}}, \quad (\lambda_1 = 0) \\ a_1 = -a_3 &= \frac{1}{\sqrt{2}}, \quad a_2 = 0, \quad (\lambda_2 = 1) \\ a_1 = a_3 &= \frac{1}{\sqrt{6}}, \quad a_2 = -\frac{1}{\sqrt{6}}, \quad (\lambda_3 = 3) \end{aligned} \tag{2-99}$$

Each mode may now be expressed in terms of a single variable q_s, defined by

$$\begin{aligned} q_1 &= \sqrt{\frac{m}{3}}(x_1 + x_2 + x_3) \\ q_2 &= \sqrt{\frac{m}{2}}(x_1 - x_3) \\ q_3 &= \sqrt{\frac{m}{6}}(x_1 - 2x_2 + x_3) \end{aligned} \tag{2-100}$$

The coefficients in (2-100) are obtained from the values of the a_r in (2-99), and the factor $\sqrt{m}$ is included for a reason which will appear below. The q_s are called *normal coordinates* and have the form

$$q_s = \sqrt{m} \sum_{r=1}^{3} a_{sr} x_r \tag{2-101}$$

where a_{sr} is the normalized amplitude associated with coordinate x_r and frequency ω_s. The potential and kinetic energies now become

$$V = \tfrac{1}{2} \sum_s \omega_s^2 q_s, \qquad T = \tfrac{1}{2} \sum_s \dot{q}_s^2 \tag{2-102}$$

so that they assume a very simple form when expressed in terms of normal coordinates. Note that cross-terms, such as $-kx_1x_2$ of (2-86), have been eliminated.

PROBLEM 2-17

Show that (2-102) follow from (2-86) and (2-88).

2-8 The planar equilateral oscillator

Consider three equal masses connected by identical springs so that the structure is an equilateral triangle at rest (Fig. 2-7). Let us assume that the masses are required to vibrate in the xy-plane. Prior to setting up and solving Lagrange's equations, we may predict the number and kinds of modes by

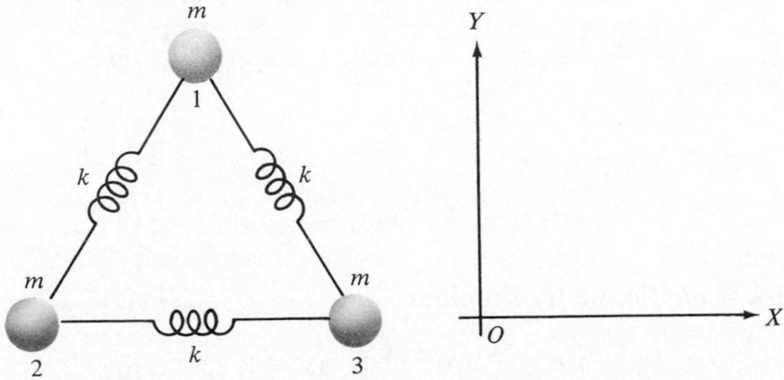

2-7 Model of an equilateral planar molecule with symmetry C_{3v}

realizing that each one of the particles is free to vibrate in each of two directions, so that there are $3 \times 2 = 6$ possible modes. However, two of these will be pure translation (one in each of the two independent directions) and a third will be a pure rotation about an axis normal to the plane. Thus, we expect one rotational, two translational, and three vibrational modes.

Orienting the axes as shown in the figure and again denoting displacements from equilibrium by x_1, y_1 for particle 1, and so on, the kinetic energy T is

$$T = \left(\frac{m}{2}\right)(\dot{x}_1^2 + \dot{y}_1^2 + \dot{x}_2^2 + \dot{y}_2^2 + \dot{x}_3^2 + \dot{y}_3^2) \tag{2-103}$$

The potential energy V may be calculated from the formula

$$V = \tfrac{1}{2}kx^2 \tag{2-85}$$

where x is the net change in length of a given spring.

PROBLEM 2-18

Show that the potential energy for the system of Fig. 2-7 is given by

$$\begin{aligned} V = \frac{k}{2}\Big[(x_3 - x_2)^2 &+ \left\{-\frac{1}{2}(x_1 - x_3) + \frac{\sqrt{3}}{2}(y_1 - y_3)\right\}^2 \\ &+ \left\{\frac{1}{2}(x_2 - x_1) + \frac{\sqrt{3}}{2}(y_2 - y_1)\right\}^2\Big] \end{aligned} \tag{2-104}$$

Expanding this expression gives

$$\begin{aligned} V = \frac{k}{2}\Big[&\frac{1}{2}x_1^2 + \frac{3}{2}y_1^2 + \frac{5}{4}x_2^2 + \frac{3}{4}y_2^2 + \frac{5}{4}x_3^2 + \frac{3}{4}y_3^2 \\ &+ \frac{\sqrt{3}}{2}x_3y_1 + \frac{\sqrt{3}}{2}x_2y_2 - 2x_2x_3 - \frac{1}{2}x_2x_1 \\ &- \frac{\sqrt{3}}{2}x_2y_1 - \frac{\sqrt{3}}{2}y_2x_1 - \frac{3}{2}y_2y_1 - \frac{1}{2}x_3x_1 \\ &- \frac{\sqrt{3}}{2}x_3y_3 + \frac{\sqrt{3}}{2}y_3x_1 - \frac{3}{2}y_3y_1\Big] \end{aligned} \tag{2-105}$$

Letting $\lambda = m\omega^2/k$, we try solutions

$$r_s = a_s e^{i\omega t} = a_s e^{i\sqrt{k\lambda/m}\,t}, \quad (s = 1, 2, \ldots, 6)$$

where $$r_1 = x_1, \quad r_2 = y_1, \quad r_3 = x_2, \quad \ldots, \quad r_6 = y_3 \tag{2-106}$$

in the Lagrange equations

$$\frac{d}{dt}\frac{\partial T}{\partial \dot{r}_s} + \frac{\partial V}{\partial r_s} = 0 \tag{2-89}$$

This results in the secular equation

$$\begin{vmatrix} \frac{1}{2} - \lambda & 0 & -\frac{1}{4} & -\frac{\sqrt{3}}{4} & -\frac{1}{4} & \frac{\sqrt{3}}{4} \\ 0 & \frac{3}{2} - \lambda & -\frac{\sqrt{3}}{4} & -\frac{3}{4} & \frac{\sqrt{3}}{4} & -\frac{3}{4} \\ -\frac{1}{4} & -\frac{\sqrt{3}}{4} & \frac{5}{4} - \lambda & \frac{\sqrt{3}}{4} & -1 & 0 \\ -\frac{\sqrt{3}}{4} & -\frac{3}{4} & \frac{\sqrt{3}}{4} & \frac{3}{4} - \lambda & 0 & 0 \\ -\frac{1}{4} & \frac{\sqrt{3}}{4} & -1 & 0 & \frac{5}{4} - \lambda & -\frac{\sqrt{3}}{4} \\ \frac{\sqrt{3}}{4} & -\frac{3}{4} & 0 & 0 & -\frac{\sqrt{3}}{4} & \frac{3}{4} - \lambda \end{vmatrix} = 0 \tag{2-107}$$

This is a sixth-order algebraic equation and it cannot be solved in any simple fashion. To determine the roots we try a series of values of λ and calculate the corresponding value D of the determinant. Plotting one against the other, the roots are given by the intersection points of the curve and the

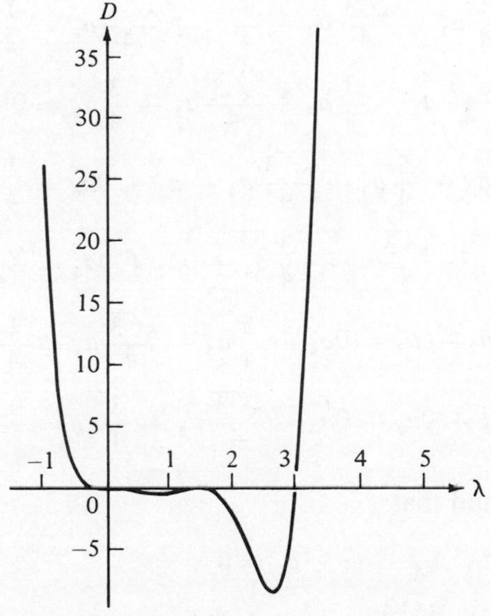

2-8 Graphical solution to a sixth degree secular equation

λ axis (Fig. 2-8). In practice, this is very tedious, and is done on a computer.[2] The figure shows that the roots are

$$\lambda = 3, \tfrac{3}{2}, 0 \tag{2-108}$$

Since Eq. (2-107) is of the sixth-order, and we have only three roots, some of them must be multiple.

PROBLEM 2-19

By examination of Fig. 2-8, show that the six values of λ are

$$\lambda = 0, 0, 0, \tfrac{3}{2}, \tfrac{3}{2}, 3 \tag{2-109}$$

Having found the eigenvalues, we substitute them one at a time into the linear equations whose coefficients are given by (2-107). Solving these equations for the a_s in the case of multiple values of λ is a matter of trial and error, for Cramer's rule yields only one set of values, even though three are possible for $\lambda = 0$. To show what can be done for this eigenvalue, we realize that two possibilities exist for a_1: either it has a finite value or it is zero. If it is finite, let us arbitrarily let it be unity and solve five of the six resulting equations for $a_2, a_3, \ldots, a_6$. These equations are

$$0a_2 - \frac{1}{4}a_3 - \frac{\sqrt{3}}{4}a_4 - \frac{1}{4}a_5 + \frac{\sqrt{3}}{4}a_6 = \frac{1}{2} \tag{2-110a}$$

$$\frac{3}{2}a_2 - \frac{\sqrt{3}}{4}a_3 - \frac{3}{4}a_4 + \frac{\sqrt{3}}{4}a_5 - \frac{3}{4}a_6 = 0 \tag{2-110b}$$

$$-\frac{\sqrt{3}}{4}a_2 + \frac{5}{4}a_3 + \frac{\sqrt{3}}{4}a_4 - a_5 + 0a_6 = \frac{1}{4} \tag{2-110c}$$

$$-\frac{3}{4}a_2 + \frac{\sqrt{3}}{4}a_3 + \frac{3}{4}a_4 + 0a_5 + 0a_6 = \frac{\sqrt{3}}{4} \tag{2-110d}$$

$$\frac{\sqrt{3}}{4}a_2 - a_3 + 0a_4 + \frac{5}{4}a_5 - \frac{\sqrt{3}}{4}a_6 = \frac{1}{4} \tag{2-110e}$$

$$-\frac{3}{4}a_2 + 0a_3 + 0a_4 - \frac{\sqrt{3}}{4}a_5 + \frac{3}{4}a_6 = -\frac{\sqrt{3}}{4} \tag{2-110f}$$

Solving for a_2, we find that

$$a_2 = 0$$

[2] A program for evaluating determinants will be found in A. Nussbaum, *FORTRAN Simplified*, Belmont Press, 5142 Belmont Ave., Minneapolis, Minn., 55419.

Then Eq. (2-110d) shows that

$$a_4 = \frac{1 - a_3}{\sqrt{3}}$$

and (2-110f) shows that

$$a_6 = \frac{a_5 - 1}{\sqrt{3}}$$

Putting these three equations into (2-110b) gives

$$a_3 = a_5$$

Carrying on in this fashion, we finally obtain two sets of values for the a_s, namely

$$a_1 = a_3 = a_5 = 1, \qquad a_2 = a_4 = a_6 = 0$$

and

$$a_1 = 1, a_2 = 0, a_3 = -\frac{1}{2}, a_4 = \frac{\sqrt{3}}{2}, a_5 = -\frac{1}{2}, a_6 = -\frac{\sqrt{3}}{2}$$

Normalizing, we obtain the entries in rows 4 and 6 of Table 2-7. The other row for $\lambda = 0$ and the two rows for $\lambda = \frac{3}{2}$ are obtained in the same way; the values of a_s for $\lambda = 3$ are easily obtained in terms of a_2.

To interpret this table, consider as an example the fourth row. This indicates that the amplitudes of the x-components of the motion of the

TABLE 2-7 ***Normal Mode Amplitudes for the Planar Equilateral Molecule***

λ	a_1	a_2	a_3	a_4	a_5	a_6
3	0	$\frac{1}{\sqrt{3}}$	$-\frac{1}{2}$	$-\frac{1}{2\sqrt{3}}$	$\frac{1}{2}$	$-\frac{1}{2\sqrt{3}}$
$\frac{3}{2}$	$\frac{1}{\sqrt{3}}$	0	$-\frac{1}{2\sqrt{3}}$	$-\frac{1}{2}$	$-\frac{1}{2\sqrt{3}}$	$\frac{1}{2}$
	0	$\frac{1}{\sqrt{3}}$	$\frac{1}{2}$	$-\frac{1}{2\sqrt{3}}$	$-\frac{1}{2}$	$-\frac{1}{2\sqrt{3}}$
0	$\frac{1}{\sqrt{3}}$	0	$\frac{1}{\sqrt{3}}$	0	$-\frac{1}{\sqrt{3}}$	0
	0	$\frac{1}{\sqrt{3}}$	0	$\frac{1}{\sqrt{3}}$	0	$\frac{1}{\sqrt{3}}$
	$\frac{1}{\sqrt{3}}$	0	$-\frac{1}{2\sqrt{3}}$	$\frac{1}{2}$	$-\frac{1}{2\sqrt{3}}$	$-\frac{1}{2}$

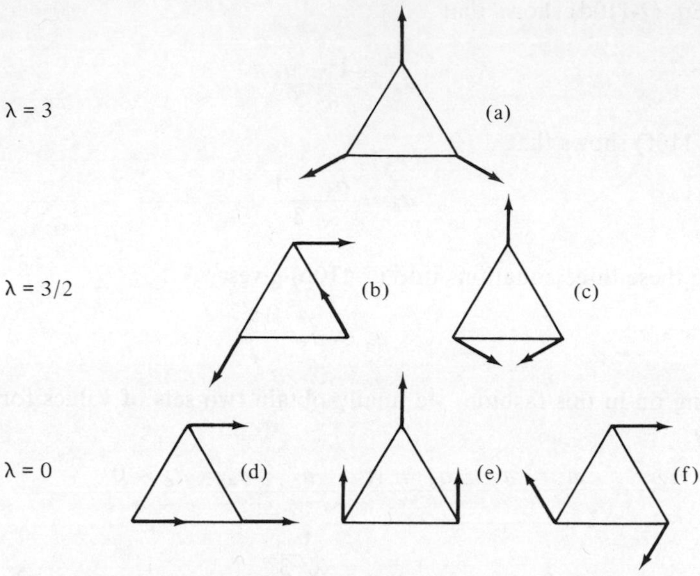

2-9 Normal modes of the planar equilateral molecule

three masses are equal to $\sqrt{3}/3$, whereas the y-components all vanish. This corresponds to a pure translation in the x-direction as indicated by Fig. 2-9(d). The next row is a translation in the y-direction [Fig. 2-9(e)], while the sixth row represents a pure rotation about an axis normal to the plane of the paper [Fig. 2-9(f)]. The expectations stated at the beginning of this section are confirmed by Fig. 2-9, which shows that the purely translational and rotational modes correspond to zero frequency.

PROBLEM 2-20

Verify the figure for the case $\lambda = 3$, which corresponds to equal amplitudes of oscillation along three directions making angles of 120° with each other.

The normal coordinates are obtained in the same way as those of Eq. (2-100), resulting in

$$q_1 = \sqrt{\frac{m}{3}}\left(y_1 - \frac{\sqrt{3}}{2}x_2 - \frac{1}{2}y_2 + \frac{\sqrt{3}}{2}x_3 - \frac{1}{2}y_3\right)$$

$$q_2 = \sqrt{\frac{m}{3}}\left(x_1 - \frac{1}{2}x_2 - \frac{\sqrt{3}}{2}y_2 - \frac{1}{2}x_3 + \frac{\sqrt{3}}{2}y_3\right)$$

$$q_3 = \sqrt{\frac{m}{3}}\left(y_1 + \frac{\sqrt{3}}{2}x_2 - \frac{1}{2}y_2 - \frac{\sqrt{3}}{2}x_3 - \frac{1}{2}y_3\right)$$

$$q_4 = \sqrt{\frac{m}{3}}(x_1 + x_2 + x_3) \qquad \textbf{(2-111)}$$

$$q_5 = \sqrt{\frac{m}{3}}(y_1 + y_2 + y_3)$$

$$q_6 = \sqrt{\frac{m}{3}}\left(x_1 - \frac{1}{2}x_2 + \frac{\sqrt{3}}{2}y_2 - \frac{1}{2}x_3 - \frac{\sqrt{3}}{2}y_3\right)$$

with inverses

$$x_1 = \frac{q_2 + q_4 + q_6}{\sqrt{3m}}$$

$$y_1 = \frac{q_1 + q_3 + q_5}{\sqrt{3m}}$$

$$x_2 = \frac{-\sqrt{3}\,q_1 - q_2 + \sqrt{3}\,q_3 + 2q_4 - q_6}{\sqrt{12m}}$$

$$y_2 = \frac{-q_1 - \sqrt{3}\,q_2 + 2q_5 + \sqrt{3}\,q_6}{\sqrt{12m}} \qquad \textbf{(2-112)}$$

$$x_3 = \frac{\sqrt{3}\,q_1 - q_2 - \sqrt{3}\,q_3 + 2q_4 - q_6}{\sqrt{12m}}$$

$$y_3 = \frac{-q_1 + \sqrt{3}\,q_2 - q_3 + 2q_5 - \sqrt{3}\,q_6}{\sqrt{12m}}$$

2-9 Projection operators

We come now to the most important concept of applied group theory: the notion of the *projection operator*. For reasons which we shall explain in great detail, it is desirable to have a method which will take the vectors used as the basis of a representation and project them along new directions which enable us to take maximum advantage of the symmetry of a given problem. It is the projection operator which performs this process. However, it is more logical to start with a simpler concept. Let us therefore study the *projection matrix*, using the approach of Hollingsworth.[3]

Consider a matrix **M** with eigenvectors $\mathbf{V}_1$, $\mathbf{V}_2$, $\mathbf{V}_3$, and associated eigenvalues λ_1, λ_2, λ_3. An arbitrary vector **V** in space may be written as the linear combination

$$\mathbf{V} = a\mathbf{V}_1 + b\mathbf{V}_2 + c\mathbf{V}_3 \qquad \textbf{(2-113)}$$

[3] C. A. Hollingsworth, *Vectors, Matrices, and Group Theory for Scientists and Engineers*, McGraw-Hill Book Company, New York, 1967.

provided the eigenvectors are noncoplanar. If $\mathbf{M}$ is symmetric, we have seen that the eigenvectors will be mutually perpendicular, so (2-113) will be valid. Now let us define three matrices $\mathbf{P}_1$, $\mathbf{P}_2$, $\mathbf{P}_3$ such that

$$\mathbf{P}_1\mathbf{V} = a\mathbf{V}_1, \qquad \mathbf{P}_2\mathbf{V} = b\mathbf{V}_2, \qquad \mathbf{P}_3\mathbf{V} = c\mathbf{V}_3 \tag{2-114}$$

That is, the matrices $\mathbf{P}_1$, $\mathbf{P}_2$, $\mathbf{P}_3$ *project* $\mathbf{V}$ along the three directions specified by the eigenvectors, and hence are *projection matrices.*

To express the $\mathbf{P}_i$ in terms of $\mathbf{M}$ and the λ_i, we form the matrix $(\mathbf{M} - \lambda_3\mathbf{I})$, where $\mathbf{I}$ is the unit matrix, as in (2-18). Then

$$(\mathbf{M} - \lambda_3\mathbf{I})\mathbf{V}_3 = \mathbf{M}\mathbf{V}_3 - \lambda_3\mathbf{I}\mathbf{V}_3 = \lambda_3\mathbf{V}_3 - \lambda_3\mathbf{V}_3 = 0$$

But

$$(\mathbf{M} - \lambda_3\mathbf{I})\mathbf{V}_2 = (\lambda_2 - \lambda_3)\mathbf{V}_2$$

$$(\mathbf{M} - \lambda_3\mathbf{I})\mathbf{V}_1 = (\lambda_1 - \lambda_3)\mathbf{V}_1$$

Further

$$(\mathbf{M} - \lambda_2\mathbf{I})(\mathbf{M} - \lambda_3\mathbf{I})\mathbf{V}_3 = 0 \tag{2-115}$$

$$(\mathbf{M} - \lambda_2\mathbf{I})(\mathbf{M} - \lambda_3\mathbf{I})\mathbf{V}_2 = 0 \tag{2-116}$$

and

$$(\mathbf{M} - \lambda_2\mathbf{I})(\mathbf{M} - \lambda_3\mathbf{I})\mathbf{V}_1 = (\lambda_1 - \lambda_2)(\lambda_1 - \lambda_3)\mathbf{V}_1 \tag{2-117}$$

These relations show that

$$\mathbf{P}_1 = \frac{(\mathbf{M} - \lambda_2\mathbf{I})(\mathbf{M} - \lambda_3\mathbf{I})}{(\lambda_1 - \lambda_2)(\lambda_1 - \lambda_3)} \tag{2-118}$$

for by (2-117)

$$\mathbf{P}_1\mathbf{V}_1 = \mathbf{V}_1$$

and by (2-115) and (2-116)

$$\mathbf{P}_1\mathbf{V}_2 = \mathbf{P}_1\mathbf{V}_3 = 0$$

Thus

$$\mathbf{P}_1\mathbf{V} = a\mathbf{P}_1\mathbf{V}_1 + b\mathbf{P}_1\mathbf{V}_2 + c\mathbf{P}_1\mathbf{V}_3 = a\mathbf{V}_1$$

as required by (2-114). Expressions for $\mathbf{P}_2$ and $\mathbf{P}_3$ are obtained from (2-118) by cyclic permutation of the subscripts.

Note that Eq. (2-118) specifically assumes that

$$\lambda_1 \neq \lambda_2 \neq \lambda_3 \tag{2-119}$$

If this is so, the eigenvalues are said to be *nondegenerate.*

Projection operators also have the property that

$$\mathbf{P}_1(\mathbf{P}_1\mathbf{V}) = \mathbf{P}_1 a\mathbf{V}_1 = a\mathbf{V}_1$$

or, in general,

$$\mathbf{P}_i\mathbf{P}_i = \mathbf{P}_i \tag{2-120}$$

Because of this, the $\mathbf{P}_i$ are known as *idempotent* matrices. Also

$$\mathbf{P}_2(\mathbf{P}_1\mathbf{V}) = \mathbf{P}_2 a\mathbf{V}_1 = 0$$

or
$$\mathbf{P}_i\mathbf{P}_j = 0, \quad (i \neq j) \tag{2-121}$$

This is the *orthogonality* property.

A matrix is said to be *degenerate* if two (or more) of the eigenvalues are identical. If $\lambda_2 = \lambda_3$, for example, then $\mathbf{P}_2$ and $\mathbf{P}_3$ cannot be defined by an expression like (2-118), for the term $(\lambda_2 - \lambda_3)$ would appear in both denominators.

Returning to the derivation of (2-118), we now have

$$(\mathbf{M} - \lambda_3\mathbf{I})\mathbf{V}_3 = (\mathbf{M} - \lambda_2\mathbf{I})\mathbf{V}_3 = 0$$
$$(\mathbf{M} - \lambda_3\mathbf{I})\mathbf{V}_2 = (\mathbf{M} - \lambda_2\mathbf{I})\mathbf{V}_2 = (\lambda_2 - \lambda_3)\mathbf{V}_2 = 0$$

and
$$(\mathbf{M} - \lambda_3\mathbf{I})\mathbf{V}_1 = (\lambda_1 - \lambda_3)\mathbf{V}_1$$

or
$$(\mathbf{M} - \lambda_2\mathbf{I})\mathbf{V}_1 = (\lambda_1 - \lambda_2)\mathbf{V}_1$$

from which
$$\mathbf{P}_1 = \frac{(\mathbf{M} - \lambda_2\mathbf{I})}{(\lambda_1 - \lambda_2)} \tag{2-122}$$

since
$$\mathbf{P}_1\mathbf{V}_1 = \mathbf{V}_1, \qquad \mathbf{P}_1\mathbf{V}_2 = \mathbf{P}_1\mathbf{V}_3 = 0$$

However, we notice that

$$(\mathbf{M} - \lambda_1\mathbf{I})\mathbf{V}_3 = (\lambda_2 - \lambda_1)\mathbf{V}_3$$
$$(\mathbf{M} - \lambda_1\mathbf{I})\mathbf{V}_2 = (\lambda_2 - \lambda_1)\mathbf{V}_2$$
$$(\mathbf{M} - \lambda_1\mathbf{I})\mathbf{V}_1 = 0$$

so that the only other matrix of the form $(\mathbf{M} - \lambda_i\mathbf{I})$ gives

$$(\mathbf{M} - \lambda_1\mathbf{I})\mathbf{V} = a(0) + b(\lambda_2 - \lambda_1)\mathbf{V}_2 + c(\lambda_2 - \lambda_1)\mathbf{V}_3$$

Hence, we define the projection matrix associated with the degenerate eigenvalues as

$$\mathbf{P}_{22} = \frac{(\mathbf{M} - \lambda_1\mathbf{I})}{(\lambda_2 - \lambda_1)} \tag{2-123}$$

This matrix, when applied to $\mathbf{V}$, gives

$$\mathbf{P}_{22}\mathbf{V} = b\mathbf{V}_2 + c\mathbf{V}_3 \tag{2-124}$$

That is, it projects $\mathbf{V}$ onto the plane defined by $\mathbf{V}_2$ and $\mathbf{V}_3$. We may think of this plane, in fact, as a *two-dimensional degenerate space*.

PROBLEM 2-21

Show that

$$\mathbf{P}_1 + \mathbf{P}_2 + \mathbf{P}_3 = \mathbf{I} \quad \text{and} \quad \mathbf{P}_1 + \mathbf{P}_{22} = \mathbf{I} \tag{2-125}$$

The projection matrices introduced above accomplish what we would like them to do; they take an arbitrary vector and annihilate all its components except those in a specified direction. However, they are not very useful because their structure involves a knowledge of the eigenvalues and eigenvectors of a matrix **M**. To find the λ_i, we generally have to solve a complicated secular determinant, such as the one in Eq. (2-107), and this is the problem which led us to projection methods in the first place. That is, we are going around in circles.

All is not lost, however, for the form of the projection matrix will lead us to the desired projection operator. We note that the three $\mathbf{P}_i$ in the nondegenerate case, above, may be written as

$$\mathbf{P}_i = \prod_{\substack{j=1 \\ i \neq j}}^{m} \frac{\mathbf{M} - \lambda_j \mathbf{I}}{\lambda_i - \lambda_j} \tag{2-126}$$

PROBLEM 2-22

Use the identity $$\frac{\mathbf{M} - \lambda_j \mathbf{I}}{\lambda_i - \lambda_j} = \mathbf{I} + \frac{\mathbf{M} - \lambda_i \mathbf{I}}{\lambda_i - \lambda_j}$$

to show that $\mathbf{P}_i$ is idempotent.

To apply (2-126) to the problem under consideration, we shall let **M** be a set of matrices associated with the class sums of the regular representation of a group. For example, we saw that the class sums for C_{3v} are

$$\begin{aligned} \mathscr{C}_1 &= E \\ \mathscr{C}_2 &= C_3 + C_3^2 \\ \mathscr{C}_3 &= \sigma_v^a + \sigma_v^b + \sigma_v^c \end{aligned} \tag{2-68}$$

and that the regular representation for σ_v^a is

$$\sigma_v^a = \begin{pmatrix} 0 & 0 & 0 & 1 & 0 & 0 \\ 0 & 0 & 0 & 0 & 0 & 1 \\ 0 & 0 & 0 & 0 & 1 & 0 \\ 1 & 0 & 0 & 0 & 0 & 0 \\ 0 & 0 & 1 & 0 & 0 & 0 \\ 0 & 1 & 0 & 0 & 0 & 0 \end{pmatrix} \tag{2-47}$$

We may obtain the matrices in the regular representation for σ_v^b and σ_v^c from

Table 2-4, and using these shows that the matrix $\mathbf{C}_3$ for the class sum $\mathscr{C}_3$ is

$$\mathbf{C}_3 = \begin{pmatrix} 0 & 0 & 0 & 1 & 1 & 1 \\ 0 & 0 & 0 & 1 & 1 & 1 \\ 0 & 0 & 0 & 1 & 1 & 1 \\ 1 & 1 & 1 & 0 & 0 & 0 \\ 1 & 1 & 1 & 0 & 0 & 0 \\ 1 & 1 & 1 & 0 & 0 & 0 \end{pmatrix} \tag{2-127}$$

To diagonalize this matrix, we use the method of Sec. 2-2, obtaining the eigenvalue equation

$$\begin{vmatrix} -\lambda & 0 & 0 & 1 & 1 & 1 \\ 0 & -\lambda & 0 & 1 & 1 & 1 \\ 0 & 0 & -\lambda & 1 & 1 & 1 \\ 1 & 1 & 1 & -\lambda & 0 & 0 \\ 1 & 1 & 1 & 0 & -\lambda & 0 \\ 1 & 1 & 1 & 0 & 0 & -\lambda \end{vmatrix} = 0 \tag{2-128}$$

which is equivalent to

$$\lambda^4(\lambda^2 - 9) = 0 \tag{2-129}$$

PROBLEM 2-23

Prove without actually expanding the 6×6 determinant that (2-128) reduces to (2-129).

The roots of (2-129) are

$$\lambda = 0, 0, 0, 0, 3, -3 \tag{2-130}$$

and we shall evaluate (2-126) for each distinct eigenvalue.

For $\lambda = 0$, we obtain

$$\mathbf{P}_0 = \frac{(\mathbf{C}_3 - 3\mathbf{I})}{(3 - 0)} \frac{(\mathbf{C}_3 + 3\mathbf{I})}{(-3 - 0)}$$

$$= -\tfrac{1}{9}(\mathbf{C}_3^2 - 9\mathbf{I})$$

But $$\mathscr{C}_3^2 = 3(E + C_3 + C_3^2) = 3(\mathscr{C}_1 + \mathscr{C}_2)$$

with a corresponding relation for the matrices. Hence

$$\begin{aligned}\mathbf{P}_0 &= -\tfrac{1}{3}(\mathbf{C}_1 + \mathbf{C}_2) + \mathbf{I} \\ &= \tfrac{2}{3}\mathbf{I} - \tfrac{1}{3}\mathbf{C}_2\end{aligned} \tag{2-131}$$

If we convert this from a matrix form to an operator form, then the projection operator P_0 corresponding to $\lambda = 0$ should be defined by the equation

$$P_0 = \tfrac{1}{3}(2E - C_3 - C_3^2) \tag{2-132}$$

In the same way, the eigenvalue $\lambda = 3$ gives

$$\begin{aligned}\mathbf{P}_3 &= \frac{(\mathbf{C}_3 - 0\mathbf{I})(\mathbf{C}_3 + 3\mathbf{I})}{(3 - 0)\{3 - (-3)\}} \\ &= \tfrac{1}{18}(\mathbf{C}_3^2 + 3\mathbf{C}_3)\end{aligned}$$

or

$$P_3 = \tfrac{1}{6}(E + C_3 + C_3^2 + \sigma_v^a + \sigma_v^b + \sigma_v^c) \tag{2-133}$$

and

$$\begin{aligned}\mathbf{P}_{-3} &= \frac{(\mathbf{C}_3 - 0\mathbf{I})(\mathbf{C}_3 - 3\mathbf{I})}{(-3 - 0)(-3 - 3)} \\ &= \tfrac{1}{18}(\mathbf{C}_3^2 - 3\mathbf{C}_3)\end{aligned}$$

or

$$P_{-3} = \tfrac{1}{6}(E + C_3 + C_3^2 - \sigma_v^a - \sigma_v^b - \sigma_v^c) \tag{2-134}$$

Next, consider the matrix $\mathbf{C}_2$ for the class sum $\mathscr{C}_2$. This leads to the eigenvalue equation

$$\begin{vmatrix} -\lambda & 1 & 1 & 0 & 0 & 0 \\ 1 & -\lambda & 1 & 0 & 0 & 0 \\ 1 & 1 & -\lambda & 0 & 0 & 0 \\ 0 & 0 & 0 & -\lambda & 1 & 1 \\ 0 & 0 & 0 & 1 & -\lambda & 1 \\ 0 & 0 & 0 & 1 & 1 & -\lambda \end{vmatrix} = 0$$

or

$$(-\lambda^3 + 3\lambda + 2)^2 = [(\lambda - 2)(\lambda + 1)^2]^2 = 0$$

with roots

$$\lambda = 2, 2, -1, -1, -1, -1$$

Then we find that

$$P_2 = \tfrac{1}{3}(E + C_3 + C_3^2) \tag{2-135}$$

and

$$P_{-1} = -\tfrac{1}{3}(2E - C_3 - C_3^2) \tag{2-136}$$

The secular equation for $\mathscr{C}_1$ is simply

$$(1 - \lambda)^6 = 0$$

and since this eigenvalue is six-fold degenerate, there is no projection operator. The procedure given above is based on a discussion by McIntosh.[4]

To interpret these results, we realize that only the eigenvalues $\lambda = 3$ and $\lambda = -3$ are nondegenerate, so that the true projection operators are P_3 and P_{-3} of (2-133) and (2-134), respectively. All the others project in some arbitrary direction in a degenerate subspace, like $\mathbf{P}_{22}$ of (2-123), rather than along an explicit direction. Hence, we use these two as a model on which to base a general definition of projection operators.

We note first that (2-133) may be written as

$$P_3 = \sum_{l=1}^{6} \chi_1(G_l)G_l$$

where G_l denotes the six operations of C_{3v}, the quantities $\chi_1(G_l)$ are the characters of the irreducible representation $\mathbf{\Gamma}_1$ of Table 2-5, and the factor $\frac{1}{6}$ is ignored. In the same way (2-134) gives

$$P_{-3} = \sum_{l=1}^{6} \chi_2(G_l)G_l$$

where the $\chi_2(G_l)$ are the characters for $\mathbf{\Gamma}_2$. These two expressions lead naturally to the definition of $P^{(k)}$, the projection operator belonging to the kth irreducible representation, as

$$P^{(k)} = \sum_{l=1}^{g} \chi^{(k)}(G_l)G_l \tag{2-137}$$

where g is the order of the group. We shall use the term *character projection operators* to describe the $P^{(k)}$. Although many authors[5] on group theory use (2-137) as their basic definition, we have seen that it leads to ambiguities when the eigenvalues are degenerate. We note, incidentally, that P_0 of (2-132) fits the character projection operator definition, since this is what we obtain when (2-137) is applied to $\mathbf{\Gamma}_3$ of Table 2-5.

In order to produce a projection operator definition of maximum usefulness, we return to (2-133) and (2-134), and utilize the fact that one-dimensional representations consist of 1×1 matrices. Then we may write (2-133) as

$$P_3 = \sum_{l=1}^{g} D_{11}^{(1)}(R_l)R_l$$

where the symbol $D^{(k)}(R_l)$ denotes the matrix which serves as the representation of the operator R_l in the kth irreducible representation, so that $D_{11}^{(1)}(R_l)$ is the 11-element (actually, the only element) of the six matrices associated

[4] H. V. McIntosh, *Notes on the Physical Applications of Group Theory*, Cornell University, 1954.

[5] J. W. Leech and D. W. Newman, *How to Use Group Theory*, Methuen, London. 1969.

with $\mathbf{\Gamma}_1$. Similarly, (2-134) is

$$P_{-3} = \sum_{l=1}^{6} D_{11}^{(2)}(R_l)R_l$$

For the 2×2 matrices associated with $\mathbf{\Gamma}_3$, it is necessary to perform summations like this on four sets of elements, and it is logical to define the projection operator $P_{ij}^{(k)}$ associated with the ij elements of the kth irreducible representation as

$$P_{ij}^{(k)} = \sum_{l=1}^{g} D_{ij}^{(k)}(R_l)R_l$$

However, in order for this definition to agree numerically with previous ones, we modify it to obtain

$$P_{ij}^{(k)} = \frac{d_k}{g} \sum_{l=1}^{g} D_{ij}^{(k)}(R_l)R_l \tag{2-138}$$

where d_k is the dimension of the representation. *This result is perhaps the single most important equation in this book.* It defines the *matrix-element projection operators*, which we shall call simply projection operators, since the character projection operators of (2-137) are not nearly so useful.

To see how (2-138) is used, we consider a simple geometric example of Schonland.[6] A basis for the group C_{3v} can be established using three mutually perpendicular vectors $\mathbf{d}_1, \mathbf{d}_2, \mathbf{d}_3$ of unit length, shown in Fig. 2-10. The point

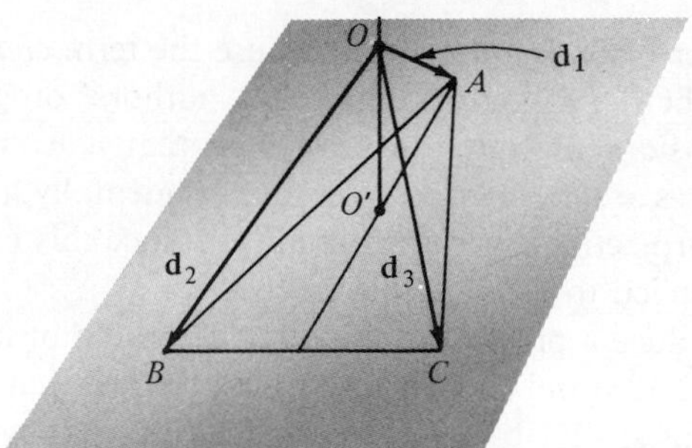

2-10 A reducible basis for the equilateral triangle

O' is the center of the equilateral triangle ABC, and OO' is perpendicular to the triangle. The effect of each operation of C_{3v} on the basis vectors is given in Table 2-8, and the 3×3 matrices comprising the representation $\mathbf{\Gamma}_6$ are simply the upper left-hand (or lower right-hand) corners of the matrices in Table 2-3.

[6] D. S. Schonland, *Molecular Symmetry*, D. Van Nostrand Co., Inc., London, 1965.

TABLE 2-8 ***Effect of C_{3v} on the Basis of Fig. 2-10***

	E	C_3	C_3^2	σ_v^a	σ_v^b	σ_v^c
$\mathbf{d}_1$	$\mathbf{d}_1$	$\mathbf{d}_2$	$\mathbf{d}_3$	$\mathbf{d}_1$	$\mathbf{d}_3$	$\mathbf{d}_2$
$\mathbf{d}_2$	$\mathbf{d}_2$	$\mathbf{d}_3$	$\mathbf{d}_1$	$\mathbf{d}_3$	$\mathbf{d}_2$	$\mathbf{d}_1$
$\mathbf{d}_3$	$\mathbf{d}_3$	$\mathbf{d}_1$	$\mathbf{d}_2$	$\mathbf{d}_2$	$\mathbf{d}_1$	$\mathbf{d}_3$

Returning to Table 2-2, we observe that there are six matrix elements $\Gamma_{ij}^{(k)}(G_l)$ corresponding to each member G_l of the group, where the group has g members $G_1, G_2, \ldots, G_g$. For example, there are three matrices under σ_v^a; two are one-dimensional and one is two-dimensional. The elements are:

for $\mathbf{\Gamma}_1$ $\qquad \mathbf{\Gamma}_{11}^{(1)}(\sigma_v^a) = 1$

for $\mathbf{\Gamma}_2$ $\qquad \mathbf{\Gamma}_{11}^{(2)}(\sigma_v^a) = -1$

and for $\mathbf{\Gamma}_3$ $\qquad \mathbf{\Gamma}_{11}^{(3)}(\sigma_v^a) = -1$

$$\mathbf{\Gamma}_{12}^{(3)}(\sigma_v^a) = \mathbf{\Gamma}_{21}^{(3)}(\sigma_v^a) = 0$$

$$\mathbf{\Gamma}_{22}^{(3)}(\sigma_v^a) = 1$$

Using (2-138) to form the projection operators, we obtain

$$P_{11}^{(1)} = (1)E + (1)C_3 + (1)C_3^2 + (1)\sigma_v^a + (1)\sigma_v^b + (1)\sigma_v^c$$

as we had in (2-133) (ignoring the numerical coefficient for the moment). Similarly, $P_{11}^{(2)}$ would agree with (2-134). However, the four operators $P_{ij}^{(3)}$ are new, and we have, for example,

$$P_{12}^{(3)} = (0)E + \frac{\sqrt{3}}{2}C_3 + \frac{\sqrt{3}}{2}C_3^2 + 0\sigma_v^a + \frac{\sqrt{3}}{2}\sigma_v^b - \frac{\sqrt{3}}{2}\sigma_v^c$$

Table 2-8 then shows that these two operators acting on $\mathbf{d}_1$, say, give

$$P_{11}^{(1)}\mathbf{d}_1 = \mathbf{d}_1 + \mathbf{d}_2 + \mathbf{d}_3 + \mathbf{d}_1 + \mathbf{d}_3 + \mathbf{d}_2 = 2(\mathbf{d}_1 + \mathbf{d}_2 + \mathbf{d}_3)$$

and

$$P_{12}^{(3)}\mathbf{d}_1 = \frac{-\sqrt{3}}{2}\mathbf{d}_2 + \frac{\sqrt{3}}{2}\mathbf{d}_3 + \frac{\sqrt{3}}{2}\mathbf{d}_3 + \frac{-\sqrt{3}}{2}\mathbf{d}_2 = \sqrt{3}(\mathbf{d}_3 - \mathbf{d}_2)$$

Using this procedure for the effect of all six projection operators on the three vectors of the basis, we obtain Table 2-9. The results have been expressed in terms of the linear combinations $(\mathbf{d}_1 + \mathbf{d}_2 + \mathbf{d}_3)$, $(2\mathbf{d}_1 - \mathbf{d}_2 - \mathbf{d}_3)$, and $(\mathbf{d}_2 - \mathbf{d}_3)$.

TABLE 2-9 ***Projections of the Basis Vectors of Fig. 2-10***

	$\mathbf{d}_1$	$\mathbf{d}_2$	$\mathbf{d}_3$
$P_{11}^{(1)}$	$2(\mathbf{d}_1 + \mathbf{d}_2 + \mathbf{d}_3)$	$2(\mathbf{d}_1 + \mathbf{d}_2 + \mathbf{d}_3)$	$2(\mathbf{d}_1 + \mathbf{d}_2 + \mathbf{d}_3)$
$P_{11}^{(2)}$	0	0	0
$P_{11}^{(3)}$	0	$\frac{3}{2}(\mathbf{d}_2 - \mathbf{d}_3)$	$\frac{3}{2}(\mathbf{d}_3 - \mathbf{d}_2)$
$P_{12}^{(3)}$	$\frac{\sqrt{3}}{2}(\mathbf{d}_3 - \mathbf{d}_2)$	$\frac{\sqrt{3}}{2}(\mathbf{d}_2 - \mathbf{d}_3)$	$\frac{\sqrt{3}}{2}(\mathbf{d}_2 - \mathbf{d}_3)$
$P_{21}^{(3)}$	0	$\frac{\sqrt{3}}{2}(-2\mathbf{d}_1 + \mathbf{d}_2 + \mathbf{d}_3)$	$\frac{\sqrt{3}}{2}(2\mathbf{d}_1 - \mathbf{d}_2 - \mathbf{d}_3)$
$P_{22}^{(3)}$	$2\mathbf{d}_1 - \mathbf{d}_2 - \mathbf{d}_3$	$\frac{1}{2}(-2\mathbf{d}_1 + \mathbf{d}_2 + \mathbf{d}_3)$	$\frac{1}{2}(-2\mathbf{d}_1 + \mathbf{d}_2 + \mathbf{d}_3)$

PROBLEM 2-24

(a) Verify the entries for $P_{11}^{(2)}$ and $P_{22}^{(3)}$.

(b) Verify from Table 2-9 that some projection operators have the orthogonality and idempotent properties that we found for projection matrices.

In performing the calculations called for by Problem 2-24(b), it will be found that $P_{11}^{(1)}, P_{11}^{(2)}, P_{11}^{(3)}$, and $P_{22}^{(3)}$ have the idempotent property but that $P_{12}^{(3)}$ and $P_{21}^{(3)}$ do not. For this reason, some authors do not regard $P_{ij}^{(k)}$ for $i \neq j$ as a projection operator and Hollingsworth[3] uses the term *shift operator*. We shall not bother with this distinction.

Let us ignore the coefficients in Table 2-9, such as the factor of 2 in front of $(\mathbf{d}_1 + \mathbf{d}_2 + \mathbf{d}_3)$, and define a new basis in terms of the normalized combinations. Let

$$\mathbf{e}_1 = \frac{1}{\sqrt{3}}(\mathbf{d}_1 + \mathbf{d}_2 + \mathbf{d}_3)$$

$$\mathbf{e}_2 = \frac{1}{\sqrt{6}}(2\mathbf{d}_1 - \mathbf{d}_2 - \mathbf{d}_3) \tag{2-139}$$

$$\mathbf{e}_3 = \frac{1}{\sqrt{2}}(\mathbf{d}_2 - \mathbf{d}_3)$$

Since the vectors $\mathbf{d}_1, \mathbf{d}_2, \mathbf{d}_3$ are orthogonal, it is easy to see that (2-139) de-

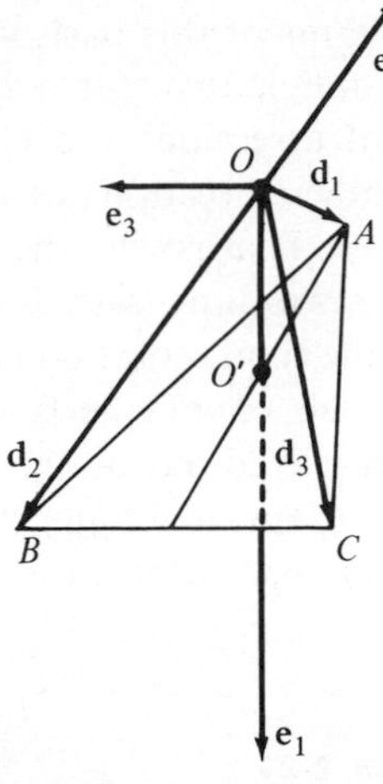

2-11 The basis generated from the previous figure by the projection operator method

fines unit vectors $\mathbf{e}_1, \mathbf{e}_2, \mathbf{e}_3$. The vector $(\mathbf{d}_1 + \mathbf{d}_2 + \mathbf{d}_3)$ lies along the normal to the plane of the triangle ABC of Fig. 2-10 and points downward. The vectors $(2\mathbf{d}_1 - \mathbf{d}_2 - \mathbf{d}_3)$ and $(\mathbf{d}_2 - \mathbf{d}_3)$ lie parallel to this plane and point in the direction shown in Fig. 2-11.

PROBLEM 2-25

Verify these statements by expressing $\mathbf{d}_1, \mathbf{d}_2$, and $\mathbf{d}_3$ in rectangular coordinates.

To check the orthogonality of $\mathbf{e}_1, \mathbf{e}_2, \mathbf{e}_3$, we compute scalar products. For example,

$$\mathbf{e}_1 \cdot \mathbf{e}_2 = \frac{1}{\sqrt{18}}(2d_1^2 - d_2^2 - d_3^2 + \mathbf{d}_1 \cdot \mathbf{d}_2 - 2\mathbf{d}_2 \cdot \mathbf{d}_3 + \mathbf{d}_1 \cdot \mathbf{d}_3) = 0$$

where we have used the fact that the basis $\mathbf{d}_1, \mathbf{d}_2, \mathbf{d}_3$ is orthogonal. Hence, the new basis is *orthonormal*; the vectors $\mathbf{e}_1, \mathbf{e}_2, \mathbf{e}_3$ are both orthogonal and normalized.

Aside from the geometry, the difference between the two bases lies in their relation to C_{3v}. In fact, if we consider the effect of each symmetry operation on the $\mathbf{e}_i$, it is clear that $\mathbf{e}_1$ is unaffected by any member R_i of the group, whereas $\mathbf{e}_2$ and $\mathbf{e}_3$ are converted into linear combinations of themselves. The projection operators have thus reduced the 3×3 basis $\mathbf{\Gamma}_6$; this is indicated by the relation

$$\mathbf{\Gamma}_6(R_i) = \begin{pmatrix} \mathbf{\Gamma}_1(R_i) & 0 & 0 \\ 0 & \multicolumn{2}{c}{\multirow{2}{*}{$\mathbf{\Gamma}_3(R_i)$}} \\ 0 & & \end{pmatrix}, \quad (i = 1, \ldots, 6) \tag{2-140}$$

and any secular equation expressed in terms of this basis would be similarly reduced. This result should have been anticipated, for a system with threefold symmetry should have one special direction: the C_3-axis. The plane normal to this axis then involves two other directions of equal importance, and they must enter the problem together. That is why the representation $\mathbf{\Gamma}_3$ of Table 2-2 is associated with the two-dimensional degenerate space specified by the plane of $\mathbf{e}_2$ and $\mathbf{e}_3$, and the representation is said to be degenerate as well. Further, $\mathbf{e}_2$ and $\mathbf{e}_3$ as a basis cannot be further separated or reduced. On the other hand, $\mathbf{e}_1$ has been picked out by the reduction process and it should belong to $\mathbf{\Gamma}_1$, since it is invariant under all six symmetry operations. We note, however, that under these operations, $\mathbf{e}_2$ and $\mathbf{e}_3$ continue to *mix*.

PROBLEM 2-26

Suppose that we had chosen $\mathbf{d}_1, \mathbf{d}_2, \mathbf{d}_3$ as identical to $\mathbf{r}_1, \mathbf{r}_2, \mathbf{r}_3$, respectively, of Fig. 2-3. Determine the new basis generated by the projection operators and explain why it fails to produce the same results as $\mathbf{e}_1, \mathbf{e}_2, \mathbf{e}_3$.

2-10 Projection operators applied to the equilateral molecule

Returning to the vibration problem of Sec. 2-8, we shall use the projection operator method to simplify the secular determinant of Eq. (2-107). Using the previous section as a guide, we shall choose the six vectors $\mathbf{e}_1, \ldots, \mathbf{e}_6$ of Fig. 2-12 as a reducible basis. That is, the symmetry of the problem is incorporated into the arbitrary initial choice.

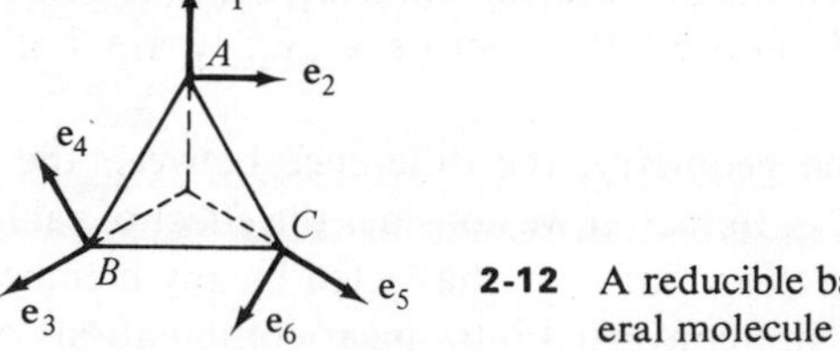

2-12 A reducible basis for the equilateral molecule

PROBLEM 2-27

Show that typical members of each class in the reducible representation $\mathbf{\Gamma}_7$ corresponding to Fig. 2-12 are

$$E = \begin{pmatrix} 1 & 0 & 0 & 0 & 0 & 0 \\ 0 & 1 & 0 & 0 & 0 & 0 \\ 0 & 0 & 1 & 0 & 0 & 0 \\ 0 & 0 & 0 & 1 & 0 & 0 \\ 0 & 0 & 0 & 0 & 1 & 0 \\ 0 & 0 & 0 & 0 & 0 & 1 \end{pmatrix}, \quad C_3 = \begin{pmatrix} 0 & 0 & 1 & 0 & 0 & 0 \\ 0 & 0 & 0 & 1 & 0 & 0 \\ 0 & 0 & 0 & 0 & 1 & 0 \\ 0 & 0 & 0 & 0 & 0 & 1 \\ 1 & 0 & 0 & 0 & 0 & 0 \\ 0 & 1 & 0 & 0 & 0 & 0 \end{pmatrix}$$

$$\sigma_v^a = \begin{pmatrix} 1 & 0 & 0 & 0 & 0 & 0 \\ 0 & -1 & 0 & 0 & 0 & 0 \\ 0 & 0 & 0 & 0 & 1 & 0 \\ 0 & 0 & 0 & 0 & 0 & -1 \\ 0 & 0 & 1 & 0 & 0 & 0 \\ 0 & 0 & 0 & -1 & 0 & 0 \end{pmatrix}$$

The corresponding characters, as well as the irreducible ones from Table 2-6, are given in Table 2-10.

TABLE 2-10 ***Character System for C_{3v} and for a Reducible Representation***

	E	$2C_3^2$	$3\sigma_v$
Γ_1	1	1	1
Γ_2	1	1	−1
Γ_3	2	−1	0
Γ_7	6	0	0

Note that the headings indicate the class members in the brief way common in the literature; for example, C_3 and C_3^2 are denoted as $2C_3$. Using (2-63), we find that

$$\begin{aligned} n_1 &= \frac{6(1) + 2(0)(1) + 3(0)(1)}{6} = 1 \\ n_2 &= \frac{6(1) + 2(0)(1) + 3(0)(-1)}{6} = 1 \\ n_3 &= \frac{6(2) + 2(0)(-1) + 3(0)(0)}{6} = 2 \end{aligned} \tag{2-141}$$

so that the reduced representation Γ_7' will contain the 1×1 matrix Γ_1, the 1×1 matrix Γ_2, and two 2×2 matrices Γ_3 on the main diagonal.

To generate the six projection operators, we should know the effect of each symmetry operation on the basis vectors $\mathbf{e}_i$, and this information is given in Table 2-11. From this table, we find as examples that

TABLE 2-11 *Effect of C_{3v} on the Basis of Fig. 2-12*

	E	C_3	C_3^2	σ_v^a	σ_v^b	σ_v^c
$\mathbf{e}_1$	$\mathbf{e}_1$	$\mathbf{e}_3$	$\mathbf{e}_5$	$\mathbf{e}_1$	$\mathbf{e}_5$	$\mathbf{e}_3$
$\mathbf{e}_2$	$\mathbf{e}_2$	$\mathbf{e}_4$	$\mathbf{e}_6$	$-\mathbf{e}_2$	$-\mathbf{e}_6$	$-\mathbf{e}_4$
$\mathbf{e}_3$	$\mathbf{e}_3$	$\mathbf{e}_5$	$\mathbf{e}_1$	$\mathbf{e}_5$	$\mathbf{e}_3$	$\mathbf{e}_1$
$\mathbf{e}_4$	$\mathbf{e}_4$	$\mathbf{e}_6$	$\mathbf{e}_2$	$-\mathbf{e}_6$	$-\mathbf{e}_4$	$-\mathbf{e}_2$
$\mathbf{e}_5$	$\mathbf{e}_5$	$\mathbf{e}_1$	$\mathbf{e}_3$	$\mathbf{e}_3$	$\mathbf{e}_1$	$\mathbf{e}_5$
$\mathbf{e}_6$	$\mathbf{e}_6$	$\mathbf{e}_2$	$\mathbf{e}_4$	$-\mathbf{e}_4$	$-\mathbf{e}_2$	$-\mathbf{e}_6$

$$
\begin{aligned}
P_{11}^{(1)}\mathbf{e}_1 &= (1)E\mathbf{e}_1 + (1)C_3\mathbf{e}_1 + (1)C_3^2\mathbf{e}_1 + (1)\sigma_v^a\mathbf{e}_1 + (1)\sigma_v^b\mathbf{e}_1 + (1)\sigma_v^c\mathbf{e}_1 \\
&= \mathbf{e}_1 + \mathbf{e}_3 + \mathbf{e}_5 + \mathbf{e}_1 + \mathbf{e}_3 + \mathbf{e}_5 = 2(\mathbf{e}_1 + \mathbf{e}_3 + \mathbf{e}_5)
\end{aligned}
$$

and
$$
\begin{aligned}
P_{12}^{(3)}\mathbf{e}_1 &= (0)E\mathbf{e}_1 + \left(-\frac{\sqrt{3}}{2}\right)C_3\mathbf{e}_1 + \frac{\sqrt{3}}{2}C_3^2\mathbf{e}_1 \\
&\quad + (0)\sigma_v^a\mathbf{e}_1 + \frac{\sqrt{3}}{2}\sigma_v^b\mathbf{e}_1 + \left(-\frac{\sqrt{3}}{2}\right)\sigma_v^c\mathbf{e}_1 \\
&= -\frac{\sqrt{3}}{2}\mathbf{e}_3 + \frac{\sqrt{3}}{2}\mathbf{e}_5 + \frac{\sqrt{3}}{2}(-\mathbf{e}_3) - \frac{\sqrt{3}}{2}(-\mathbf{e}_5) \\
&= \sqrt{3}\,(\mathbf{e}_5 - \mathbf{e}_3)
\end{aligned}
$$

In a similar way, we may generate all 36 entries of Table 2-12.

The projection operators have produced a new basis which we normalize and denote as

$$
\begin{aligned}
\mathbf{g}_1 &= \frac{1}{\sqrt{3}}(\mathbf{e}_1 + \mathbf{e}_3 + \mathbf{e}_5) \\
\mathbf{g}_2 &= \frac{1}{\sqrt{3}}(\mathbf{e}_2 + \mathbf{e}_4 + \mathbf{e}_6) \\
\mathbf{g}_3 &= \frac{1}{\sqrt{6}}(2\mathbf{e}_2 - \mathbf{e}_4 - \mathbf{e}_6) \\
\mathbf{g}_4 &= \frac{1}{\sqrt{2}}(\mathbf{e}_3 - \mathbf{e}_5) \\
\mathbf{g}_5 &= \frac{1}{\sqrt{6}}(2\mathbf{e}_1 - \mathbf{e}_3 - \mathbf{e}_5) \\
\mathbf{g}_6 &= \frac{1}{\sqrt{2}}(\mathbf{e}_4 - \mathbf{e}_6)
\end{aligned}
$$

TABLE 2-12 *Projections of the Basis Vectors of Fig. 2-12*

	$\mathbf{e}_1$	$\mathbf{e}_2$	$\mathbf{e}_3$	$\mathbf{e}_4$	$\mathbf{e}_5$	$\mathbf{e}_6$
$P_{11}^{(1)}$	$2(\mathbf{e}_1 + \mathbf{e}_2 + \mathbf{e}_3)$	0	$2(\mathbf{e}_1 + \mathbf{e}_2 + \mathbf{e}_3)$	0	$2(\mathbf{e}_1 + \mathbf{e}_2 + \mathbf{e}_3)$	0
$P_{11}^{(2)}$	0	$2(\mathbf{e}_2 + \mathbf{e}_4 + \mathbf{e}_6)$	0	$2(\mathbf{e}_2 + \mathbf{e}_4 + \mathbf{e}_6)$	0	$2(\mathbf{e}_2 + \mathbf{e}_4 + \mathbf{e}_6)$
$P_{11}^{(3)}$	0	$2\mathbf{e}_2 - \mathbf{e}_4 - \mathbf{e}_6$	$\frac{3}{2}(\mathbf{e}_3 - \mathbf{e}_5)$	$\frac{1}{2}(\mathbf{e}_4 + \mathbf{e}_6 - 2\mathbf{e}_2)$	$\frac{3}{2}(\mathbf{e}_5 - \mathbf{e}_3)$	$\frac{1}{2}(\mathbf{e}_4 + \mathbf{e}_6 - 2\mathbf{e}_2)$
$P_{12}^{(3)}$	$\sqrt{3}\,(\mathbf{e}_3 - \mathbf{e}_5)$	0	$\frac{\sqrt{3}}{2}(\mathbf{e}_3 - \mathbf{e}_5)$	$\frac{\sqrt{3}}{2}(-2\mathbf{e}_2 + \mathbf{e}_4 + \mathbf{e}_6)$	$\frac{\sqrt{3}}{2}(\mathbf{e}_3 - \mathbf{e}_5)$	$\frac{\sqrt{3}}{2}(2\mathbf{e}_2 - \mathbf{e}_4 - \mathbf{e}_6)$
$P_{21}^{(3)}$	0	$\frac{\sqrt{3}}{2}(\mathbf{e}_6 - \mathbf{e}_4)$	$\frac{\sqrt{3}}{2}(-2\mathbf{e}_1 + \mathbf{e}_3 + \mathbf{e}_5)$	$\frac{\sqrt{3}}{2}(\mathbf{e}_4 - \mathbf{e}_6)$	$\frac{\sqrt{3}}{2}(-2\mathbf{e}_1 + \mathbf{e}_3 + \mathbf{e}_5)$	$\frac{\sqrt{3}}{2}(\mathbf{e}_4 - \mathbf{e}_6)$
$P_{22}^{(3)}$	$2\mathbf{e}_1 - \mathbf{e}_3 - \mathbf{e}_5$	0	$\frac{1}{2}(-2\mathbf{e}_1 + \mathbf{e}_3 + \mathbf{e}_5)$	$\frac{3}{2}(\mathbf{e}_4 - \mathbf{e}_6)$	$\frac{1}{2}(-2\mathbf{e}_1 + \mathbf{e}_3 + \mathbf{e}_5)$	$\frac{3}{2}(\mathbf{e}_6 - \mathbf{e}_4)$

PROBLEM 2-28

Show that the vectors $\mathbf{g}_i$ are orthogonal.

Although it now appears as if we have six coplanar, orthonormal vectors, we realize that the combination $\mathbf{e}_1 + \mathbf{e}_3 + \mathbf{e}_5$ is equivalent to zero, so that $\mathbf{g}_1$ is not a vector in the two-dimensional space of the molecular plane; rather, it is defined in the six-dimensional normal-coordinate space. Let us multiply this combination by a quantity $X_1/\sqrt{3}$ to obtain $X_1(\mathbf{e}_1 + \mathbf{e}_3 + \mathbf{e}_5)/\sqrt{3}$.

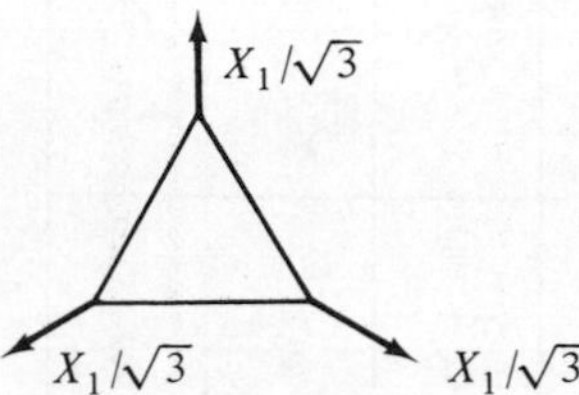

2-13 Normal mode generated by projection operator $P_{11}^{(1)}$

We may visualize this result as the sum of the three displacements of Fig. 2-13. Each of these displacements has a component $(X_1/\sqrt{3})\cos 30° = X_1/2$ along each side, so that the total stretching of a side is $2(X_1/2) = X_1$. Hence, the potential energy of the system is

$$V = \frac{k}{2}(X_1^2 + X_1^2 + X_1^2) = \frac{3k}{2}X_1^2$$

In addition, the kinetic energy is

$$T = \frac{m}{2}\left(\frac{\dot{X}_1^2}{3} + \frac{\dot{X}_1^2}{3} + \frac{\dot{X}_1^2}{3}\right) = \frac{m\dot{X}_1^2}{2}$$

since each mass is moving a distance $X_1/\sqrt{3}$. The Lagrange equation (2-89) becomes

$$m\ddot{X}_1 + 3kX_1 = 0$$

and trying a solution

$$X_1 = A_1 e^{i\omega t} \tag{2-142}$$

gives

$$-m\omega^2 + 3k = 0 \quad \text{or} \quad \omega^2 = \frac{3k}{m}$$

Thus the representation Γ_1 generates the vibrational mode of Fig. 2-13 with a frequency $\sqrt{3k/m}$, agreeing with Fig. 2-9(a).

The only linear combination produced by $P_{11}^{(2)}$ is $(\mathbf{e}_2 + \mathbf{e}_4 + \mathbf{e}_6)$, and we use it to define a set of displacements $X_2(\mathbf{e}_2 + \mathbf{e}_4 + \mathbf{e}_6)/\sqrt{3}$, as shown in Fig. 2-14. Any pair of displacements at each end of a spring produces a zero net change in length (as may easily be seen geometrically), so that

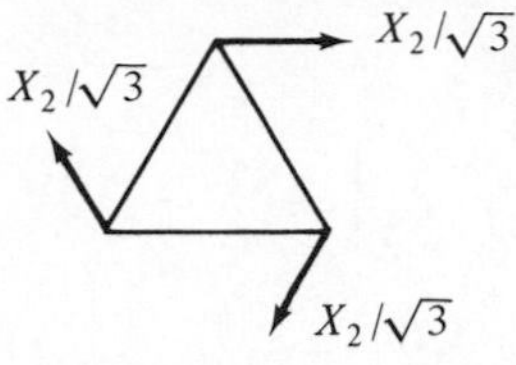

2-14 Normal mode corresponding to projection operator $P_{11}^{(2)}$

$$V = 0$$

and we obtain

$$\omega^2 = 0$$

This is the mode corresponding to pure rotation, which we now see belongs to $\boldsymbol{\Gamma}_2$.

For the section of the table associated with $P_{11}^{(3)}$ we have what appears to be several linear combinations, but some of them are the same except for sign (again ignoring coefficients) and only two are actually independent. Consider then the displacements specified by

$$X_3\frac{(2\mathbf{e}_2 - \mathbf{e}_4 - \mathbf{e}_6)}{\sqrt{6}} + X_4\frac{(\mathbf{e}_5 - \mathbf{e}_3)}{\sqrt{2}}$$

where each term is normalized, the quantities X_3, X_4 are to be determined, and the selection of linear combinations from the second and fifth columns is arbitrary. This situation is shown in Fig. 2-15, and again we may show geometrically that

$$V = \frac{k}{2}\left[\frac{3}{4}X_3^2 - \frac{3}{2}X_3X_4 + \frac{3}{4}X_4^2\right]$$
$$T = \frac{m}{2}[\dot{X}_3^2 + \dot{X}_4^2] \qquad \textbf{(2-143)}$$

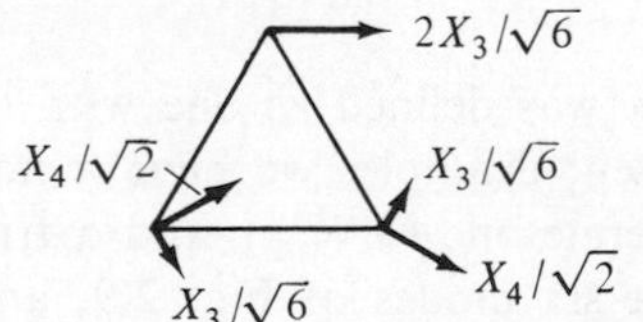

2-15 One of the linear combinations corresponding to $P_{11}^{(3)}$

PROBLEM 2-29

Verify (2-143).

The simultaneous equations for the amplitudes A_3, A_4 are

$$(-\lambda + \tfrac{3}{4})A_3 - \tfrac{3}{4}A_4 = 0$$
$$-\tfrac{3}{4}A_3 + (-\lambda + \tfrac{3}{4})A_4 = 0$$

giving the secular equation

$$\begin{vmatrix} \frac{3}{4} - \lambda & -\frac{3}{4} \\ -\frac{3}{4} & \frac{3}{4} - \lambda \end{vmatrix} = 0$$

and the eigenvalues $\lambda = \frac{3}{2}, 0$

The amplitudes then obey the relations

$$A_3 = A_4, \quad (\lambda = 0)$$
$$A_3 = -A_4, \quad (\lambda = \tfrac{3}{2})$$

It follows that $X_3 = X_4$ for $\lambda = 0$, so that the vectors at the base of the triangle combine to give the resultants of Fig. 2-16. For $\lambda = \frac{3}{2}$, we obtain Fig. 2-17 in the same way. These will be recognized as the modes of Fig. 2-9-(b) and (d), and belong to $\mathbf{\Gamma}_3$. The final pair of modes comes from the combination

$$\frac{X_5}{\sqrt{6}}(2\mathbf{e}_1 - \mathbf{e}_3 - \mathbf{e}_5) + \frac{X_6}{\sqrt{2}}(\mathbf{e}_4 - \mathbf{e}_6)$$

which produces the remainder of Fig. 2-9 as expected.

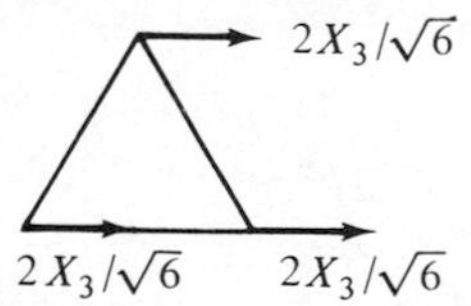

2-16 Normal mode obtained from previous figure

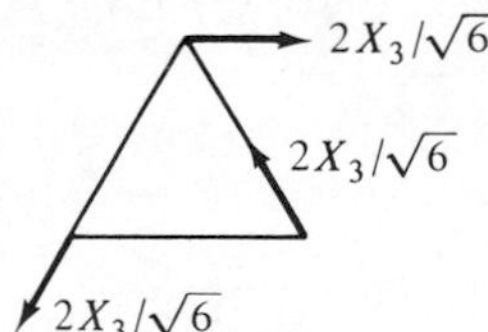

2-17 The other normal mode obtained from $P_{11}^{(3)}$

In Sec. 2-9, a degenerate matrix was defined as one with two or more identical eigenvalues. For our present example, we have a nondegenerate eigenvalue ($\lambda = 3$), a doubly-degenerate one ($\lambda = \frac{3}{2}$), and a triply-degenerate one ($\lambda = 0$). If we return to the six modes of Fig. 2-9, and sort them according to both frequency and representation, we obtain the arrangement of Fig. 2-18. Hence, we see that the dimensionality of a representation indicates the degree of degeneracy, for the nondegenerate representations $\mathbf{\Gamma}_1$ and $\mathbf{\Gamma}_2$ are one-dimensional whereas the two-dimensional representation $\mathbf{\Gamma}_3$ is doubly-degenerate. We stated above that the eigenvalue $\lambda = 0$ appears triply-degenerate, but we realize that this is a consequence of the fact that pure translation and pure rotation both correspond to $\omega = 0$.

Group theory has thus given us the frequencies and amplitudes of the normal modes without the necessity of solving a high order secular determi-

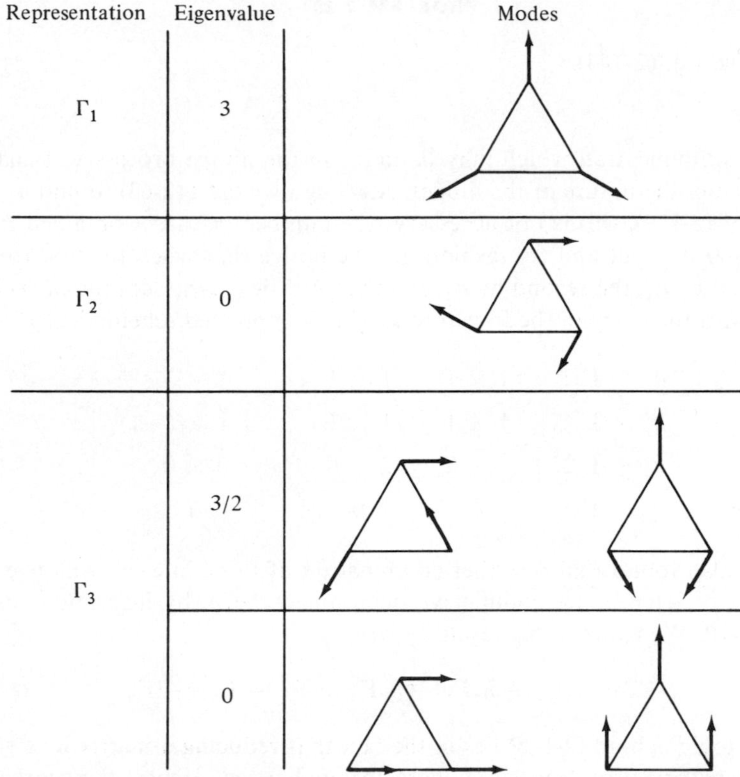

2-18 Identification of normal modes with irreducible representations

nant. What has happened, in fact, is that the sixth-order determinant itself has been reduced, for we may combine the individual results above in the following way:

$$\begin{array}{c} \Gamma_1 \\ \Gamma_2 \\ \\ \Gamma_3 \\ \\ \\ \end{array}\left|\begin{array}{cccccc} 3-\lambda & 0 & 0 & 0 & 0 & 0 \\ 0 & 0-\lambda & 0 & 0 & 0 & 0 \\ 0 & 0 & \frac{3}{4}-\lambda & -\frac{3}{4} & 0 & 0 \\ 0 & 0 & -\frac{3}{4} & \frac{3}{4}-\lambda & 0 & 0 \\ 0 & 0 & 0 & 0 & \frac{3}{4}-\lambda & -\frac{3}{4} \\ 0 & 0 & 0 & 0 & -\frac{3}{4} & \frac{3}{4}-\lambda \end{array}\right| = 0 \qquad \text{(2-144)}$$

PROBLEM 2-30

Derive Eq. (2-144).

One simplification which may be made in the above process will enable us to predict the nature of the modes. Although we used (2-63) to find n_1, n_2, and n_3 of (2-129), this is not necessary; the numbers n_i can be obtained from Table 2-80 in a cut and try fashion, for we notice that when the first row is multiplied by n_1, the second by n_2, and the third by n_3, the sum for each class agrees with the entry in the fourth row. This is expressed schematically by

$n_1 \times \Gamma_1$:	1×1	1×1	1×1
$n_2 \times \Gamma_2$:	1×1	1×1	$1 \times (-1)$
$n_3 \times \Gamma_3$:	2×2	$2 \times (-1)$	2×0
Γ_7:	6	0	0

It is also found that no other combination of three integers will give the same result. Hence, we could have determined the n_i by inspection, using Table 2-10. We express this result by writing

$$\Gamma_7' = n_1\Gamma_1 + n_2\Gamma_2 + n_3\Gamma_3 = \Gamma_1 + \Gamma_2 + 2\Gamma_3 \tag{2-145}$$

The logic behind (2-145) lies in the fact that reducing a matrix by a similarity transformation leaves the character unchanged. Hence, the characters of the diagonal matrices in the reduced representation must add up to that of the original character for the reducible one.

PROBLEM 2-31

(a) If $\mathbf{P}$ and $\mathbf{Q}$ are two $n \times n$ matrices, show using (2-11) that

$$\chi(\mathbf{PQ}) = \chi(\mathbf{QP}) \tag{2-146}$$

where $\chi(\mathbf{PQ})$ is the character of the product $\mathbf{PQ}$.

(b) Use (2-146) to show that if

$$\mathbf{B} = \Theta^{-1}\mathbf{A}\Theta$$

then

$$\chi(\mathbf{B}) = \chi(\mathbf{A}) \tag{2-147}$$

Thus, while a matrix may change under a similarity transformation, its character is invariant.

PROBLEM 2-32

As shown by Schonland,[6] it is possible to apply group theory to the equilateral molecule by using a less convenient basis. Consider six vectors $\mathbf{g}_1, \ldots,$ $\mathbf{g}_6$ oriented so that $\mathbf{g}_1, \mathbf{g}_3, \mathbf{g}_5$ are along the x-axis at the points A, B, C, respectively, while $\mathbf{g}_2, \mathbf{g}_4, \mathbf{g}_6$ are along the y-axis.

(a) Make a table showing the effect of each symmetry operation on the $\mathbf{g}_i$.

(b) Use (2-138) to obtain the linear combinations which reduce the 3×3 representation, and identify the associated irreducible representation.

One thing we should note in connection with projection operator methods is that we need only work with a single operator $P_{ij}^{(k)}$ in each row. That is, both $P_{11}^{(3)}$ and $P_{12}^{(3)}$ of Table 2-9 produce the combination $\mathbf{d}_2 - \mathbf{d}_3$ and both $P_{21}^{(3)}$ and $P_{22}^{(3)}$ produce $2\mathbf{d}_1 - \mathbf{d}_2 - \mathbf{d}_3$. In the same way, the first row of $\mathbf{\Gamma}_3$ (Table 2-12) generates $\mathbf{d}_3 - \mathbf{d}_5$ and $2\mathbf{d}_2 - \mathbf{d}_4 - \mathbf{d}_6$; the second row of $\mathbf{\Gamma}_3$ generates $\mathbf{d}_4 - \mathbf{d}_6$ and $2\mathbf{d}_1 - \mathbf{d}_3 - \mathbf{d}_5$. A general proof of this property of projection operators and the others that we have used is given by Hamermesh.[7]

Another interesting aspect of projection operator methods concerns the use of (2-137) rather than (2-138). Many authors advocate this approach, since it is easier to calculate character tables than it is to determine complete representations. To compare the usefulness of the two types of operators, consider $P^{(3)}$ belonging to $\mathbf{\Gamma}_3$, which has the form

$$\begin{aligned} P^{(3)} &= \sum_{l=1}^{6} \chi^{(3)}(G_l)G_l \\ &= 2E - 1C_3 - 1C_3^2 + 0\sigma_v^a + 0\sigma_v^b + 0\sigma_v^c \\ &= 2E - C_3 - C_3^2 \end{aligned} \tag{2-148}$$

Then $P^{(3)}$ acting on $\mathbf{e}_1, \ldots, \mathbf{e}_6$ of Table 2-11 gives

$$\begin{aligned} P^{(3)}\mathbf{e}_1 &= 2\mathbf{e}_1 - \mathbf{e}_3 - \mathbf{e}_5 \\ P^{(3)}\mathbf{e}_2 &= 2\mathbf{e}_2 - \mathbf{e}_4 - \mathbf{e}_6 \\ P^{(3)}\mathbf{e}_3 &= 2\mathbf{e}_3 - \mathbf{e}_5 - \mathbf{e}_1 \\ P^{(3)}\mathbf{e}_4 &= 2\mathbf{e}_4 - \mathbf{e}_6 - \mathbf{e}_2 \\ P^{(3)}\mathbf{e}_5 &= 2\mathbf{e}_5 - \mathbf{e}_1 - \mathbf{e}_3 \\ P^{(3)}\mathbf{e}_6 &= 2\mathbf{e}_6 - \mathbf{e}_2 - \mathbf{e}_4 \end{aligned} \tag{2-149}$$

[7] M. Hamermesh, *Group Theory*, Addison-Wesley Publishing Co., Inc., Reading, Mass., 1962.

The first two of these equations produce combinations which we have already found for $\mathbf{\Gamma}_3$. The next four appear unfamiliar; let us therefore write the identity

$$(2\mathbf{e}_2 - \mathbf{e}_4 - \mathbf{e}_6) = -(2\mathbf{e}_4 - \mathbf{e}_6 - \mathbf{e}_2) - (2\mathbf{e}_6 - \mathbf{e}_2 - \mathbf{e}_4)$$

That is, a linear combination of $P^{(3)}\mathbf{e}_4$ and $P^{(3)}\mathbf{e}_6$ is equivalent to $P^{(3)}\mathbf{e}_2$. Hence, this approach does not provide the complete basis $\mathbf{g}_1, \ldots, \mathbf{g}_6$ needed to reduce $\mathbf{\Gamma}_7$.

Finally, we should mention that the basis provided by the projection operators is said to be *symmetry adapted* and the vectors $\mathbf{g}_i$ are called *partners* in the basis.

2-11 The theory of projection operators

Let us consider in more detail the relation

$$P_{ij}^{(k)} = \frac{d_k}{g} \sum_{l=1}^{g} D_{ij}^{(k)}(R_l)R_l \tag{2-138}$$

which defines the projection operator $P_{ij}^{(k)}$. We start by realizing that the vectors of a reducible basis are generally converted into linear combinations of one another by the symmetry operations R_l of the group.

PROBLEM 2-33

Show that this is the case for the basis vectors of Fig. 2-7.

The basis $\mathbf{d}_1, \mathbf{d}_2, \mathbf{d}_3$ of Fig. 2-10 gives "linear combinations" involving only a single term, but in the general case we would expect the effect of an operation R_j on a basis vector $\mathbf{d}_i$ to be expressible as

$$R_j\mathbf{d}_i = \sum_{k=1}^{n} a_{kij}\mathbf{d}_k, \quad (i = 1, 2, \ldots, n; j = 1, 2, \ldots, g) \tag{2-150}$$

for a basis with n vectors. The coefficients d_{kij} require three subscripts since they differ for each operation R_j and each vector $\mathbf{d}_i$.

Let us now show that these coefficients may be regarded as the elements of a representation of the group. We first replace a_{kij} with a new symbol $D_{ki}(R_j)$ which specifies the k, i-element in the matrix $D(R_j)$. Then (2-150) becomes

$$R_j\mathbf{d}_i = \sum_{k=1}^{n} D_{ki}(R_j)\mathbf{d}_k \tag{2-151}$$

Then we operate on both sides of this with another element R_m, obtaining

$$\begin{aligned} R_m R_j \mathbf{d}_i &= R_m \sum_k D_{ki}(R_j)\mathbf{d}_k \\ &= \sum_k [\sum_{k'} D_{k'k}(R_m)\mathbf{d}_{k'}] D_{ki}(R_j) \\ &= \sum_{k'} [\sum_k D_{k'k}(R_m) D_{ki}(R_j)]\mathbf{d}_{k'} \\ &= \sum_{k'} [D(R_m)D(R_j)]_{k'i}\mathbf{d}_{k'} \end{aligned} \tag{2-152}$$

Thus, the matrices obey the Cayley table, so that they constitute a representation.

PROBLEM 2-34

Show that the representation $\mathbf{\Gamma}_4$ of Table 2-3 is based on (2-151).

Next, we use the fact (first brought out at the end of Sec. 2-4) that irreducible representations are equivalent to orthogonal vectors. That is, the matrix elements $D_{ij}^{(k)}$ of C_{3v} used in (2-138) define six orthogonal vectors $\mathbf{a}_k$ as

$$\mathbf{a}_1 = \begin{pmatrix} 1 \\ 1 \\ 1 \\ 1 \\ 1 \\ 1 \end{pmatrix}, \quad \mathbf{a}_2 = \begin{pmatrix} 1 \\ 1 \\ 1 \\ -1 \\ -1 \\ -1 \end{pmatrix}, \quad \mathbf{a}_3 = \begin{pmatrix} 1 \\ -\frac{1}{2} \\ -\frac{1}{2} \\ -1 \\ \frac{1}{2} \\ \frac{1}{2} \end{pmatrix}$$

$$\mathbf{a}_4 = \begin{pmatrix} 0 \\ -\sqrt{3}/2 \\ \sqrt{3}/2 \\ 0 \\ \sqrt{3}/2 \\ -\sqrt{3}/2 \end{pmatrix}, \quad \mathbf{a}_5 = \begin{pmatrix} 0 \\ \sqrt{3}/2 \\ -\sqrt{3}/2 \\ 0 \\ \sqrt{3}/2 \\ -\sqrt{3}/2 \end{pmatrix}, \quad \mathbf{a}_6 = \begin{pmatrix} 1 \\ -\frac{1}{2} \\ -\frac{1}{2} \\ 1 \\ -\frac{1}{2} \\ -\frac{1}{2} \end{pmatrix}$$

where

$$\mathbf{a}_i \cdot \mathbf{a}_j = 0$$

for $i \neq j$. More generally,

$$\sum_l D_{ij}^{(\mu)}(R_l) D_{pq}^{(\nu)}(R_l) = \left(\frac{g}{d_\mu}\right)\delta_{\mu\nu}\delta_{ip}\delta_{jq} \tag{2-153}$$

where δ_{ij} is the Kronecker delta ($\delta_{ij} = 1$ if $i = j$, $\delta_{ij} = 0$ if $i \neq j$), and the other symbols are the same as those of (2-138). Equation (2-153) is often called the *great orthogonality theorem.*

Returning to (2-139), we may solve these equations for the $\mathbf{d}_i$ in terms of the $\mathbf{e}_j$, obtaining

$$\begin{aligned}
\mathbf{d}_1 &= \frac{1}{\sqrt{3}}\mathbf{e}_1 + \frac{2}{\sqrt{6}}\mathbf{e}_2 + 0\mathbf{e}_3 \\
\mathbf{d}_2 &= \frac{1}{\sqrt{3}}\mathbf{e}_1 - \frac{1}{\sqrt{6}}\mathbf{e}_2 + \frac{1}{\sqrt{2}}\mathbf{e}_3 \\
\mathbf{d}_3 &= \frac{1}{\sqrt{3}}\mathbf{e}_1 - \frac{1}{\sqrt{6}}\mathbf{e}_2 - \frac{1}{\sqrt{2}}\mathbf{e}_3
\end{aligned} \tag{2-154}$$

These equations indicate that if we operate on one of the $\mathbf{d}_i$ with a symmetry operator R_j, we generate linear combinations of the $\mathbf{e}_k$; these combinations do *not* mix. For example,

$$\begin{aligned}
\sigma_v^a \mathbf{d}_2 &= \sigma_v^a\left(\frac{\mathbf{e}_1}{\sqrt{3}} - \frac{\mathbf{e}_2}{\sqrt{6}} + \frac{\mathbf{e}_3}{\sqrt{2}}\right) \\
&= \frac{1}{\sqrt{3}}\mathbf{e}_1 - \frac{1}{\sqrt{6}}\mathbf{e}_2 - \frac{1}{\sqrt{2}}\mathbf{e}_3 = \mathbf{d}_3
\end{aligned}$$

Relations of this kind may be written, in general, as

$$R_j \mathbf{d}_i = \sum_{k=1}^{d_1} D_{ki}^{(1)}(R_j)\mathbf{e}_k^{(1)} + \sum_{k=1}^{d_2} D_{ki}^{(2)}(R_j)\mathbf{e}^{(2)} + \ldots \tag{2-155}$$

where the sums are taken over each irreducible representation, with dimension $d_1, d_2, \ldots, d_\nu, \ldots$, respectively. Multiplying both sides of (2-155) by $D_{nm}^{(\mu)}(R_j)$ and summing over j, a typical term on the right would be

$$\begin{aligned}
\sum_{j=1}^{g} \sum_{k=1}^{d_\nu} D_{ki}^{(\nu)}(R_j) D_{nm}^{(\mu)}(R_j)\mathbf{e}_k^{(\nu)} &= \sum_{k=1}^{d_\nu}\left(\frac{g}{d_\mu}\right)\delta_{kn}\delta_{im}\delta_{\mu\nu}\mathbf{e}_k^{(\nu)} \\
&= \sum_{k=1}^{d_\mu}\left(\frac{g}{d_\mu}\right)\delta_{kn}\delta_{im}\mathbf{e}_k^{(\mu)}
\end{aligned} \tag{2-156}$$

Hence, only one sum on the right of (2-155) contributes to (2-156), and equating this to what we obtain from the left-hand side gives

$$\sum_j D_{nm}^{(\mu)}(R_j) R_j \mathbf{d}_i = \sum_{k=1}^{d_\mu}\left(\frac{g}{d_\mu}\right)\delta_{kn}\delta_{im}\mathbf{e}_k^{(\mu)} \tag{2-157}$$

which we recognize as the definition of the projection operator, given in (2-138).

We note further that since the $\mathbf{d}_i$ with which we start are arbitrary; there is no reason why they cannot, in fact, be the $\mathbf{e}_i$. Letting $\mathbf{d}_i$ in (2-157) be $\mathbf{e}_m^{(\mu)}$ gives

$$\sum_j D_{nm}^{(\mu)}(R_j)R_j\mathbf{e}_m^{(\mu)} = \sum_{k=1}^{d_\mu}\left(\frac{g}{d_\mu}\right)\delta_{kn}\delta_{im}\mathbf{e}_k^{(\mu)}$$

For the case $n = m$, this simplifies to

$$\sum_j D_{mm}^{(\mu)}(R_j)R_j\mathbf{e}_m^{(\mu)} = \frac{g}{d_\mu}\mathbf{e}_m^{(\mu)} \tag{2-158}$$

PROBLEM 2-35

Verify (2-158) for the basis of Table 2-12.

2-12 *Internal coordinates*

We have seen that the determination of the normal modes by setting up and solving the Lagrange equations of motion generates the translational and rotational modes as well as the purely vibrational ones. It would seem logical to simplify vibrational problems by considering only relative motions of one atom with respect to another. This would eliminate the necessity of dealing with pure translations and rotations, which we may regard as motions for which the molecule acts as a rigid body. Hence, we shall introduce *relative* or *internal coordinates* for this purpose.

As a simple example, consider the diatomic molecule of Fig. 2-19, composed of one X atom and one Y atom. Denoting the displacement from equilibrium of the X atom as x_1 and of the Y atom as x_2, the internal coordinate S_1—which we may take as the bond extension—is defined as

$$S_1 = x_1 - x_2 \tag{2-159}$$

The potential energy is

$$V = \tfrac{1}{2}kS_1^2 \tag{2-160}$$

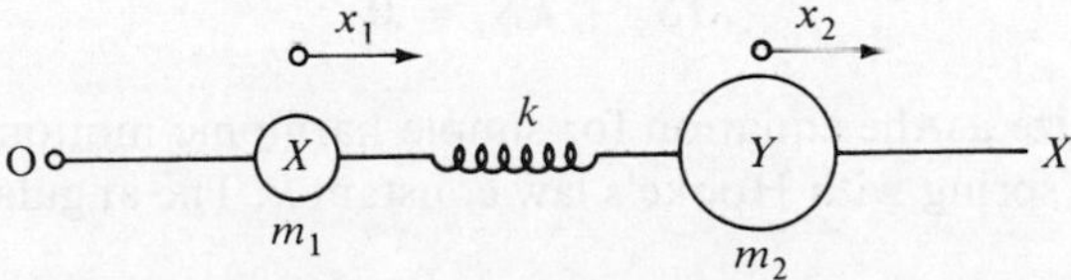

2-19 An XY molecule

and the kinetic energy is $$T = \frac{m_1}{2}\dot{x}_1^2 + \frac{m_2}{2}\dot{x}_2^2 \tag{2-161}$$

Following Brand and Speakman,[8] it is convenient to introduce *mass-weighted coordinates* by

$$q_1 = \sqrt{m_1}\,x_1, \qquad q_2 = \sqrt{m_2}\,x_2 \tag{2-162}$$

and (2-161) becomes $$2T = \dot{q}_1^2 + \dot{q}_2^2 \tag{2-163}$$

This expression for the kinetic energy can be converted into one involving S_1 of (2-159) by making use of the fact that a pair of masses vibrating at the ends of a spring has no total linear momentum if the system does not undergo translation. That is,

$$p_1 = -p_2 \quad \text{or} \quad m_1\dot{x}_1 = -m_2\dot{x}_2 \tag{2-164}$$

PROBLEM 2-36

Prove that $$m_1\dot{x}_1^2 + m_2\dot{x}_2^2 = \frac{m_1 m_2}{m_1 + m_2}(\dot{x}_1^2 - \dot{x}_2^2) \tag{2-165}$$

Combining (2-160) and (2-165) shows that

$$2T = M\dot{S}_1^2 \tag{2-166}$$

where M, called the *reduced mass* of the system, is defined by

$$\frac{1}{M} = \frac{1}{m_1} + \frac{1}{m_2} \tag{2-167}$$

The Lagrange equation (2-89) for the symmetry coordinate S_1 becomes

$$\frac{d}{dt}\frac{\partial T}{\partial \dot{S}_1} + \frac{\partial V}{\partial S_1} = 0 \tag{2-168}$$

or by (2-160) and (2-166)

$$M\ddot{S}_1 + kS_1 = 0 \tag{2-169}$$

This we recognize as the equation for simple harmonic motion of a mass M on the end of a spring with Hooke's law constant k. The angular frequency is of course

[8] J. C. D. Brand and J. C. Speakman, *Molecular Structure*, Edward Arnold, London, 1960.

$$\omega = \sqrt{k/M} \tag{2-170}$$

and the secular equation which comes from (2-168) is

$$-M\omega^2 + k = 0 \tag{2-171}$$

or

$$-\lambda + 1 = 0 \tag{2-172}$$

where

$$\lambda = \frac{\omega^2 M}{k} \tag{2-173}$$

Let us recast the secular equation into a form used by Wilson, Decius, and Cross.[9] They write (2-159) as

$$S_t = \sum_{i=1}^{3N} B_{ti} x_i \tag{2-174}$$

where

$$B_{11} = 1, \qquad B_{12} = -1 \tag{2-175}$$

in this example. They then define a matrix G in terms of the coefficients B_{mn} as

$$G_{tt'} = \sum_{i=1}^{3M} \frac{1}{m_i} B_{ti} B_{t'i} \tag{2-176}$$

For the present example, $t = t' = 1$ and $i = 1, 2$, so that

$$G_{11} = \frac{B_{11}^2}{m_1} + \frac{B_{12}^2}{m_2} = \frac{1}{m_1} + \frac{1}{m_2} \tag{2-177}$$

Finally, they define a matrix F whose elements are the Hooke's law constants (called *force constants* by chemists). This definition comes from (2-160), which is written as

$$2V = \sum_{t,t'} F_{tt'} S_t S_{t'} \tag{2-178}$$

where

$$F_{11} = k \tag{2-179}$$

We can now show that the secular equation has the form

$$|\mathbf{GF} - \omega^2 \mathbf{I}| = 0 \tag{2-180}$$

where $\mathbf{I}$ is the unit matrix of Eq. (2-18). By (2-177) and (2-179), (2-180) becomes

$$\left| \frac{k}{M} - \omega^2 \right| = 0 \tag{2-181}$$

[9]E. B. Wilson, Jr., J. C. Decius, and P. C. Cross, *Molecular Vibrations*, McGraw-Hill Book Company, New York, 1955.

which is identical to (2-171). A general proof of the Wilson, Decius, and Cross equation is given in their book;[9] we shall be content with this simple justification.

2-13 Internal symmetry coordinates

Let us next consider a situation for which we can not only use internal coordinates but choose them in such a way as to take advantage of the symmetry. As an example, consider the H_2O molecule of Fig. 2-20. It is known

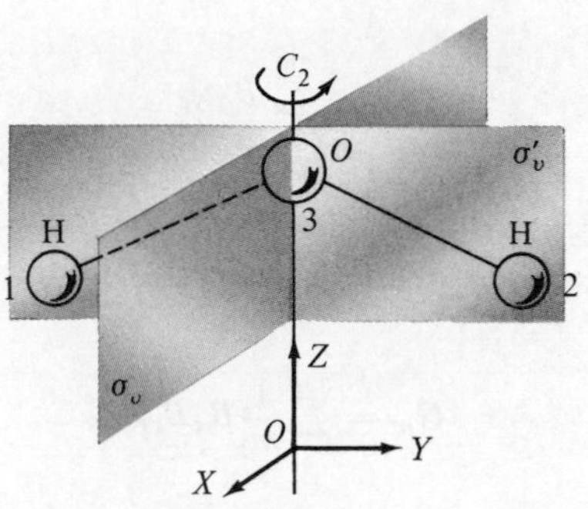

2-20 The water molecule

that the equilibrium value of the interbond angle α is 104.5°, and its point group is C_{2v}, with the axes and planes as shown. The symmetry operations for a molecule of this configuration are seen to be

E, the identity
C_2, a two-fold axis (chosen along OZ)
σ_v, a reflection plane (chosen as XOZ)
σ'_v, a reflection plane (chosen as YOZ)

The character system, determined as described earlier, is given in Table 2-13. In this table, we have changed the designation of the representations

TABLE 2-13 ***Character System and the Basis Functions of the Group C_{2v}***

	E	C_2	σ_v	$\sigma_{v'}$	
A_1	1	1	1	1	z, x^2, y^2, z^2
A_2	1	1	−1	−1	xy, R_z
B_1	1	−1	1	−1	x, R_y, xz
B_2	1	−1	−1	1	y, R_x, yz

from the arbitrary $\Gamma_1, \Gamma_2, \Gamma_3, \Gamma_4$ used by physicists to the notation favored by chemists, since in this chapter we are primarily interested in spectroscopic problems. The logic behind this notation (due to R.S. Mulliken), as given by Colthup, Daly, and Wiberly[10] is

A, symmetric with respect to the principal axis of symmetry
B, antisymmetric with respect to the principal axis of symmetry
E, doubly-degenerate representation
F, triply-degenerate representation (some authors use *T*)

Subscripts or primes are added to these symbols in the following way

g and *u*,
symmetric or antisymmetric with respect to a center of symmetry, where *g* comes from the German *gerade* (even) and *u* from *ungerade* (odd).

1 and 2,
symmetric or antisymmetric with respect to a rotation axis (C_n) or rotation-reflection axis (S_n) other than the principal axis or (in those point groups with only one symmetry axis) with respect to a plane of symmetry

prime and double prime,
symmetric or antisymmetric with respect to a plane of symmetry

The representation A_1, then, is symmetric with respect to the *OZ*-axis, since the character of C_2 in this representation is $+1$. The characters of both σ_v and σ'_v in A_1 are also $+1$, explaining the subscript; on the other hand, these operations have characters of -1 in A_2. In the same way, C_2 has a character of -1 in both B_1 and B_2, whereas σ_v and σ'_v follow the same subscript pattern as for A_1 and A_2. (The symbols at the right-hand end of Table 2-13 will be explained shortly.) For the group C_{3v}, the representations of Table 2-14 indicate that A_1 should have characters of $+1$ for both C_3 and σ_v, that A_2 should have $+1$ for C_3 but -1 for σ_v, and that E is doubly-degenerate.

TABLE 2-14 ***Character System and Basis Functions of*** C_{3v}

	E	$2C_3$	$3\sigma_v$		
A_1	1	1	1	$z, x^2 + y^2, z^2$	
A_2	1	1	-1	R_z	
E	2	-1	0	$\begin{cases}(x, y)\\(R_x, R_y)\end{cases}$	$\begin{cases}(x^2 - y^2, xy)\\(xz, yz)\end{cases}$

[10]E. B. Colthup, L. E. Daly, and S. E. Wiberly, *Introduction to Infrared and Raman Spectroscopy*, Academic Press Inc., New York, 1964.

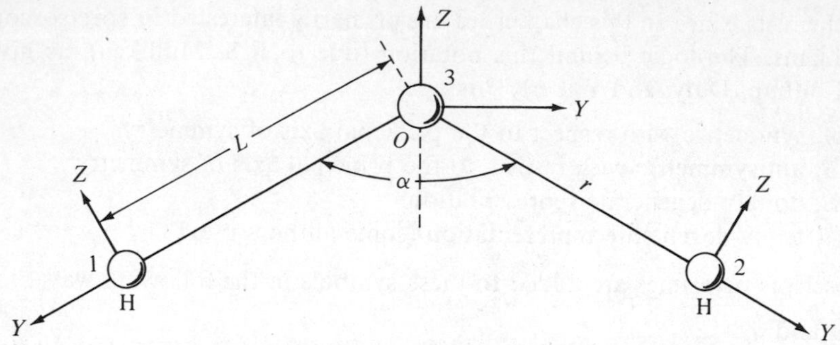

2-21 Internal coordinates for H_2O

Returning to the H_2O molecule of Fig. 2-20, we assume that the vibrations remain in the plane of the molecule. Using the coordinate system of Fig. 2-21, we choose as relative coordinates the stretching of the two oxygen-hydrogen bonds and the vibration of the bonds with respect to each other. The relative or internal coordinate S_{31} for atom 3 with respect to atom 1 is

$$S_{31} = y_1 + y_3 \sin \frac{\alpha}{2} + z_3 \cos \frac{\alpha}{2}$$

Similarly, for the other bond

$$S_{23} = y_2 - y_3 \sin \frac{\alpha}{2} + z_3 \cos \frac{\alpha}{2} \tag{2-182}$$

The relative angle θ may be expressed

$$S_{12} = L\theta \tag{2-183}$$

where L is the equilibrium bond length. For the left side of the figure, for example, changes in y_1 and y_3 do not affect θ, but to a first approximation, $z_1 - z_3 \sin(\alpha/2)$ will equal $L\theta$ if atom 2 is stationary. Adding a similar contribution from the right, we obtain

$$S_{12} = z_1 + z_2 - 2z_3 \sin \frac{\alpha}{2} \tag{2-184}$$

Now we convert to *internal symmetry coordinates* by using the projection operator method of Sec. 2-10. In Table 2-15, we show the effect of each operation on each coordinate.

Forming the projection operators from Table 2-13 and working out the linear combinations, we obtain the combinations of Table 2-16. Thus,

TABLE 2-15 ***Effect of C_{2v} on the Internal Coordinates of the Water Molecule***

	E	C_2	σ_v	$\sigma_{v'}$
S_{31}	S_{31}	S_{23}	S_{23}	S_{31}
S_{23}	S_{23}	S_{31}	S_{31}	S_{23}
S_{12}	S_{12}	S_{12}	S_{12}	S_{12}

TABLE 2-16 ***Projections of the Coordinates of Table 2-15***

Representation	Coordinate	Linear combination
A_1	S_{31} S_{23} S_{12}	$2(S_{31} + S_{23})$ $2(S_{23} + S_{31})$ $4S_{12}$
A_2	S_{31} S_{23} S_{12}	0 0 0
B_1	S_{31} S_{23} S_{12}	0 0 0
B_2	S_{31} S_{23} S_{12}	$2(S_{31} - S_{23})$ $2(S_{23} - S_{31})$ 0

group theory gives the following normalized linear combinations

$$\begin{aligned} S_1 &= \frac{1}{\sqrt{2}}(S_{31} + S_{23}) \\ S_2 &= S_{12} \\ S_3 &= \frac{1}{\sqrt{2}}(S_{31} - S_{23}) \end{aligned} \tag{2-185}$$

This we might have expected, since S_{12} is invariant under all operations of the group, whereas S_{13} and S_{23} are converted into one another by some of the operations. We realize that the "symmetry" part of the name for the coordinates S_1, S_2, and S_3 is due to the fact that we have taken the original internal coordinates S_{31}, S_{23}, S_{12} and used the projection operator method to generate a symmetric basis.

To use the Wilson, Decius, and Cross determinant we combine (2-185) with (2-182), (2-183), and (2-184) to obtain

$$S_1 = \frac{y_1 + y_2 + 2z_3 \cos\frac{\alpha}{2}}{\sqrt{2}}$$

$$S_2 = z_1 + z_2 - 2z_3 \sin\frac{\alpha}{2} \tag{2-186}$$

$$S_3 = \frac{y_1 - y_2 + 2y_3 \sin\frac{\alpha}{2}}{\sqrt{2}}$$

The elements of G are then

$$G_{11} = \frac{1}{m_H} + \frac{1}{m_O}(1 + \cos\alpha)$$

$$G_{12} = G_{21} = -\frac{\sin\alpha}{\sqrt{2}\, m_O}$$

$$G_{13} = G_{31} = 0$$

$$G_{22} = 2\left[\frac{1}{m_H} + \frac{1 - \cos\alpha}{m_O}\right] \tag{2-187}$$

$$G_{23} = G_{32} = 0$$

$$G_{33} = \frac{1}{m_H} + \frac{1}{m_O}(1 - \cos\alpha)$$

where we have used obvious symbols for the masses.

PROBLEM 2-37

Verify (2-187) using

$$1 + \cos\alpha = 2\cos^2\frac{\alpha}{2}, \qquad 1 - \cos\alpha = 2\sin^2\frac{\alpha}{2}$$

The potential energy, in terms of internal coordinates, is

$$2V = k_r S_{31}^2 + k_r S_{23}^2 + k_\theta S_{12}^2$$

or from (2-185)

$$2V = k_r(S_1^2 + S_3^2) + k_\theta S_3^2 \tag{2-188}$$

Hence, the nonzero elements of the matrix **F** are

$$F_{11} = F_{33} = k_r, \qquad F_{22} = k_\theta \tag{2-189}$$

The secular equation then becomes

$$\left|\begin{pmatrix} G_{11} & G_{12} & 0 \\ G_{12} & G_{22} & 0 \\ 0 & 0 & G_{33} \end{pmatrix}\begin{pmatrix} F_{11} & 0 & 0 \\ 0 & F_{22} & 0 \\ 0 & 0 & F_{11} \end{pmatrix} - \omega^2\begin{pmatrix} 1 & 0 & 0 \\ 0 & 1 & 0 \\ 0 & 0 & 1 \end{pmatrix}\right| = 0$$

or

$$\begin{vmatrix} G_{11}F_{11} - \omega^2 & G_{12}F_{22} & 0 \\ G_{12}F_{11} & G_{22}F_{22} - \omega^2 & 0 \\ 0 & 0 & G_{33}F_{11} - \omega^2 \end{vmatrix} = 0 \quad \text{(2-190)}$$

Thus, the use of internal symmetry coordinates has automatically reduced the secular determinant to a 2×2 and a 1×1 block. The normal frequencies for the 1×1 block are then given by

$$\omega_3^2 = k_r\left(\frac{1}{m_H} + \frac{1 - \cos\alpha}{m_O}\right) \quad \text{(2-191)}$$

For the 2×2 block, the algebra is rather a nuisance, and it is simpler to insert numerical values. Brand and Speakman[8] give the following values for the force constants

$$k_r = 7.6 \times 10^2 \text{ newt/m}, \qquad k_\theta = 0.70 \times 10^2 \text{ newt/m}$$

from which they obtain

$$\omega_1^2 = 4.736 \times 10^{29}/\text{sec}^2, \qquad \omega_2^2 = 0.901 \times 10^{29}/\text{sec}^2,$$
$$\omega_3^2 = 5.008 \times 10^{29}/\text{sec}^2$$

PROBLEM 2-38

(a) Show that the off-diagonal terms in (2-176) are almost negligible with respect to the diagonal terms.

(b) If the off-diagonal terms actually vanish, describe the normal modes.

PROBLEM 2-39

Discuss the nature of the normal modes when $\alpha = 180°$.

One aspect of the H_2O molecule which appears puzzling at first sight is the structure of the secular determinant in (2-190). There is a 2×2 block in the upper left-hand corner belonging to A_1 of Table 2-13 and a 1×1 block in the lower right-hand corner belonging to B_2. This is in contrast to Eq.

(2-144), in which 1×1 blocks correspond to one-dimensional representations Γ_1 and Γ_2, while 2×2 blocks correspond to the two-dimensional representation Γ_3. For (2-190), there is a 2×2 block corresponding to A_1. Although we might initially think that the presence of a degenerate block belonging to a nondegenerate representation is a contradiction, we realize that it is actually a nondegenerate situation. There are two linear combinations, namely $(S_{31} + S_{23})$ and S_{12}, belonging to A_1, each of which has a distinct frequency associated with it. We have thus discovered a feature of the use of group theory to reduce a secular determinant which is often overlooked: the dimensions of the irreducible representations are a lower limit to the capabilities of the reduction process, but there is no guarantee that we shall achieve this limit. We were lucky in the case of the planar equilateral molecule; we shall not be so lucky with the example in Sec. 2-15. Nevertheless, group theory is worth the trouble because of the insights it provides.

2-14 The qualitative treatment of normal modes

Sometimes the analysis of spectroscopic data merely requires a knowledge of the number of pure vibrational modes and of the representations to which they belong. Let us see how group theory produces this information with very little trouble, before we proceed to consider examples which are more involved than H_2O or the equilateral molecule. We saw in Sec. 2-10 that the irreducible representation Γ_3 of the group C_{3v} could be identified with the translational mode, the representation Γ_2 with the rotational mode, and both Γ_1 and Γ_3 with the vibrational modes. We indicate this symbolically by writing

$$\Gamma_{\text{tot}} = \Gamma_{\text{trans}} + \Gamma_{\text{rot}} + \Gamma_{\text{vib}} \tag{2-192}$$

where the subscripts refer, respectively, to the total representation, and to the translational, rotational, and vibrational ones.

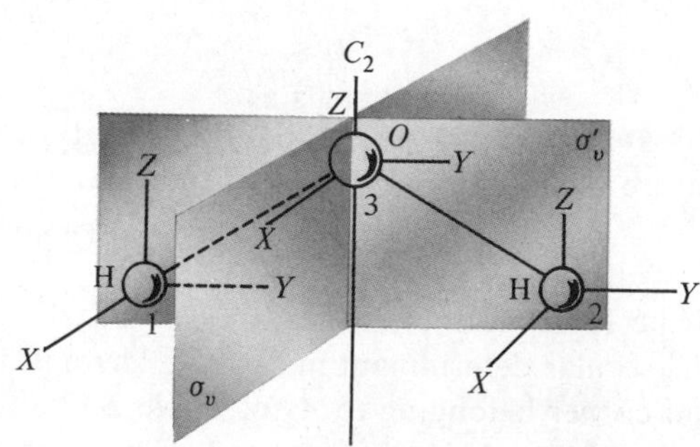

2-22 A reducible basis for H_2O

To put this equation on a firmer foundation, let us consider a reducible representation for the H_2O molecule using as a basis the coordinate system of Fig. 2-22, which differs from that of 2-21. Using coordinates $x_1, x_2, x_3, \ldots, z_3$, there will be a 9×9 reducible representation for C_{2v}. We can determine the characters χ_{tot} of the representation Γ_{tot} in the following simple way:

1. E

Since $x_1, y_1, \ldots, z_3$ are unaffected by E, the matrix equation is

$$\begin{pmatrix} 1 & 0 & 0 & 0 & 0 & 0 & 0 & 0 & 0 \\ 0 & 1 & 0 & 0 & 0 & 0 & 0 & 0 & 0 \\ 0 & 0 & 1 & 0 & 0 & 0 & 0 & 0 & 0 \\ 0 & 0 & 0 & 1 & 0 & 0 & 0 & 0 & \cdot \\ \cdot & \cdot & \cdot & \cdot & \cdot & \cdot & \cdot & \cdot & \cdot \\ 0 & 0 & 0 & 0 & 0 & 0 & 0 & 0 & 1 \end{pmatrix} \begin{pmatrix} x_1 \\ x_2 \\ \cdot \\ \cdot \\ \cdot \\ z_3 \end{pmatrix} = \begin{pmatrix} x_1 \\ x_2 \\ \cdot \\ \cdot \\ \cdot \\ z_3 \end{pmatrix} \tag{2-193}$$

and the character is

$$\chi_{\text{tot}}(E) = 9$$

2. C_2

Only the oxygen atom is unshifted. For hydrogen 1, for example, we see that (using an arrow to denote "is shifted to")

$$x_1 \longrightarrow -x_2, \qquad y_1 \longrightarrow -y_2, \qquad z_1 \longrightarrow z_2$$

whereas

$$x_3 \longrightarrow -x_3, \qquad y_3 \longrightarrow -y_3, \qquad z_3 \longrightarrow z_3$$

The matrix equation is then

$$\begin{pmatrix} 0 & 0 & 0 & -1 & 0 & 0 & 0 & 0 & 0 \\ 0 & 0 & 0 & 0 & -1 & 0 & 0 & 0 & 0 \\ 0 & 0 & 0 & 0 & 0 & 1 & 0 & 0 & 0 \\ -1 & 0 & 0 & 0 & 0 & 0 & 0 & 0 & 0 \\ 0 & -1 & 0 & 0 & 0 & 0 & 0 & 0 & 0 \\ 0 & 0 & 1 & 0 & 0 & 0 & 0 & 0 & 0 \\ 0 & 0 & 0 & 0 & 0 & 0 & -1 & 0 & 0 \\ 0 & 0 & 0 & 0 & 0 & 0 & 0 & -1 & 0 \\ 0 & 0 & 0 & 0 & 0 & 0 & 0 & 0 & 1 \end{pmatrix} \begin{pmatrix} x_1 \\ y_1 \\ z_1 \\ x_2 \\ y_2 \\ z_2 \\ x_3 \\ y_3 \\ z_3 \end{pmatrix} = \begin{pmatrix} -x_2 \\ -y_2 \\ z_2 \\ -x_1 \\ -y_1 \\ z_1 \\ -x_3 \\ -y_3 \\ z_3 \end{pmatrix} \tag{2-194}$$

and the character is $\chi_{\text{tot}}(C_2) = -1$

3. σ_v

Again, only the oxygen atom is unshifted, and the coordinates change in accordance with

$$x_3 \longrightarrow x_3, \qquad y_3 \longrightarrow -y_3, \qquad z_3 \longrightarrow z_3$$

Using the same arguments as above, we can see without even writing out the matrix that

$$\chi_{\text{tot}}(\sigma_v) = 1$$

4. σ'_v

All three atoms are unshifted, so that

$$\begin{aligned} x_1 &\longrightarrow -x_1, & x_2 &\longrightarrow -x_2, & x_3 &\longrightarrow -x_3 \\ y_1 &\longrightarrow y_1, & y_2 &\longrightarrow y_2, & y_3 &\longrightarrow y_3 \\ z_1 &\longrightarrow z_1, & z_2 &\longrightarrow z_2, & z_3 &\longrightarrow z_3 \end{aligned}$$

and

$$\chi_{\text{tot}}(\sigma'_v) = 3$$

The character system for $\mathbf{\Gamma}_{\text{tot}}$ is thus

	E	C_3	σ_v	σ'_v
Γ_{tot}	9	-1	1	3

(2-195)

We can now use (2-63) to find how the irreducible representations combine to form $\mathbf{\Gamma}_{\text{tot}}$, but in this case it is just as easy to do this empirically. We find that the only combination of representations from Table 2-13 that add up to give $\mathbf{\Gamma}_{\text{tot}}$ is

$$\mathbf{\Gamma}_{\text{tot}} = 3A_1 + A_2 + 2B_1 + 3B_2 \tag{2-196}$$

Since there are $3N - 6 = 3$ vibrational modes for H_2O (using a three-dimensional basis as we have done), six of the representations in (2-196) must be associated with translations or rotations.

Let us first identify the translations. A general translation $\mathbf{r} = x\mathbf{i} + y\mathbf{j} + z\mathbf{k}$ will have contributions along each of the three axes of Fig. 2-22. To find the association between coordinates and representations, we again use the projection operator method. For x, for example, the effect of each operation in C_{2v} is

$$Ex = x, \qquad C_2x = -x, \qquad \sigma_v x = x, \qquad \sigma'_v x = -x$$

Hence, by (2-138) we have for A_1,

$$P^{(1)}x = 1(x) + 1(-x) + 1(x) + 1(-x) = 0$$

and similarly for A_2,

$$P^{(2)}x = 1(x) + 1-(x) - 1(x) - 1(-x) = 0$$

For B_1, $$P^{(3)}x = 1(x) - 1(-x) + 1(x) - 1(-x) = 4x$$

and for B_2 the result is the same as for A_1 and A_2. Hence, x belongs to B_1. In the same way, we find that y belongs to B_2 and z to A_1, as indicated at the far right of Table 2-13. Thus, we may write the translational part of (2-192) as

$$\Gamma_{\text{trans}} = A_1 + B_1 + B_2 \tag{2-197}$$

For rotations, consider the three operations indicated in Fig. 2-23(a). Each rotation R_x, R_y, R_z is right-handed; if we apply a two-fold rotation

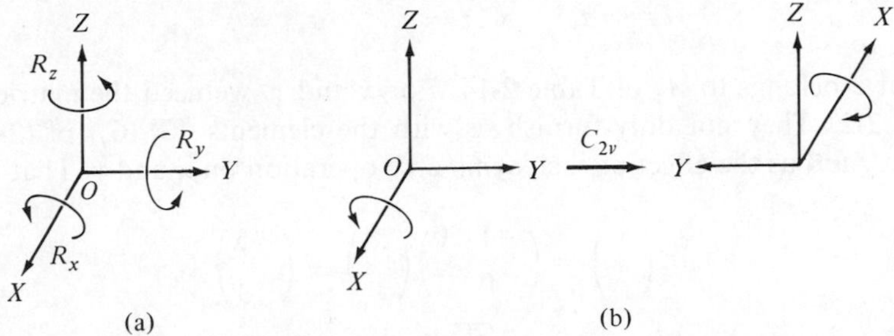

2-23 Identification of rotations with representations

about OZ to this system, then R_x and R_y, for example, reverse in sense, as indicated in Fig. 2-23(b). Hence, the associated characters are -1. Similar arguments for the other operations enable us to identify R_x with B_2 of Table 2-13, and using the entries for R_y and R_z as well, we have

$$\Gamma_{\text{rot}} = A_2 + B_1 + B_2 \tag{2-198}$$

Then subtracting Γ_{trans} and Γ_{rot} from (2-192) leaves finally

$$\Gamma_{\text{vib}} = 2A_1 + B_2 \tag{2-199}$$

exactly as we have already determined in Table 2-16, where the combinations $(S_{13} + S_{23})$ and S_{12} belong to A_1, while $(S_{13} - S_{23})$ belongs to B_2.

PROBLEM 2-40

Verify that the second-degree functions $x^2, y^2, z^2, xy, xz, yz$ belong to the representations indicated in Table 2-13.

PROBLEM 2-41

(a) Show for the molecule of Sec. 2-10 that the characters of Γ_{tot} in (2-192) are

$$\chi_{\text{tot}}(E) = 9, \qquad \chi_{\text{tot}}(C_3) = 0, \qquad \chi_{\text{tot}}(\sigma_v) = 1$$

(b) Show also that

$$\Gamma_{\text{tot}} = 2A_1 + A_2 + 3E$$

For groups such as C_{3v}, the identifications made above are more of a nuisance, since we get linear combinations from the projection operators. Placing the coordinate system XYZ with C_3 along OZ, we see that

$$Ez = z, \qquad C_3 z = z, \qquad \sigma_v z = z$$

so that z belongs to A_1 of Table 2-14. For x and y, we need the matrices in Table 2-2. They not only furnish us with the elements $\Gamma_{ij}^{(k)}(G_l)$ of (2-138), but they tell us the effect of each symmetry operation on x and y. That is,

$$\sigma_v^a\begin{pmatrix} x \\ y \end{pmatrix} = \begin{pmatrix} -1 & 0 \\ 0 & 1 \end{pmatrix}\begin{pmatrix} x \\ y \end{pmatrix} = \begin{pmatrix} -x \\ y \end{pmatrix}$$

but
$$C_3\begin{pmatrix} x \\ \\ y \end{pmatrix} = \begin{pmatrix} -\frac{1}{2} & -\frac{\sqrt{3}}{2} \\ \frac{\sqrt{3}}{2} & \frac{1}{2} \end{pmatrix}\begin{pmatrix} x \\ \\ y \end{pmatrix} = \begin{pmatrix} -\frac{x}{2} - \frac{\sqrt{3}\,y}{2} \\ \frac{\sqrt{3}\,x}{2} + \frac{y}{2} \end{pmatrix}$$

and x or y are converted into a linear combination of both variables. We then find, as examples, that for A_1

$$\begin{aligned} P_{11}^{(1)}x &= 1(Ex) + 1(C_3x) + \dots + 1(\sigma_v^c x) \\ &= x + \left(-\frac{x}{2} - \frac{\sqrt{3}\,y}{2}\right) + \dots + \left(\frac{x}{2} - \frac{\sqrt{3}\,y}{2}\right) = 0 \end{aligned}$$

with a similar result for A_2. However, for E

$$\begin{aligned} P_{11}^{(3)}x = 1(x) &- \frac{1}{2}\left(-\frac{x}{2} - \frac{\sqrt{3}\,y}{2}\right) - \frac{1}{2}\left(-\frac{x}{2} + \frac{\sqrt{3}\,y}{2}\right) \\ &- 1(-x) + \frac{1}{2}\left(\frac{x}{2} + \frac{\sqrt{3}\,y}{2}\right) + \frac{1}{2}\left(\frac{x}{2} - \frac{\sqrt{3}\,y}{2}\right) = 3x \end{aligned}$$

with corresponding values for y, and for the other three operators. That is, x

and y together belong to the degenerate representation E. They are partners, and we indicate this by the symbol (x, y) in Table 2-14. As we may anticipate, a similar situation exists for the partner rotations (R_x, R_y). Although this procedure we have just used is tedious, we need do it only once and then tabulate the results for each group. Complete sets of character tables and basis functions will be found in many books.[11,12,13]

PROBLEM 2-42

Finish the analysis of the molecule begun in Problem 2-41.

Having shown qualitatively how to predict the symmetries of the normal modes, we next ask if we can do anything similar with regard to actually predicting the nature of these mode. As we shall see, for nondegenerate representations, this is fairly easy. Consider the three normal modes of H_2O in Fig. 2-24, as taken from Herzberg.[13] The mode with frequency ν_1 has to

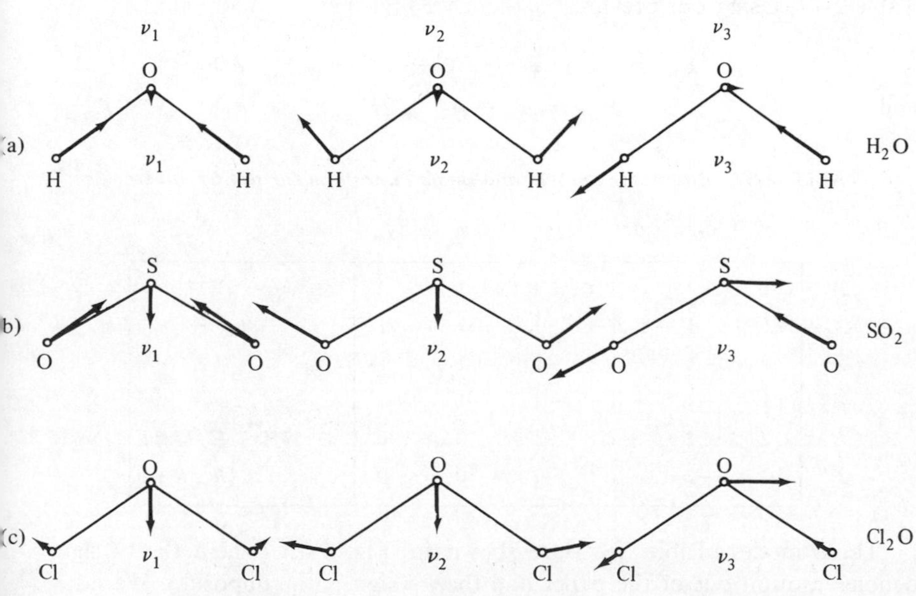

2-24 The normal modes for some typical XY_2 molecules

[11] F. A. Cotton, *Chemical Applications of Group Theory*, Interscience, New York, 1963.

[12] R. M. Hochstrasser, *Molecular Aspects of Symmetry*, W. A. Benjamin, New York, 1966.

[13] G. Herzberg, *Molecular Structure and Molecular Spectra, II. Infrared and Raman Structure of Polyatomic Molecules*, D. Van Nostrand Co., Inc., Princeton, N. J., 1945.

be symmetrical under all operations of the group, since it belongs to A_1, as can be seen from Fig. 2-24 and Table 2-13. It consists mostly of stretching of the bonds, with a small amount of bending. Since we predicted two A_2 modes in (2-199), the second one—also symmetrical—might be expected to be mostly bending. Finally, we need a mode which is invariant under σ'_v, but reverses sign for C_3 or σ_v. Hence, the stretchings should be related as shown for the third mode, and the oxygen atom then moves as shown in order to keep the center of mass constant. Figures 2-24(b) and (c) show the normal modes for two other molecules similar to H_2O but having different atomic masses. We note that the light H atoms in H_2O have the largest amplitude, but the corresponding Cl atoms in Cl_2O, which are heavy, have very small amplitudes. Although these amplitude differences exist, all three molecules show the same mode symmetries.

Turning now to the molecule of Fig. 2-18, we certainly expect that the vibrational mode belonging to A_1 (or Γ_1) will be completely symmetrical. The degenerate modes, however, are not as straightforward to predict. To see why, let us consider an example given by Cotton.[11] The planar XY_3 molecule will belong to the group D_{3h}, with the character system as given by Table 2-17. Using our previous method, we find that

$$\Gamma_{\text{tot}} = A'_1 + A'_2 + 2A''_2 + 3E' + E''$$

and

$$\Gamma_{\text{vib}} = A'_1 + A''_2 + 2E'$$

TABLE 2-17 ***Character System and Basis Functions for the Group D_{3h}***

	E	σ_h	$2C_3$	$2S_3$	$3C'_2$	$3\sigma_v$	
A'_1	1	1	1	1	1	1	$x^2 + y^2, z^2$
A'_2	1	1	1	1	−1	−1	R_z
A''_1	1	−1	1	−1	1	−1	
A''_2	1	−1	1	−1	−1	1	z
E'	2	2	−1	−1	0	0	$(x, y), (x^2 - y^2, xy)$
E''	2	−2	−1	1	0	0	$(xz, yz), (R_x, R_y)$

These modes (Table 2-17) are shown in Fig. 2-25 (where the + sign denotes motion out of the paper and the − sign is the opposite). We note that the two nondegenerate modes, denoted $A'_1(1)$ and $A''_2(2)$, fit their respective representations as demanded by the characters. The left-hand mode of the first degenerate pair $E'(3a)$ and $E'(3b)$ may be converted into a new configuration by the operation C_3, as shown in Fig. 2-26(a). Then, as shown in Fig. 2-26(b), this new mode is actually the sum of $-\frac{1}{2}E'(3a)$ and $\frac{1}{2}E'(3b)$. It can similarly be shown that the right-hand member of this pair is the

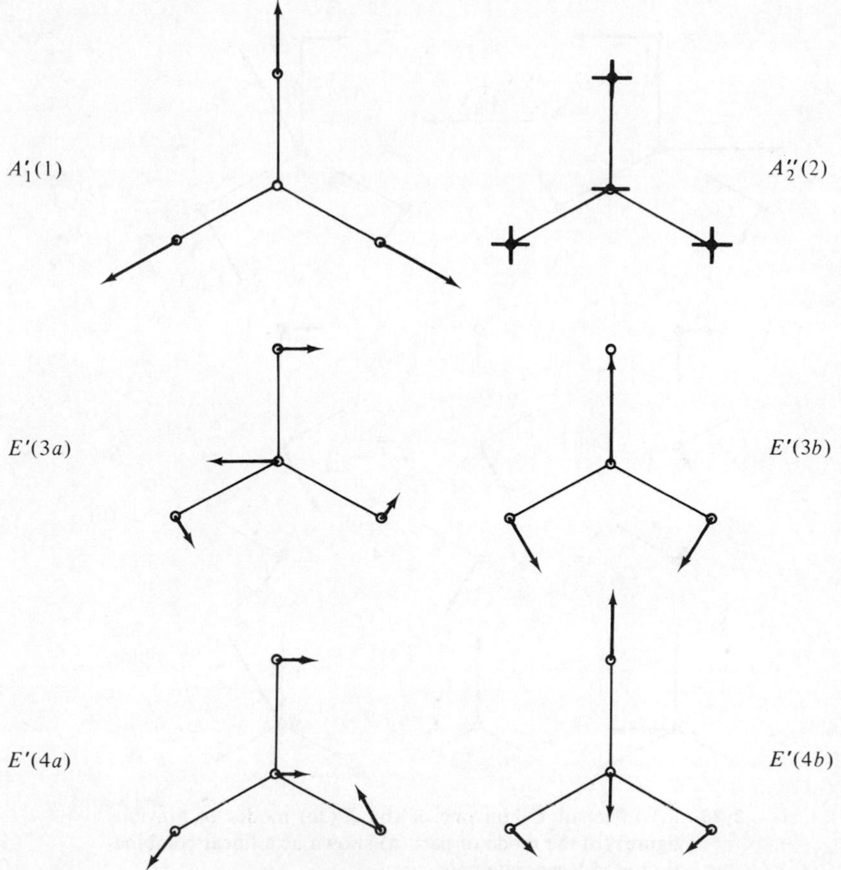

2-25 Normal modes of the planar XY_3 molecule

sum of $-\frac{3}{2}E'(3a)$ and $-\frac{1}{2}E'(3b)$. The matrix of the coefficients of these two combinations is

$$\begin{pmatrix} -\frac{1}{2} & \frac{1}{2} \\ -\frac{3}{2} & -\frac{1}{2} \end{pmatrix}$$

whose character is -1, which is the character of C_3 in the representation E'. Similarly, it may be shown that $C_2'E'(3a) = -E'(3a)$ and $C_2'E'(3b) = E'(3b)$, giving the matrix

$$\begin{pmatrix} -1 & 0 \\ 0 & 1 \end{pmatrix}$$

whose character is 0. Continuing in this way, we would find that this pair matches E' completely.

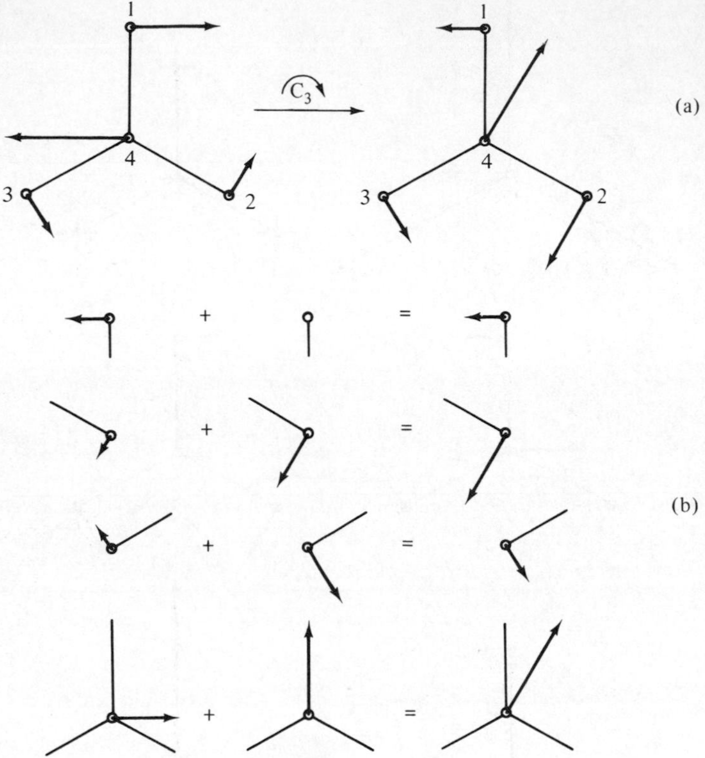

2-26 (a) Effect of C_3 on one of the $E'(3a)$ modes of previous figure; (b) the mode of part (a) shown as a linear combination of a degenerate pair

PROBLEM 2-43

Show that the sum of the second degenerate mode in Fig. 2-18 and of this mode when subjected to C_3 will be the first degenerate mode.

2-15 An extension to a more intricate example: the chloroform molecule

The $CHCl_3$ molecule, as discussed by Colthup, Daly, and Wiberly[5] and by Wilde,[14] is a good example of how involved a normal mode calculation for a five-atom molecule can be, even with the help of group theory. Figure

[14] R. E. Wilde, *Am. J. Phys.*, **32**, 45 (1964).

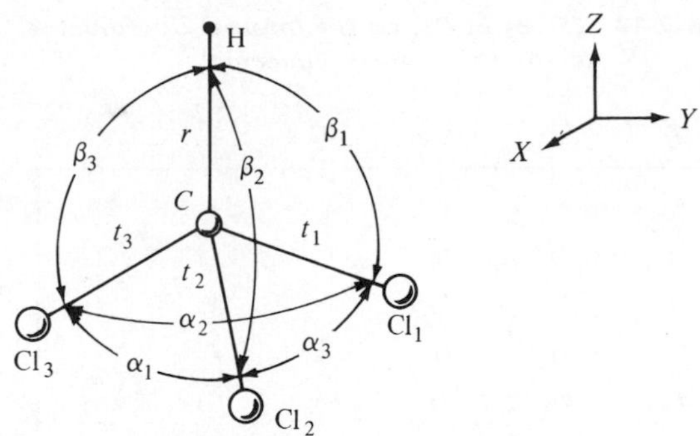

2-27 Internal coordinates for the chloroform molecule

2-27 shows the structure, with the internal coordinates chosen as the C—Cl bond length changes (t_1, t_2, t_3), the C—H bond length change (r), the Cl—C—Cl angle changes (α_1, α_2, α_3), and the H—C—H angle changes (β_1, β_2, β_3).

Although these ten coordinates seem like an obvious choice, one of them is actually redundant, for the number of vibrational modes is

$$3N - 6 = 15 - 6 = 9$$

The total representation has characters

$$\chi_{\text{tot}}(E) = 15, \qquad \chi_{\text{tot}}(C_3) = 0, \qquad \chi_{\text{tot}}(\sigma_v) = 3 \tag{2-200}$$

The result for C_3 may be a little surprising; although the Cl atoms all change places under C_3 and C_3^2, the C, and the H atoms have invariant z values. However that portion of the 15×15 matrix representation of C_3 which applies to these atoms is actually two 3×3 matrices, identical to

$$C_3 = \begin{pmatrix} -1/2 & -\sqrt{3}/2 & 0 \\ \sqrt{3}/2 & -1/2 & 0 \\ 0 & 0 & 1 \end{pmatrix}$$

with $\chi = 0$, so that the total character of the 15×15 matrix is also zero. Comparing (2-200) with Table 2-14 for group C_{3v} or $3m$, to which $CHCl_3$ belongs, we see that

$$\Gamma_{\text{tot}} = 4A_1 + A_2 + 5E$$

and that $\qquad \Gamma_{\text{trans}} = A_1 + E, \qquad \Gamma_{\text{rot}} = A_2 + E$

so that $\qquad \Gamma_{\text{vib}} = 3A_1 + 0A_2 + 3E \qquad$ (2-201)

The effect of each symmetry operation on the ten internal coordinates is

TABLE 2-18 ***Effect of C_{3v} on the Internal Coordinates of the Chloroform Molecule***

	E	C_3	C_3^2	σ_v^a	σ_v^b	σ_v^c
r	r	r	r	r	r	r
t_1	t_1	t_3	t_2	t_1	t_3	t_2
t_2	t_2	t_1	t_3	t_3	t_2	t_1
t_3	t_3	t_2	t_1	t_2	t_1	t_3
α_1	α_1	α_3	α_2	α_1	α_3	α_2
α_2	α_2	α_1	α_3	α_3	α_2	α_1
α_3	α_3	α_2	α_1	α_2	α_1	α_3
β_1	β_1	β_3	β_2	β_1	β_3	β_2
β_2	β_2	β_1	β_3	β_3	β_2	β_1
β_3	β_3	β_2	β_1	β_2	β_1	β_3

TABLE 2-19 ***Projections of the Coordinates of Table 2-18***

Representation	Symmetric combination
A_1	$S_1 = r$
A_1	$S_2 = \dfrac{t_1 + t_2 + t_3}{\sqrt{3}}$
A_1	$S_3 = \dfrac{\alpha_1 + \alpha_2 + \alpha_3}{\sqrt{3}}$
A_1	$S_4 = \dfrac{\beta_1 + \beta_2 + \beta_3}{\sqrt{3}}$
E	$S_5 = \dfrac{2t_1 - t_2 - t_3}{\sqrt{6}}$
E	$S_6 = \dfrac{t_2 - t_3}{\sqrt{2}}$
E	$S_7 = \dfrac{2\alpha_1 - \alpha_2 - \alpha_3}{\sqrt{6}}$
E	$S_8 = \dfrac{\alpha_2 - \alpha_3}{\sqrt{2}}$
E	$S_9 = \dfrac{2\beta_1 - \beta_2 - \beta_3}{\sqrt{6}}$
E	$S_{10} = \dfrac{\beta_2 - \beta_3}{\sqrt{2}}$

given in Table 2-18. Using the projection operators of Table 2-9, we obtain linear combinations of the same form, as listed in Table 2-19 (still carrying a redundant combination). At this point in the analysis of H_2O we determined the force constants $F_{tt'}$ of (2-178) from a knowledge of the internal symmetry coordinates S_i. In Table 2-19, we can label the linear combinations S_1 through S_{10} and proceed as before. However, the algebra is far more tedious in the present example, because of the difficulty in generating the expression corresponding to (2-188). First, we need the potential energy in terms of the ten internal coordinates we have chosen. This has the form

$$\begin{aligned}
2V = {} & f_r(r^2) + f_t(t_1^2 + t_2^2 + t_3^2) + d^2 f_\alpha(\alpha_1^2 + \alpha_2^2 + \alpha_3^2) \\
& + d^2 f_\beta(\beta_1^2 + \beta_2^2 + \beta_3^2) + 2f_{rt}(rt_1 + rt_2 + rt_3) \\
& + 2f_{tt}(t_1t_2 + t_1t_3 + t_2t_3) + 2d^2 f_{\alpha\alpha}(\alpha_1\alpha_2 + \alpha_1\alpha_3 + \alpha_2\alpha_3) \\
& + 2d^2 f_{\beta\beta}(\beta_1\beta_2 + \beta_1\beta_3 + \beta_2\beta_3) + 2df_{r\alpha}(r\alpha_1 + r\alpha_2 + r\alpha_3) \\
& + 2df_{r\beta}(r\beta_1 + r\beta_2 + r\beta_3) + 2df'_{t\alpha}(t_1\alpha_1 + t_2\alpha_2 + t_3\alpha_3) \\
& + 2df_{t\alpha}(t_1\alpha_2 + t_1\alpha_3 + t_2\alpha_1 + t_2\alpha_3 + t_3\alpha_1 + t_3\alpha_2) \\
& + 2df_{t\beta}(t_1\beta_1 + t_2\beta_2 + t_3\beta_3) + 2df'_{t\beta}(t_1\beta_2 + t_1\beta_3 + t_2\beta_1 \\
& + t_2\beta_3 + t_3\beta_1 + t_3\beta_2) + 2d^2 f'_{\alpha\beta}(\alpha_1\beta_1 + \alpha_2\beta_2 + \alpha_3\beta_3) \\
& + 2d^2 f_{\alpha\beta}(\alpha_1\beta_2 + \alpha_1\beta_3 + \alpha_2\beta_1 + \alpha_2\beta_3 + \alpha_3\beta_1 + \alpha_3\beta_2)
\end{aligned} \tag{2-202}$$

where the meaning of the subscripts on the force constants is obvious and where terms involving angular displacements have been multiplied by d or d^2 (d is an appropriate distance) so that all units are expressed as newt/m. Solving the 10 equations

$$\begin{aligned}
S_1 &= r \\
S_2 &= \frac{t_1 + t_2 + t_3}{\sqrt{3}} \\
&\cdots\cdots\cdots\cdots \\
S_{10} &= \frac{\beta_2 - \beta_3}{\sqrt{2}}
\end{aligned} \tag{2-203}$$

for $r, t_1, t_2, \ldots, \beta_2, \beta_3$ and substituting into (2-201) would give the desired expression for V in terms of the S_i.

It is quite obvious that a more systematic procedure is needed here and what is done is to define a transformation matrix $\mathbf{U}$ connecting the internal coordinates $r_1, r_2, \ldots, r_{10}$ with the internal symmetry coordinates $S_1, S_2, \ldots, S_{10}$ through the equation

$$\mathbf{S} = \mathbf{Ur} \tag{2-204}$$

where $\quad r_1 = r, \quad r_2 = t_1, \quad r_3 = t_2, \quad \ldots, \quad r_{10} = \beta_3$

Referring back to Eq. (2-2), we have regarded y as a 3×1 *column matrix.* We may also treat y as a *row matrix* by putting (2-2) in the form

$$\begin{pmatrix} z_1 \\ z_2 \\ z_3 \end{pmatrix} = (y_1 \quad y_2 \quad y_3) \begin{pmatrix} a_{11} & a_{12} & a_{13} \\ a_{21} & a_{22} & a_{23} \\ a_{31} & a_{32} & a_{33} \end{pmatrix}$$

and the rule for matrix multiplication leads to Eqs. (2-1), as before.

It is interesting to note that the well-known definition of the scalar product

$$\mathbf{A} \cdot \mathbf{A} = A_x^2 + A_y^2 + A_z^2$$

may be expressed in matrix form by writing **A** as a row matrix and as a column matrix, so that

$$\mathbf{A} \cdot \mathbf{A} = (A_x \quad A_y \quad A_z) \begin{pmatrix} A_x \\ A_y \\ A_z \end{pmatrix}$$

Using **r** in both row and column form, Eq. (2-202) can be simplified to

$$2V = \mathbf{r}^{-1}\mathbf{f}\mathbf{r} \tag{2-205}$$

provided $\mathbf{r} = \mathbf{r}^{-1}$ or $r^2 = 1$, which will be true if we incorporate the proper numerical factor. Since (2-178) is equivalent to

$$2V = \mathbf{S}^{-1}\mathbf{F}\mathbf{S} \tag{2-206}$$

then

$$\mathbf{r}^{-1}\mathbf{f}\mathbf{r} = \mathbf{S}^{-1}\mathbf{F}\mathbf{S} \tag{2-207}$$

From (2-204)

$$\mathbf{r}^{-1}\mathbf{f}\mathbf{r} = (\mathbf{U}\mathbf{r})^{-1}\mathbf{F}(\mathbf{U}\mathbf{r}) = \mathbf{r}^{-1}\mathbf{U}^{-1}\mathbf{F}\mathbf{U}\mathbf{r}$$

or

$$\mathbf{F} = \mathbf{U}^{-1}\mathbf{f}\mathbf{U} \tag{2-208}$$

where we have used the relation

$$(\mathbf{A}\mathbf{B})^{-1} = \mathbf{B}^{-1}\mathbf{A}^{-1} \tag{2-209}$$

PROBLEM 2-44

Prove (2-209).

The structure of the matrix **U**, as determined by inspection from Table 2-19, is

$$\mathbf{U} = \begin{pmatrix} 1 & 0 & 0 & 0 & 0 & 0 & 0 & 0 & 0 & 0 \\ 0 & 1/a & 1/a & 1/a & 0 & 0 & 0 & 0 & 0 & 0 \\ 0 & 0 & 0 & 0 & 1/a & 1/a & 1/a & 0 & 0 & 0 \\ 0 & 0 & 0 & 0 & 0 & 0 & 0 & 1/a & 1/a & 1/a \\ 2/b & -1/b & -1/b & 0 & 0 & 0 & 0 & 0 & 0 & 0 \\ 0 & 1/c & 1/c & 0 & 0 & 0 & 0 & 0 & 0 & 0 \\ 0 & 0 & 0 & 2/b & -1/b & -1/b & 0 & 0 & 0 & 0 \\ 0 & 0 & 0 & 0 & 1/c & -1/c & 0 & 0 & 0 & 0 \\ 0 & 0 & 0 & 0 & 0 & 0 & 0 & 2/b & -1/b & -1/b \\ 0 & 0 & 0 & 0 & 0 & 0 & 0 & 0 & 1/c & -1/c \end{pmatrix}$$

where $a = \sqrt{3}, \quad b = \sqrt{6}, \quad c = \sqrt{2}$ **(2-210)**

Rather than work with this entire matrix, let us break it up into parts. For example, the portion which belongs to S_6, S_8, and S_{10} is

$$\mathbf{U}_{6,8,10}^{-1} = \begin{pmatrix} 0 & 1/c & -1/c & 0 & 0 & 0 & 0 & 0 & 0 & 0 \\ 0 & 0 & 0 & 1/c & -1/c & 0 & 0 & 0 & 0 & 0 \\ 0 & 0 & 0 & 0 & 0 & 1/c & -1/c & 0 & 0 & 0 \end{pmatrix} \qquad \textbf{(2-211)}$$

where the symbol $\mathbf{U}_{6,8,10}^{-1}$ denotes a row matrix which will be used as the left-hand factor in (2-208). The matrix **f** as specified by (2-202) is

$$\mathbf{f} = \begin{pmatrix} f_r & f_{rt} & f_{rt} & f_{rt} & df_{r\alpha} & df_{r\alpha} & df_{r\alpha} & df_{r\beta} & df_{r\beta} & df_{r\beta} \\ & f_t & f_{tt} & f_{tt} & df'_{t\alpha} & df_{t\alpha} & df_{t\alpha} & df_{t\beta} & df'_{t\beta} & df'_{t\beta} \\ & & f_t & f_{tt} & df_{t\alpha} & df'_{t\alpha} & df_{t\alpha} & df'_{t\beta} & df_{t\beta} & df'_{t\beta} \\ & & & f_t & df_{t\alpha} & df_{t\alpha} & df'_{t\alpha} & df'_{t\beta} & df'_{t\beta} & df_{t\beta} \\ & & & & d^2f_\alpha & d^2f_{\alpha\alpha} & d^2f_{\alpha\alpha} & d^2f'_{\alpha\beta} & d^2f_{\alpha\beta} & d^2f_{\alpha\beta} \\ & & & & & d^2f_\alpha & d^2f_{\alpha\alpha} & d^2f_{\alpha\beta} & d^2f'_{\alpha\beta} & d^2f_{\alpha\beta} \\ & & & & & & d^2f_\alpha & d^2f_{\alpha\beta} & d^2f_{\alpha\beta} & d^2f'_{\alpha\beta} \\ & & & & & & & d^2f_\beta & d^2f_{\beta\beta} & d^2f_{\beta\beta} \\ & & & & & & & & d^2f_\beta & d^2f_{\beta\beta} \\ & & & & & & & & & d^2f_\beta \end{pmatrix} \qquad \textbf{(2-212)}$$

where only half of the symmetric matrix is shown. Then, obtaining $\mathbf{U}_{6,8,10}$ by reversing rows and columns in (2-211), we find that the corresponding part of **F** simplifies to a 3 × 3 matrix

$$\mathbf{F}_{6,8,10} = \begin{pmatrix} f_t - f_{tt} & df'_{t\alpha} - df_{t\alpha} & df_{t\beta} - df'_{t\beta} \\ df'_{t\alpha} - df_{t\alpha} & d^2 f_\alpha - d^2 f_{\alpha\alpha} & d^2 f'_{\alpha\beta} - d^2 f_{\alpha\beta} \\ df_{t\beta} - df'_{t\beta} & d^2 f'_{\alpha\beta} - d^2 f_{\alpha\beta} & d^2 f_\beta - d^2 f_{\beta\beta} \end{pmatrix} \tag{2-213}$$

It is found that the same 3 × 3 matrix is obtained by using the portion of **U** belonging to S_5, S_7, and S_9—as we would expect—for this set of six coordinates belongs to the doubly-degenerate representation E.

The matrix **G**, as we saw in Eq. (2-187), has elements which depend on the bond angles and the individual masses of the atoms forming the molecule. Again, the algebra is tedious, but simplifying formulas will be found in Appendix 6 of Wilson, Decius, and Cross.[9] Using this approach, or the straightforward method of Sec. 2-13, we find that the matrix **G**, which we shall denote by **g** when expressed in terms of the original internal coordinates $r_1, r_2, \ldots, r_{12}$, has a form similar to (2-212). However, many of the elements turn out to be identical and the matrix simplifies to

$$\mathbf{g} = \begin{pmatrix} g_r & g_{rt} & g_{rt} & g_{rt} & g_{r\alpha} & g_{r\alpha} & g_{r\alpha} & g_{r\beta} & g_{r\beta} & g_{r\beta} \\ & g_t & g_{tt} & g_{tt} & g'_{t\alpha} & g_{t\alpha} & g_{t\alpha} & g_{t\beta} & g'_{t\beta} & g'_{t\beta} \\ & & g_t & g_{tt} & g_{t\alpha} & g'_{t\alpha} & g_{t\alpha} & g'_{t\beta} & g_{t\beta} & g'_{t\beta} \\ & & & g_t & g_{t\alpha} & g_{t\alpha} & g'_{t\alpha} & g'_{t\beta} & g'_{t\beta} & g_{t\beta} \\ & & & & g_\alpha & g_{\alpha\alpha} & g_{\alpha\alpha} & g'_{\alpha\beta} & g_{\alpha\beta} & g_{\alpha\beta} \\ & & & & & g_\alpha & g_{\alpha\alpha} & g_{\alpha\beta} & g'_{\alpha\beta} & g_{\alpha\beta} \\ & & & & & & g_\alpha & g_{\alpha\beta} & g_{\alpha\beta} & g'_{\alpha\beta} \\ & & & & & & & g_\beta & g_{\beta\beta} & g_{\beta\beta} \\ & & & & & & & & g_\beta & g_{\beta\beta} \\ & & & & & & & & & g_\beta \end{pmatrix} \tag{2-214}$$

The explicit expressions for these elements, as given by Colthup, Daly, and Wiberly[10] are

$$g_{r\alpha} = \frac{2\sqrt{2}}{3}\rho_{\mathrm{CCl}}\mu_{\mathrm{C}}, \qquad g_{t\alpha} = -\frac{2\sqrt{2}}{3}\rho_{\mathrm{CCl}}\mu_{\mathrm{C}},$$

$$g_{r\beta} = -\frac{2\sqrt{2}}{3}\rho_{\mathrm{CCl}}\mu_{\mathrm{C}}, \qquad g_{t\beta} = -\frac{2\sqrt{2}}{3}\rho_{\mathrm{CH}}\mu_{\mathrm{C}},$$

$$g_t = \mu_{\mathrm{C}} + \mu_{\mathrm{Cl}}, \qquad g'_{t\beta} = \frac{\sqrt{2}}{3}(\rho_{\mathrm{CH}} + \rho_{\mathrm{CCl}})\mu_{\mathrm{C}},$$

$$g_{tt} = -\frac{1}{3}\mu_{\mathrm{C}}, \qquad g_\alpha = \frac{8}{3}\rho_{\mathrm{CCl}}^2\mu_{\mathrm{C}} + 2\rho_{\mathrm{CCl}}^2\mu_{\mathrm{Cl}},$$

$$g'_{t\alpha} = \frac{2\sqrt{2}}{3}\rho_{\mathrm{CCl}}\mu_{\mathrm{C}},$$

$$g_{\alpha\alpha} = -\frac{1}{2}\rho_{\text{CCl}}^2\mu_{\text{Cl}}, \tag{2-215}$$

$$g'_{\alpha\beta} = -\frac{4}{3}(\rho_{\text{CCl}}^2 + \rho_{\text{CH}}\rho_{\text{CCl}})\mu_{\text{C}},$$

$$g_{\alpha\beta} = -\left(\frac{2}{3}\rho_{\text{CCl}}^2 - \frac{2}{3}\rho_{\text{CH}}\rho_{\text{CCl}}\right)\mu_{\text{C}} - \frac{1}{2}\rho_{\text{CCl}}^2\mu_{\text{Cl}},$$

$$g_{\beta} = \rho_{\text{CH}}^2\mu_{\text{H}} + \rho_{\text{CCl}}^2\mu_{\text{Cl}} + \left(\rho_{\text{CH}}^2 + \rho_{\text{CCl}}^2 + \frac{2}{3}\rho_{\text{CH}}\rho_{\text{CCl}}\right)\mu_{\text{C}},$$

$$g_{\beta\beta} = -\frac{1}{2}\rho_{\text{CH}}^2\mu_{\text{H}} - \frac{1}{6}(3\rho_{\text{CH}}^2 + 2\rho_{\text{CH}}\rho_{\text{CCl}} - 5\rho_{\text{CCl}}^2)\mu_{\text{C}}$$

where μ_{H}, μ_{C}, and μ_{Cl} are the reciprocals of the atomic masses of hydrogen, carbon, and chlorine and ρ_{CH} and ρ_{CCl} are the reciprocals of the C—H and C—Cl bond distances, respectively. Diagonalizing this with $\mathbf{U}^{-1}$ and $\mathbf{U}$ as we did for $\mathbf{f}$, we obtain the reduced matrix $\mathbf{G}$. The 3×3 sections belonging to S_6, S_8, S_{10} and to S_5, S_7, S_9 are

$$\mathbf{G}_{6,8,10} = \begin{pmatrix} \frac{4}{3}\mu_{\text{C}} + \mu_{\text{Cl}} & \frac{4\sqrt{2}}{3}\rho_{\text{CCl}}\mu_{\text{C}} & -\left(\sqrt{2}\,\rho_{\text{CH}} + \frac{\sqrt{2}}{3}\rho_{\text{CCl}}\right)\mu_{\text{C}} \\ \frac{4\sqrt{2}}{3}\rho_{\text{CCl}}\mu_{\text{C}} & \frac{8}{3}\rho_{\text{CCl}}^2\mu_{\text{C}} + \frac{5}{2}\rho_{\text{CCl}}^2\mu_{\text{Cl}} & -\frac{2}{3}(\rho_{\text{CCl}}^2 + \rho_{\text{CH}}\rho_{\text{CCl}})\mu_{\text{C}} + \frac{1}{2}\rho_{\text{CCl}}^2\mu_{\text{Cl}} \\ -\left(\sqrt{2}\,\rho_{\text{CH}} + \frac{\sqrt{2}}{3}\rho_{\text{CCl}}\right)\mu_{\text{C}} & -\frac{2}{3}(\rho_{\text{CCl}}^2 + \rho_{\text{CH}}\rho_{\text{CCl}})\mu_{\text{C}} + \frac{1}{2}\rho_{\text{CCl}}^2\mu_{\text{Cl}} & \frac{3}{2}\rho_{\text{CH}}^2\mu_{\text{H}} + \rho_{\text{CCl}}^2\mu_{\text{Cl}} + \left(\frac{3}{2}\rho_{\text{CH}}^2 + \frac{1}{6}\rho_{\text{CCl}}^2 + \rho_{\text{CH}}\rho_{\text{CCl}}\right)\mu_{\text{C}} \end{pmatrix} \tag{2-216}$$

As before, it is better to work with numerical quantities from here on. The force constants as given by Zeitlow[15] are listed in Table 2-20, where $f'_{\alpha\beta}$ is taken as zero because it is unknown. Using these values, the third-order secular equation becomes

$$\begin{vmatrix} 19.1267 \times 10^{27} - \omega^2 & 2.2781 \times 10^{19} & -1.7657 \times 10^{19} \\ 10.3437 \times 10^{35} & 4.1782 \times 10^{27} - \omega^2 & -2.1083 \times 10^{27} \\ 20.5809 \times 10^{35} & -15.2280 \times 10^{27} & 51.6481 \times 10^{27} - \omega^2 \end{vmatrix} = 0$$

from which

$$\omega_4^2 = 5.13 \times 10^{28}\ \text{sec}^{-2}$$
$$\omega_5^2 = 2.12 \times 10^{28}\ \text{sec}^{-2}$$
$$\omega_6^2 = 2.41 \times 10^{27}\ \text{sec}^{-2}$$

[15] J. P. Zeitlow, M. S. Thesis, Illinois Institute of Technology, Chicago, 1949.

TABLE 2-20 ***Force Constants for*** $CHCl_3$

f_{ij}	Value (dynes/cm)	df_{ij} or d^2f_{ij} ($d = 1.77 \times 10^{-8}$ cm)	Value (ergs)
f_r	4.8540×10^5		
f_t	3.4782×10^5		
f_α	0.51275×10^5	d^2f_α	1.6064×10^{-11}
f_β	0.20834×10^5	d^2f_β	0.6525×10^{-11}
f_{rt}	0.0880×10^5		
f_{tt}	0.3526×10^5		
$f_{r\alpha} - f_{r\beta}$	0.31208×10^5	$df_{r\alpha} - df_{r\beta}$	0.5524×10^{-3}
$f_{t\alpha}$	0.23378×10^5	$df_{t\alpha}$	0.4136×10^{-3}
$f_{t\beta}$	0.18319×10^5	$df_{t\beta}$	0.3243×10^{-3}
$f'_{t\alpha}$	-0.11123×10^5	$df'_{t\alpha}$	-0.1968×10^{-3}
$f'_{t\beta}$	-0.0927×10^5	$df'_{t\beta}$	-0.1641×10^{-3}
$f_{\beta\beta}$	0.14287×10^5	$d^2f_{\alpha\alpha}$	0.4476×10^{-11}
$f_{\beta\beta}$	0.00287×10^5	$d^2f_{\beta\beta}$	0.00899×10^{-11}
$f'_{\alpha\beta}$	0		
$f_{\alpha\beta}$	0.049242×10^5	$d^2f_{\alpha\beta}$	0.1543×10^{-11}

PROBLEM 2-45

(a) Find the normal modes for NH_3 by setting up and solving the 12×12 secular determinant which comes from Lagrange's equations, using Herzberg[13] as a reference for bond angles.

(b) Do the same thing with group theory.

The most extensive compilation of symmetry coordinates is that of Cyvin.[16] This reference, plus Wilson, Decius, and Cross,[9] provide a thorough treatment of molecular vibration theory.

[16] S. J. Cyvin, *Molecular Vibrations and Mean Square Amplitudes*, Elseveir Publishing Co., Amsterdam, 1968.

CHAPTER THREE INFRARED AND RAMAN VIBRATIONAL SPECTRA

3-1 *The Schroedinger equation*

The first problem we wish to consider in this chapter is the determination of the equation of motion of an electron. We shall define an electron as a particle of mass m ($m = 9.1 \times 10^{-31}$ kgm) and charge e ($e = -1.6 \times 10^{-19}$ coul). In addition, we shall require it to be subject to Planck's quantum hypothesis

$$E = h\nu \tag{3-1}$$

and to the concept of wave-particle duality as expressed by de Broglie's relation

$$p = \frac{h}{\lambda} \tag{3-2}$$

where E is energy, h is Planck's constant ($h = 6.6 \times 10^{-34}$ joule-sec), ν is frequency, p is momentum, and λ is wave length. There is some controversy about the meaning of Eq. (3-2), but for our purposes here, we shall accept the traditional interpretation.

Let us consider an electric field of intensity $\mathscr{E}$ being propagated in free space with velocity c ($c = 3.0 \times 10^8$ m/sec), where

$$\nu\lambda = c \tag{3-3}$$

This wave may be expressed analytically as

$$\mathscr{E} = \mathscr{E}_m e^{i(kx-\omega t)} \tag{3-4}$$

where the propagation constant k is

$$k = \frac{2\pi}{\lambda} \tag{3-5}$$

the angular frequency is

$$\omega = 2\pi\nu \tag{3-6}$$

and the subscript m denotes the maximum value of the amplitude. (A more detailed discussion of wave motion is given in Sec. 5-1.) We see by direct substitution that $\mathscr{E}$ obeys the classical wave equation

$$\frac{\partial^2 \mathscr{E}}{\partial x^2} = \frac{1}{c^2}\frac{\partial^2 \mathscr{E}}{\partial t^2}$$

Using the Dirac constant

$$\hbar = \frac{h}{2\pi} \tag{3-7}$$

Eq. (3-1) becomes

$$E = \hbar\omega \tag{3-8}$$

and (3-2) becomes

$$p = \hbar k \tag{3-9}$$

Substituting (3-8) and (3-9) into (3-4) and replacing $\mathscr{E}$ with an undefined quantity Ψ, we obtain

$$\Psi = \Psi_m e^{i(px - Et)/\hbar} \tag{3-10}$$

as the solution to a differential equation specifying the behavior of an electron. The equation itself is generated in the following problem, and the meaning of Ψ will be considered afterwards.

PROBLEM 3-1

(a) Calculate $\partial^2\Psi/\partial x^2$ and $\partial\Psi/\partial t$.

(b) Make use of the fact that the total energy E of the electron is the sum of its kinetic energy T and its potential energy V, i.e.,

$$E = T + V$$

and combine this with (a) to show that

$$-\frac{\hbar^2}{2m}\frac{\partial^2\Psi}{\partial x^2} + V - i\hbar\frac{\partial\Psi}{\partial t} = 0 \tag{3-11}$$

This is the *time-dependent Schroedinger equation*, and it governs the motion of an electron. Since Eq. (3-11) is a partial differential equation involving x and t as independent variables, we simplify it by using the method of separation of variables. Assume that $\Psi(x, t)$ can be expressed as the product of two new unknown functions, denoted as $\psi(x)$, which depends only on x, and $T(t)$, which depends only on t. Then

$$\Psi(x, t) = \psi(x)T(t)$$

Since $\psi(x)$ is not a function of t, then

$$\frac{\partial \Psi}{\partial t} = \psi \frac{dT}{dt}$$

Similarly,
$$\frac{\partial^2 \Psi}{\partial x^2} = T \frac{d^2 \psi}{dx^2}$$

Using these, Eq. (3-11) becomes, after rearrangement

$$\frac{\hbar^2}{2m} \frac{d^2 \psi}{dx^2} + V = \frac{i\hbar}{T} \frac{dT}{dt} \tag{3-12}$$

This is an equation in which one side depends only on x and the other side depends only on t. The only way in which this is possible is for both sides to be equal to a constant C, giving two ordinary differential equations, of which one is

$$i\hbar \frac{dT}{dt} = CT$$

and this can be integrated to give

$$T = e^{Ct/i\hbar}$$

so that
$$\bar{\Psi} = \psi(x) e^{-iCt/\hbar} \tag{3-13}$$

where we have used $i = -1/i$. Comparing (3-13) with (3-10) shows that

$$E = C$$

The other separated equation then becomes

$$\frac{-\hbar^2}{2m} \frac{d^2 \psi}{dx^2} + V\psi = E\psi \tag{3-14}$$

This is called the *Schroedinger amplitude equation* or simply the Schroedinger equation, and the solution ψ is called a *wave function.*

3-2 The quantum oscillator

As discussed in Chapter 2, a harmonic force F exerted by a spring, for example, is related to the corresponding displacement x by the equation

$$F = -\kappa x$$

where we use κ for the Hooke's law constant to distinguish it from k, the propagation constant. The potential energy is then

$$V = -\int F\,dx = \tfrac{1}{2}\kappa x^2$$

Hence, Eq. (3-14) becomes

$$\frac{d^2\psi}{dx^2} + \frac{2m}{\hbar^2}\left(E - \frac{\kappa x^2}{2}\right)\psi = 0 \tag{3-15}$$

If we let $\alpha = 2mE/\hbar^2$ and $\beta^2 = m\kappa/\hbar^2$, then Eq. (3-15) can be written

$$\frac{d^2\psi}{dx^2} + (\alpha - \beta^2 x^2)\,\psi = 0 \tag{3-16}$$

and (3-16) has a solution given by

$$\psi = e^{-\beta x^2/2} \tag{3-17}$$

provided $\alpha = \beta$.

PROBLEM 3-2

Verify this solution.

This leads us to try as a solution to (3-16) the more general expression

$$\psi(x) = e^{-\beta x^2/2} f(x) \tag{3-18}$$

where $f(x)$ is a new function to be determined. Substituting Eq. (3-18) into Eq. (3-17) leads to

$$\frac{d^2f}{dx^2} - 2\beta x\frac{df}{dx} + (\alpha - \beta)f = 0 \tag{3-19}$$

This is known as the *Hermite equation* (after a nineteenth century mathematician and astronomer), and it has an infinite number of solutions, as follows:

$$\begin{aligned} f &= 1 & &\text{if } \alpha = \beta \\ f &= x & &\text{if } \alpha = 3\beta \\ f &= 2\beta x^2 - 1 & &\text{if } \alpha = 5\beta \\ &\ \ \vdots & &\quad\ \vdots \end{aligned} \tag{3-20}$$

The correctness of these solutions can be verified by substitution into Eq. (3-19). From the definitions of α and β, we have for $\alpha = \beta$

$$E = \frac{\hbar\sqrt{\kappa/m}}{2} = \frac{\hbar\omega}{2} = \frac{h\nu}{2}$$

and for $\alpha = 3\beta$, $\alpha = 5\beta$, etc., we obtain

$$E = \frac{3h\nu}{2}, \qquad E = \frac{5h\nu}{2}, \qquad \cdots$$

or, in general,

$$E_n = (n + \tfrac{1}{2})h\nu, \quad (n = 0, 1, 2, \ldots) \tag{3-21}$$

where the integer n is called the *quantum number*. The Schroedinger equation has thus shown us that the harmonic oscillator is allowed only certain energy levels, and the energy of this system is said to be *quantized*.

Now let us consider the wave functions, which by Eqs. (3-18) and (3-20), are

$$\begin{aligned} \psi_0 &= e^{-\beta x^2/2} \\ \psi_1 &= xe^{-\beta x^2/2} \\ \psi_2 &= (2\beta x^2 - 1)e^{-\beta x^2/2} \\ &\ \ \vdots \end{aligned} \tag{3-22}$$

From Eq. (3-22), we expect that the functions ψ_3, ψ_4, etc. will involve more complicated polynomials in x, which turns out to be the case. For reasons which will soon be apparent, we show a plot of the square of the function ψ_{10} in Fig. (3-1). In this figure, the solid line is the curve ψ_{10}^2 vs. x. The dotted line is the average value of ψ_{10}^2, obtained by taking the area under each "hump" and dividing by the corresponding length of the base line. The points

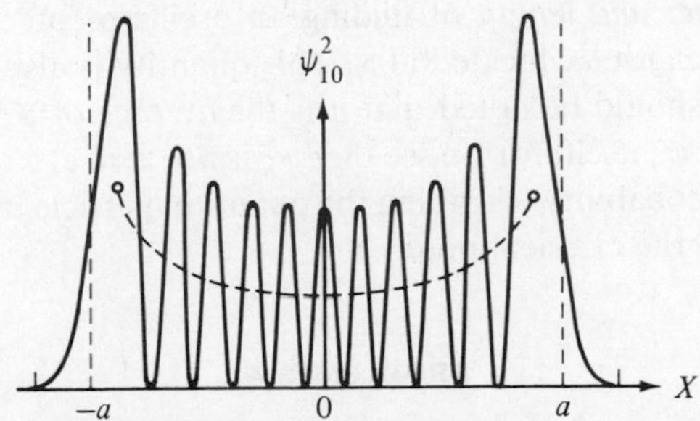

3-1 A plot of ψ_{10}^2 as a function of x (solid curve) for the quantum harmonic oscillator. The dotted curve represents the average probability.

$\pm a$ on the x-axis represent the amplitude of a classical oscillator having the same total energy.

PROBLEM 3-3

Consider a mass m attached to a spring of Hooke's law constant κ and vibrating with amplitude a. Show that the probability dP of finding the mass in a small interval dx at a distance x from its equilibrium position may be expressed as

$$dP \propto \frac{dx}{\sqrt{a^2 - x^2}} \tag{3-23}$$

A plot of $1/\sqrt{a^2 - x^2}$ is shown in Fig. 3-2. A comparison of the two figures shows that there is a close resemblance between the probability curve for the classical vibrating mass and for the curve showing the average value

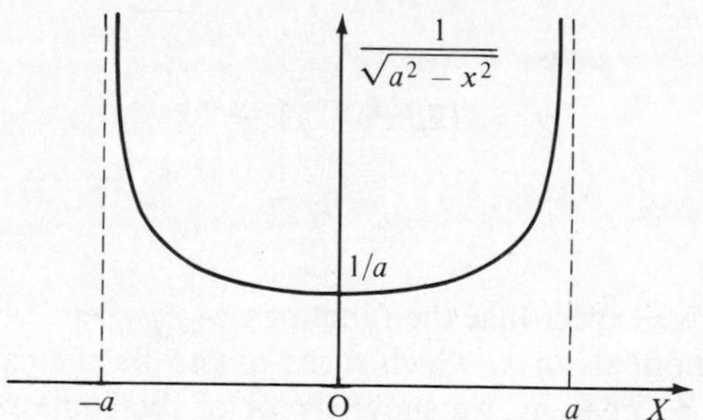

3-2 Probability curve for a classical harmonic oscillator

of ψ_{10}^2 for the particle which obeys the Schroedinger equation. It seems, therefore, very reasonable to postulate that $\psi_n^2(x)$, where $n = 0, 1, 2, \ldots$ is the *probability per unit length* of finding an oscillator of energy $E_n = (n + \frac{1}{2})hv$ in a small length dx located at x. This quantity is also called the *probability density*. It should be noted that it is the *average* of ψ_n^2 which resembles Fig. 3-2, and not ψ_n^2 itself, for we see that ψ_n^2 is not zero at $x = \pm a$, and hence there is a finite probability of finding the quantum particle in places which are not accessible to the classical one.

PROBLEM 3-4

(a) Prove that the total energy E and the amplitude a of an oscillator are related by

$$a = \sqrt{2E/\kappa}$$

(b) Use this result to show that the points $x = \pm a$ of Fig. 3-1 are located a distance $(21h/4\pi^2 mv)^{1/2}$ from the origin.

If the solutions ψ_n to the Schroedinger equation are to be interpreted in terms of a probability-density, as above, then they must obey the condition that

$$\int_{-\infty}^{\infty} \psi_n^2 \, dx = 1 \tag{3-24}$$

This equation may be interpreted as saying that the probability of finding the quantum oscillator *somewhere* on the x-axis is unity. This, in turn, means that the solutions (3-18) should be written

$$\psi_n = N_n f_n(x) e^{-\beta x^2/2}, \quad (n = 0, 1, 2, \ldots) \tag{3-25}$$

where N_n is a *normalization constant*, whose purpose is to be sure that ψ_n satisfies (3-24). For example, for ψ_1 we have

$$\int_{-\infty}^{\infty} \psi_1^2 \, dx = N_1^2 \int_{-\infty}^{\infty} x^2 e^{-\beta x^2} \, dx = \frac{1}{4}\sqrt{\frac{\pi}{\beta^3 N_1^2}} = 1$$

using standard tables. Hence

$$N_1 = \frac{2\beta^{3/4}}{\pi^{1/4}}$$

Eyring, Walter, and Kimball[1] show how to compute N_n for any n; we shall not need the general expression here.

It is also interesting to note that if we replace ψ_n^2 by $\psi_m \psi_n$ in (3-24), where $m \neq n$, then

$$\int_{-\infty}^{\infty} \psi_m \psi_n \, dx = 0 \tag{3-26}$$

Two functions ψ_m and ψ_n which obey a relation like (3-26) are said to be *orthogonal* over the interval $(-\infty, \infty)$. This is analogous to the scalar product of two orthogonal vectors, which is also zero.

PROBLEM 3-5

Verify (3-26) for $m = 0$, $n = 1$.

---

[1] H. Eyring, J. Walter, and G. E. Kimball, *Quantum Chemistry*, John Wiley & Sons, Inc., New York, 1944.

Again, we refer the reader to Eyring, Walter, and Kimball[1] for a proof of (3-26) valid for any m and n.

3-3 The Schroedinger equation and dipoles

In order to show how the normal mode theory developed in Chapter 2 applies to the interpretation of spectroscopic data, let us briefly review some elementary ideas from electrostatic theory. A pair of equal but opposite point charges, of magnitude $\pm q$ and which are arbitrarily close, form a

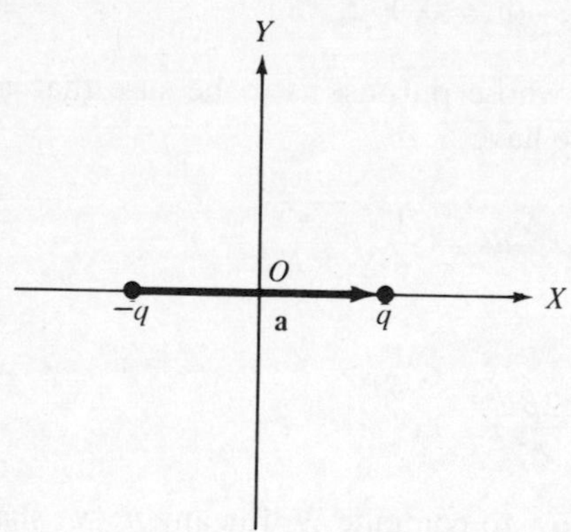

3-3 An electric dipole

dipole. If the separation of two such charges is the vector **a** of Fig. 3-3, we define the *dipole moment* **p** as the quantity

$$\mathbf{p} = q\mathbf{a} \tag{3-27}$$

Let us consider a material, such as a crystal, composed of a large number of identical atoms, each atom containing a nucleus surrounded by Z electrons, where Z is the atomic number. In many materials—said to be *nonpolar*—this situation may be indicated as shown in Fig. 3-4(a). The nucleus is essentially a point charge of magnitude Ze ($e = 1.6 \times 10^{-19}$ coul) immersed in a spherical cloud of electrons of charge $-Ze$. Since the electron cloud is equivalent to a point charge $-Ze$ centered on the nucleus, the material is everywhere neutral.

If an electric field of intensity $\mathscr{E}$ is applied as shown in Fig. 3-4(b), it causes the nucleus to shift to the right and the electron cloud to the left, generating the *induced* dipole of Fig. 3-4(c). For a linear, isotropic medium

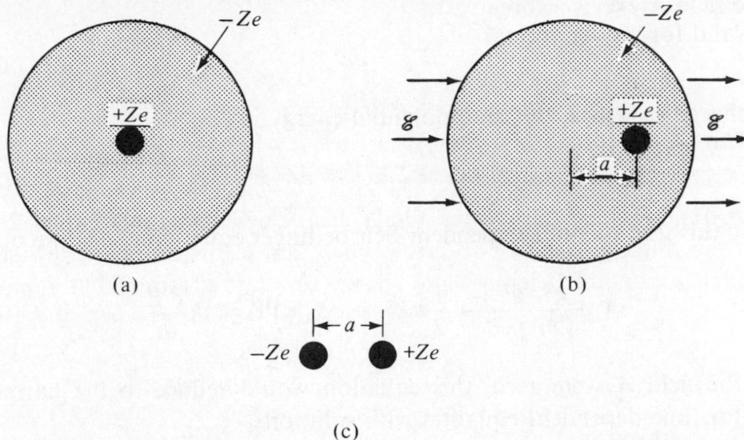

3-4 Electronic polarization, showing (a) the nucleus and the electron charge cloud; (b) the effect of an applied field; and (c) the equivalent dipole

the induced dipole moment $\mathbf{p}_{\text{ind}}$ is related to the field creating this dipole by

$$\mathbf{p}_{\text{ind}} = \alpha \mathscr{E} \tag{3-28}$$

where the proportionality constant α is called the *electronic polarizability* of the material. For anisotropic media, this equation must be generalized to

$$\mathbf{p}_{\text{ind}} = \boldsymbol{\alpha} \mathscr{E} \tag{3-29}$$

since the polarizability is a tensor when the induced dipole moment does not lie along the inducing field.[2]

Let us consider next the situation when $\mathscr{E}$ varies with time. We shall express this in the form

$$\mathscr{E}_x = \mathscr{E}_{mx} \cos 2\pi\nu t = \frac{\mathscr{E}_{mx}(e^{i2\pi\nu t} + e^{-i2\pi\nu t})}{2} \tag{3-30}$$

where $\mathscr{E}_{mx}$ is the maximum value of the x-component of $\mathscr{E}$. The dipole itself is then executing harmonic motion, but the potential energy is no longer simply

$$V(x) = \tfrac{1}{2}\kappa x^2 \tag{3-31}$$

[2]Readers unfamiliar with tensors may refer to the author's *Electromagnetic and Quantum Properties of Materials*, Prentice-Hall, Inc., Englewood Cliffs, N. J., 1966, Sec. 5-7, for a simple introduction to matrix algebra and the use of tensors.

for the field $\mathscr{E}$ exerts a force

$$F_x = e\mathscr{E}_x \tag{3-32}$$

on a charge e. Hence, the total potential energy is

$$V(x) = \tfrac{1}{2}\kappa x^2 + e\mathscr{E}_x x \tag{3-33}$$

Putting this into the time-dependent Schroedinger equation (3-11), we obtain

$$\left(-\frac{\hbar^2}{2m}\frac{\partial^2}{\partial x^2} + \frac{1}{2}\kappa x^2 + e\mathscr{E}_x x\right)\Psi = i\hbar\frac{\partial\Psi}{\partial t} \tag{3-34}$$

If the field $\mathscr{E}$ were zero, this equation would reduce to the harmonic oscillator time-dependent equation with solutions

$$\Psi_n = \psi_n(x)e^{-iE_nt/\hbar} \tag{3-35}$$

as given by (3-13), where

$$E_n = (n + \tfrac{1}{2})h\nu \tag{3-21}$$

Let us therefore try to solve (3-34) by taking a linear combination

$$\Psi = C_n\Psi_n + C_p\Psi_p \tag{3-36}$$

of two distinct solutions ψ_n and ψ_p, where C_n and C_p are functions of time. Substituting (3-36) into the right side of (3-34) gives

$$\begin{aligned} i\hbar\frac{\partial}{\partial t}(C_n\psi_n + C_p\psi_p) &= i\hbar\frac{\partial}{\partial t}(C_n\psi_n e^{-iE_nt/\hbar} + C_p\psi_p e^{-iE_pt/\hbar}) \\ &= i\hbar\left(\psi_n\frac{dC_n}{dt} + \psi_p\frac{dC_p}{dt}\right) + C_n\psi_n E_n + C_p\psi_p E_p \end{aligned} \tag{3-37}$$

The use of (3-36) in the left-hand side of (3-34) converts the first two terms into

$$\left(-\frac{\hbar^2}{\partial m}\frac{\partial^2}{\partial x^2} + \frac{1}{2}\kappa x^2\right)(C_n\Psi_n + C_p\Psi_p) = C_nE_n\Psi_n + C_pE_p\Psi_p \tag{3-38}$$

and this cancels the corresponding terms on the right of (3-37), leaving

$$e\mathscr{E}_x x\Psi = i\hbar\left(\Psi_n\frac{dC_n}{dt} + \Psi_p\frac{dC_p}{dt}\right) \tag{3-39}$$

Multiply both sides of this equation by Ψ_p^*, the complex conjugate of Ψ_p, and integrate from $-\infty$ to ∞. This results in the equation

$$\int_{-\infty}^{\infty} e\mathscr{E}_x x\Psi_p^*(C_n\Psi_n + C_p\Psi_p)\,dx = i\hbar\int_{-\infty}^{\infty} \Psi_p^*\left(\Psi_n\frac{dC_n}{dt} + \Psi_p\frac{dC_p}{dt}\right)dx \tag{3-40}$$

However, integrals involving $\Psi_p^*\Psi_n$ vanish, since

$$\int_{-\infty}^{\infty} \Psi_p^*\Psi_n\,dx = e^{i(E_p-E_n)t/\hbar}\int_{-\infty}^{\infty} \psi_p\psi_n\,dx = 0 \tag{3-41}$$

by (3-26), and those involving $\Psi_p^*\Psi_p$ reduce to unity. Hence

$$\frac{dC_p}{dt} = -\left(\frac{i}{\hbar}\right)\left(C_p\int_{-\infty}^{\infty} e\mathscr{E}_x x\Psi_p^*\Psi_p\,dx + C_n\int_{-\infty}^{\infty} e\mathscr{E}_x x\Psi_p^*\Psi_n\,dx\right) \tag{3-42}$$

By (3-30) and (3-13), this becomes

$$\frac{dC_p}{dt} = -\left(\frac{i}{\hbar}\right)\mathscr{E}_{mx}\left[C_p\left\{\frac{e^{i2\pi\nu t} + e^{-i2\pi\nu t}}{2}\right\}e\int_{-\infty}^{\infty} x\psi_p^2\,dx\right.$$
$$\left. + C_n\left\{\frac{e^{i(E_p-E_n+h\nu)t/\hbar} + e^{i(E_p-E_n-h\nu)t/\hbar}}{2}\right\}e\int_{-\infty}^{\infty} x\Psi_p\Psi_n\,dx\right]$$

We shall show below that the first integral on the right of this equation vanishes, leaving

$$\frac{dC_p}{dt} = -\left(\frac{i}{\hbar}\right)\mathscr{E}_{mx}C_n\left\{\frac{e^{i(E_p-E_n+h\nu)t/\hbar} + e^{i(E_p-E_n-h\nu)t/\hbar}}{2}\right\}e\int_{-\infty}^{\infty} x\Psi_p\Psi_n\,dx \tag{3-43}$$

PROBLEM 3-6

Show that the solution of the differential equation (3-43) subject to the initial conditions

$$C_p = 0 \quad \text{and} \quad C_n = 1 \quad \text{at} \quad t = 0$$

is

$$C_p = \left(\frac{\mathscr{E}_{mx}}{2}\right)\left(e\int_{-\infty}^{\infty} x\psi_p\psi_n\,dx\right)\left[\frac{1 - e^{i(E_p-E_n+h\nu)t/\hbar}}{E_p - E_n + h\nu} + \frac{1 - e^{i(E_p-E_n-h\nu)t/\hbar}}{E_p - E_n - h\nu}\right] \tag{3-44}$$

provided $C_n \sim 1$ for $t \neq 0$. (The reasons for the conditions on C_n and C_p will be given shortly.)

The second term on the right becomes very large if its denominator approaches zero; that is, if

$$E_p - E_n = h\nu \tag{3-45}$$

which is the well-known *Bohr frequency condition.* Thus, transitions from an energy E_n to a higher energy E_p involve the transfer of one quantum of energy. The first term will be large if the opposite process occurs, so that we can neglect one of the two terms. Keeping the second term, then, and multiplying (3-44) by its complex conjugate gives

$$C_p^* C_p = |C_p|^2 = \frac{\mathscr{E}_{mx}^2 \left[\int_{-\infty}^{\infty} ex\psi_p \psi_n \, dx\right]^2 \sin^2\left[\dfrac{(E_p - E_n - h\nu)t}{2\hbar}\right]}{(E_p - E_n - h\nu)^2} \tag{3-46}$$

Integrating over all frequencies ν from 0 to ∞, and using

$$\int_0^{\infty} \left(\frac{\sin^2 x}{x^2}\right) dx = \frac{\pi}{2}$$

we obtain

$$|C_p|^2 = \left(\frac{\mathscr{E}_{mx}}{2\hbar}\right)^2 \left[\int_{-\infty}^{\infty} ex\psi_p \psi_n \, dx\right]^2 t \tag{3-47}$$

To interpret this result, we recall that we have made the approximation that only the transition from a level of energy E_n to a higher level E_p is of importance, so that (3-47) refers to the absorption of radiation. Since C_p is the coefficient of ψ_p, the wave-function after absorption, we regard $|C_p|^2$ as the contribution to the probability that the harmonic oscillator will make the transition from E_n to E_p. The integral on the right involves the x-component ex of the dipole moment $\mathbf{p} = e\mathbf{r}$, where $\mathbf{r}$ is the separation induced by $\mathscr{E}$. Hence, we call the integral the *transition moment*, and we see that the transition probability is proportional to this integral.

Integrals of this form have another extremely important property; we may show that

$$\int_{-\infty}^{\infty} x\psi_p \psi_n \, dx \begin{cases} \neq 0 & \text{if } p = n + 1 \text{ or } n - 1 \\ = 0 & \text{otherwise} \end{cases} \tag{3-48}$$

For example, using ψ_0 and ψ_2 of (3-22) gives

$$\int_{-\infty}^{\infty} (2\beta x^3 - x)e^{-\beta x^2} \, dx = 0$$

since either $xe^{-\beta x^2}$ or $x^3 e^{-\beta x^2}$ is an odd function whose integral over $-\infty$, ∞ will vanish. Equation (3-48) is equivalent to stipulating that a transition from an energy E_n to an energy E_p is possible only if

$$E_p - E_n = (p + \tfrac{1}{2})h\nu - (n + \tfrac{1}{2})h\nu = [p - (p \pm 1)h\nu] = \pm h\nu \qquad \textbf{(3-49)}$$

and if

$$p - n = \pm 1 \quad \text{or} \quad \Delta n = \pm 1 \qquad \textbf{(3-50)}$$

for the probability of any other transition is zero. Equation (3-50) is called the *selection rule* for transitions between vibrational levels. This selection rule may not always be strictly obeyed, however; although the quantum mechanical basis is perfectly general, we must remember that molecules are not really particles on the ends of ideal springs. Hooke's law is only an approximation corresponding to the first term of a Taylor series. That is, if the force depends also on $x^2, x^3, \ldots$, these *anharmonic* terms should appear in the classical or quantum oscillator equation, and the selection rule is no longer rigorously valid.

Returning to the conditions imposed on C_n and C_p in Problem 3-6, we now realize that the specification of $C_n = 1$, or $|C_n|^2 = 1$, at $t = 0$ guarantees that the oscillator definitely possesses the energy E_n at the start. The condition $C_n \sim 1$ at $t \neq 0$ requires some explanation, however. In order to solve Eq. (3-43), we implicity assumed that the electric field is applied only for a short time from $t = 0$ to t and that C_n does not change substantially in that time. The solution we obtained, of the form (3-46), is valid for arbitrary values of t, but this involves a rather complicated proof.

3-4 Vibrational infrared spectra

The human eye responds to light with wavelengths from approximately 4×10^{-7} m (0.4 microns, 4,000 angstroms, or 0.4 micrometers in SI units) to 7×10^{-7} m, corresponding respectively to the violet and red ends of the visible spectrum. Wavelengths of 0.8 microns or longer lie in the *infrared*, and this is an area of interest both to chemists and physicists. If we consider the $CHCl_3$ molecule of Chapter 2, measurement of its infrared absorption properties results in the curve shown schematically in Fig. 3-5(a). This graph is a plot of the transparency of a sample of chloroform vapor as a function of the frequency of the incident infrared radiation. The vertical axis represents transmittance in arbitrary units and the lower horizontal scale is the frequency in units commonly used by spectroscopists. That is, the frequency ν in units of c, the velocity of light, is

$$\frac{1}{\lambda} = \frac{\nu}{c} \qquad \textbf{(3-51)}$$

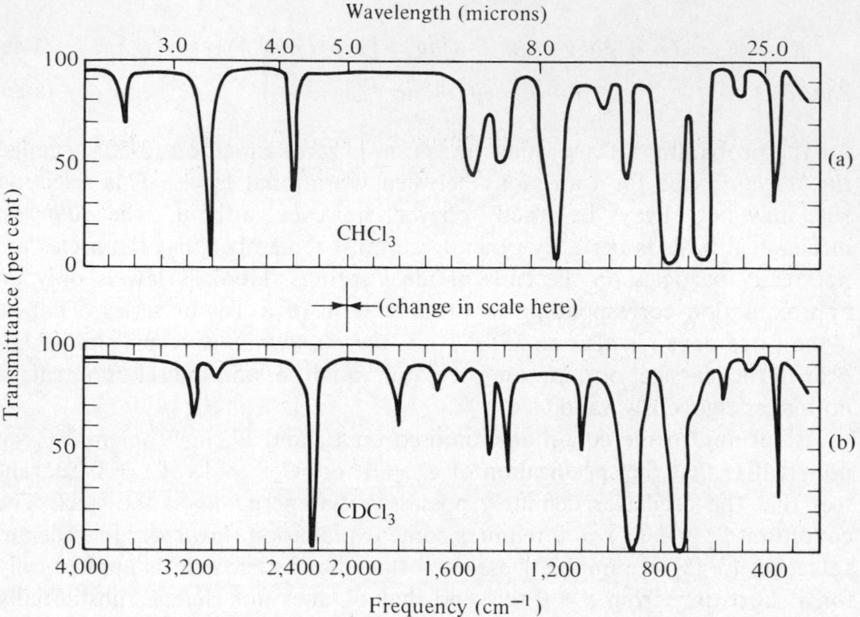

3-5 Infrared transmission properties of (a) chloroform ($CHCl_3$) and (b) deuterium-substituted chloroform ($CDCl_3$)

so that $1/\lambda$ (expressed in cm^{-1}) is a measure of the frequency. For example, the wavelength of 3 microns converts to

$$\frac{1}{\lambda} = \frac{1}{3 \times 10^{-4}} = 3{,}333.3 \text{ cm}^{-1}$$

(Note the change in horizontal scale at $1/\lambda = 2{,}000 \text{ cm}^{-1}$.)

The graph of Fig. 3-5(a) shows that incident light with wavelengths near 3, 8, 13, or 15 microns is very strongly absorbed and there are other less pronounced *absorption bands* at several more places. These bands are due in part to the activation of the vibrational and the rotational normal modes we previously calculated in Chapter 2. If we take the values of ω^2 for the E modes, as well as the three for the A_1 modes, and convert to units of $1/\lambda$, we obtain the frequencies of Table 3-1.

The frequencies ν_4, ν_5, ν_6 are determined by converting $\omega_4, \omega_5, \omega_6$, respectively, of Sec. 2-15; the other three frequencies, which must then belong to A_1 in accordance with Eq. (2-201), are taken from Colthup, Daly, and Wiberly.[3] All but the 38 micron band appears in Fig. 3-5(a), and these

[3]E. B. Colthup, L. E. Daly, and S. E. Wiberly, *Introduction to Infrared and Raman Spectroscopy*, Academic Press, Inc., New York, 1964.

TABLE 3-1 ***Fundamental Infrared Frequencies of $CHCl_3$***

Frequency Designation	Wavelength (microns)	Frequency (cm^{-1})	Representation
ν_1	3.3	3,025	A_1
ν_2	15.0	668	A_1
ν_3	26.8	373	A_1
ν_4	8.2	1,203	E
ν_5	13.2	773	E
ν_6	38.4	261	E

fundamental frequencies correspond to the strongest absorption. This may be regarded as the optical analogue of a resonating system; if we push a swing at its natural frequency, the amplitude is large.

We should remark that our calculation for $CHCl_3$ gave only the six vibrational frequencies, since internal coordinates were used. The rotational modes of a great many molecules also produce infrared bands. It is also possible, in fact, to have interaction of vibrational and rotational modes, giving more complicated spectra. Since our interest here is in the application of group theory, we shall assume that we need work only with the vibrational modes.

The figure shows absorption frequencies other than the fundamental ones listed in Table 3-1. As is the case for any resonance phenomenon, we should also expect frequencies which are of the form $n\nu_i (n = 2, 3, \ldots)$ and $\nu_i \pm n\nu_j (i \neq j)$. These are known as harmonics or *overtones* and as beat notes or *combinations*, respectively. Table 3-2 lists some of the frequencies of this type.

The identification of an overtone or a combination frequency with a representation, as given in the last column of Table 3-2, is the next thing we

TABLE 3-2 ***Overtones and Combinations for $CHCl_3$***

Frequency (cm^{-1})	Designation	Representation
2,400	$2\nu_4$	$A_1 + E$
1,521	$2\nu_5$	$A_1 + E$
1,423	$\nu_2 + \nu_5$	E
497	$\nu_5 - \nu_6$	$A_1 + A_2 + E$
230	$\nu_5 - 2\nu_6$	$A_1 + A_2 + 2E$

should consider. If we add $\nu_2 = 668\ \text{cm}^{-1}$ and $\nu_5 = 773\ \text{cm}^{-1}$, we obtain $\nu_2 + \nu_5 = 1{,}441\ \text{cm}^{-1}$, which agrees very well with the corresponding entry in Table 3-2. Since ν_2 belongs to A_1 and ν_5 belongs to E, the reason why the combination belongs to E may be puzzling. The explanation comes from considering the solutions (2-91) to the Lagrange equations (2-89). If the oscillations of a system can be described as the combination of two simultaneous normal modes of frequencies ν_i and ν_j, then a function corresponding to a frequency $(\nu_i + \nu_j)$ has the form

$$x_{i+j} = a_{i+j}e^{i(\omega_i+\omega_j)t} \tag{3-52}$$

which we may regard as being generated from the product of the two individual solutions

$$x_i = a_i e^{i\omega_i t} \tag{3-53}$$

and

$$x_j = a_j e^{i\omega_j t} \tag{3-54}$$

Hence, we get the representation associated with a combination frequency by multiplying the appropriate quantities for the individual frequencies; that is, we simply multiply the two sets of characters. This is an example of what is known as the *Kronecker product*. That is, we obtain the Kronecker product for two representations in a given group by simply multiplying the corresponding entries in the character table. Thus, Table 2-14 shows that the characters for the Kronecker product $A_1 \times E$ of the representations A_1 and E can be indicated as follows:

	E	$2C_3$	$3\sigma_v$
A_1	1	1	1
E	2	−1	0
$A_1 \times E$	2	−1	0

This result shows that the characters for $A_1 \times E$ are the same as for E, which explains why the combination of ν_2 (belonging to A_1) and ν_5 (belonging to E) gives a frequency $(\nu_2 + \nu_5)$ belonging only to E. It should be noted that there is a formal similarity between the Kronecker product and the direct product introduced in connection with Table 1-15. We emphasize that the Kronecker product applies to the combination of two representations of the *same* group; however, the same symbol is used for the two types of product.

The next example to consider is the combination $\nu_5 - \nu_6$. The results we shall obtain apply to $\nu_5 \pm \nu_6$, for two normal modes can add or subtract. Since both frequencies belong to E, we compute the Kronecker product as follows:

	E	$2C_3$	$3\sigma_v$
E	2	-1	0
E	2	-1	0
$E \times E$	4	1	0

The resulting representation must consist of 4×4 matrices, and hence is reducible. As Table 2-14 shows, it decomposes into irreducible representations according to the relation

$$E \times E = A_1 + A_2 + E$$

thus explaining the entry for $\nu_5 - \nu_6$ in Table 3-2.

Considering now the overtones, although Table 3-2 does not show any belonging to A_1, it is quite clear that all those for ν_1, ν_2, and ν_3 belong to this representation, since

$$A_1 \times A_1 = A_1$$

It is also possible to have overtones associated with A_2 (for which there are no fundamentals). Using

$$\begin{aligned} A_2 \times A_2 &= A_1 \\ A_2 \times A_2 \times A_2 &= A_2 \\ A_2 \times A_2 \times A_2 \times A_2 &= A_1 \end{aligned} \qquad \textbf{(3-55)}$$

it follows that overtones of the form $n\nu$ belong to A_1 when n is even and to A_2 when n is odd.

The final situation to be considered is that for overtones belonging to E. Two-dimensional (and higher) representations were stated to be degenerate in Chapter 2, and these are complicated to deal with. The method for finding the representations associated with overtones of degenerate frequencies is due to Tisza[4]; we shall give a simplified version of his treatment. It is based on the two-dimensional Schroedinger equation, which we would expect to have the form

$$-\frac{\hbar^2}{2m}\left(\frac{\partial^2\psi}{\partial x^2} + \frac{\partial^2\psi}{\partial y^2}\right) + V\psi = E\psi \qquad \textbf{(3-56)}$$

[4] L. Tisza, *Zeit. Physik*, **82**, 48 (1933); a translation will be found in P. H. Meijer, *Group Theory and Solid State Physics*, Gordon and Breach, New York, 1964.

for when $\psi(x, y)$ in this equation reduces to $\psi(x)$, a function of x only, then (3-56) is the same as (3-14). For a two-dimensional *isotropic oscillator*, the potential energy is

$$V = \frac{\kappa}{2}(x^2 + y^2) \tag{3-57}$$

Substituting this into (3-56) gives the partial differential equation

$$-\frac{\hbar^2}{2m}\left(\frac{\partial^2\psi}{\partial x^2} + \frac{\partial^2\psi}{\partial y^2}\right) + \frac{\kappa}{2}(x^2 + y^2)\psi = E\psi \tag{3-58}$$

To solve this, we try the method of separation of variables; let

$$\psi(x, y) = X(x)Y(y) \tag{3-59}$$

where $X(x)$ is a function which depends only on x, and $Y(y)$ depends only on y. Inserting this into (3-58) we obtain

$$-\frac{\hbar^2}{2m}\left(Y\frac{d^2X}{dx^2} + X\frac{d^2Y}{dy^2}\right) + \frac{\kappa}{2}(x^2 + y^2)XY = EXY$$

or dividing by XY and rearranging

$$-\frac{\hbar^2}{2m}\left(\frac{1}{X}\frac{d^2X}{dx^2}\right) + \frac{\kappa}{2}x^2 = \frac{\hbar^2}{2m}\left(\frac{1}{Y}\frac{d^2Y}{dy^2}\right) - \frac{\kappa}{2}x^2 + E \tag{3-60}$$

Using an arbitrary separation constant which we denote by E_x, we obtain two ordinary differential equations of the form

$$-\frac{\hbar^2}{2m}\frac{d^2X}{dx^2} + \frac{1}{2}\kappa x^2X = E_xX \tag{3-61}$$

$$-\frac{\hbar^2}{2m}\frac{d^2Y}{dy^2} + \frac{1}{2}\kappa y^2Y = E_yY \tag{3-62}$$

where

$$E_y = E - E_x \quad \text{or} \quad E = E_x + E_y \tag{3-63}$$

Each of the equations (3-61) and (3-62) will be recognized as the quantum oscillator equation (3-15) with Hermite functions as solutions. That is,

$$X = e^{-\beta x^2/2}f(x)$$

$$Y = e^{-\beta y^2/2}f(y)$$

and

$$\psi(x, y) = e^{-\beta(x^2+y^2)/2}f(x)f(y) \tag{3-64}$$

The energy levels are

$$E_x = (n_x + \tfrac{1}{2})h\nu, \quad (n_x = 0, 1, 2, \ldots)$$

$$E_y = (n_y + \tfrac{1}{2})h\nu, \quad (n_y = 0, 1, 2, \ldots)$$

and by (3-63) $$E_{xy} = (n_x + n_y + 1)h\nu$$

For $n_x = n_y = 0$, we have $E_{xy} = h\nu$, and this level is nondegenerate. For the pair of quantum numbers $n_x = 1$, $n_y = 0$ or $n_x = 0$, $n_y = 1$, we have

$$\begin{aligned} \psi_{1,0}(x, y) &= e^{-\beta(x^2+y^2)/2}x \\ \psi_{0,1}(x, y) &= e^{-\beta/(x^2+y^2)/2}y \end{aligned} \tag{3-65}$$

and $$E_{1,0} = E_{0,1} = 2h\nu$$

This solution thus has a two-fold degeneracy; there are two distinct wavefunctions corresponding to the next-to-lowest energy level $2h\nu$. The terms "degenerate" and "nondegenerate" as we use them here have the same meaning as in Chapter 2; that is, a degenerate solution means that two or more functions correspond to a given value of the energy.

The third level has an energy

$$E_{2,0} = E_{1,1} = E_{0,2} = 3h\nu$$

which is triply-degenerate, with the three functions

$$\begin{aligned} \psi_{2,0} &= e^{-\beta(x^2+y^2)/2}(2\beta x^2 - 1) \\ \psi_{1,1} &= e^{-\beta(x^2+y^2)/2}xy \\ \psi_{0,2} &= e^{-\beta(x^2+y^2)/2}(2\beta y^2 - 1) \end{aligned} \tag{3-66}$$

The only portions of this triply-degenerate set which depend on x and y individually, rather than the symmetric quantity $r^2 = x^2 + y^2$, are x^2, xy, and y^2, respectively. These quantities form a basis for generating a reducible representation corresponding to an energy change of $2h\nu$. As the Bohr condition, Eq. (3-45), shows, a change in energy by an amount $h\nu$ represents a photon of frequency ν. Hence, an energy change of $2h\nu$ corresponds to a frequency of 2ν, and this is the first harmonic or overtone.

Consider next the effect of some operation G in the symmetry group of the molecule on x and y. Let us assume for simplicity that we have chosen a coordinate system such that

$$Gx = a_1x, \qquad Gy = a_2y \tag{3-67}$$

and this may be extended to give

$$\begin{aligned} Gx^2 &= (a_1x)^2 \\ Gy^2 &= (a_2y)^2 \\ Gxy &= (a_1x)(a_2y) \end{aligned} \tag{3-68}$$

The matrix equivalent of (3-67) is

$$\begin{pmatrix} a_1 & 0 \\ 0 & a_2 \end{pmatrix}\begin{pmatrix} x \\ y \end{pmatrix} = \begin{pmatrix} a_1 x \\ a_2 y \end{pmatrix}$$

or

$$G = \begin{pmatrix} a_1 & 0 \\ 0 & a_2 \end{pmatrix} \tag{3-69}$$

Similarly, (3-68) in matrix form leads to

$$G^{(1)} = \begin{pmatrix} a_1^2 & 0 & 0 \\ 0 & a_1 a_2 & 0 \\ 0 & 0 & a_2^2 \end{pmatrix} \tag{3-70}$$

where the superscript denotes the connection with the first overtone through the basis functions of (3-66). We have further that

$$G^2 = \begin{pmatrix} a_1^2 & 0 \\ 0 & a_2^2 \end{pmatrix} \tag{3-71}$$

Computing the characters of G, $G^{(1)}$, and G^2 gives

$$\begin{aligned} \chi(G) &= a_1 + a_2 \\ \chi(G^2) &= a_1^2 + a_2^2 \\ \chi(G^{(1)}) &= a_1^2 + a_1 a_2 + a_2^2 \end{aligned} \tag{3-72}$$

from which we see that

$$\chi(G^{(1)}) = \tfrac{1}{2}[\chi^2(G) + \chi(G^2)] \tag{3-73}$$

Similar formulas for the overtones of triply-degenerate frequencies and for higher overtones are given by Wilson, Decius, and Cross.[6]

Let us apply (3-73) to the $CHCl_3$ molecule. The character system for $\chi_E^2(G)$ is

	E	C_3	σ_v
$\chi_E^2(G)$	4	1	0

The squares of the operations are

$$EE = E, \qquad C_3 C_3 = C_3^2, \qquad \sigma_v \sigma_v = E$$

and the corresponding characters are

	E^2	C_3^2	σ_v^2
$\chi_E(G^2)$	2	−1	2

Equation (3-73) then gives the first overtone character system as

	E	C_3	σ_v
$\chi_E(G^{(1)})$	3	0	1

from which we identify the irreducible representations as

$$\Gamma^{(1)} = A_1 + E \tag{3-74}$$

and this is indicated for the first two entries in Table 3-2.

PROBLEM 3-7

Explain the last entry in Table 3-2.

3-5 Selection rules

Let us put the considerations of the previous section on firmer foundations. To start with a familiar concept we know that an *odd function* $f(x)$ in the region $(L, -L)$ is one for which

$$f(x) = -f(-x) \tag{3-75}$$

and that

$$\int_{-L}^{L} f(x)\,dx = 0 \tag{3-76}$$

For example, $f(x) = x$ is such a function, and

$$\int_{-L}^{L} x\,dx = \frac{x^2}{2}\bigg|_{-L}^{L} = 0$$

In group-theoretical terms, we say that the integral vanishes because x is antisymmetric, i.e., it has inversion symmetry but not reflection symmetry.

Another way of saying this is to realize that x cannot belong to the completely symmetric representation A_1, or 1 1 1 ... 1, of any of the groups we have studied.

Extending this to any integral

$$\int f_A(x, y, z) f_B(x, y, z)\, dx\, dy\, dz$$

involving the product of two functions, a value of zero will be obtained unless at least one term in the product $f_A f_B$ (assuming each function is the sum of several terms) is a basis for A_1. We can examine this question by finding the irreducible representations $\mathbf{\Gamma}_A$ and $\mathbf{\Gamma}_B$ associated with f_A and f_B and forming the Kronecker product. We shall show that the decomposition of the representation $\mathbf{\Gamma}_{A\times B}$ for the Kronecker product $A \times B$ will contain A_1 *only* if

$$\mathbf{\Gamma}_A = \mathbf{\Gamma}_B \tag{3-77}$$

We apply the relation

$$n_k = \frac{1}{g} \sum \chi(R) \chi_k(R) \tag{2-63}$$

with the irreducible representation taken as A_1. Hence,

$$\chi_k = 1$$

and

$$n_1 = \frac{\sum_R \chi_{A\times B}(R)}{g} \tag{3-78}$$

for $\mathbf{\Gamma}_{A\times B}$. Since

$$\chi_{A\times B}(R) = \chi_A(R) \chi_B(R)$$

then (3-78) becomes

$$n_1 = \frac{\sum_R \chi_A(R) \chi_B(R)}{g} \tag{3-79}$$

But we have seen that characters of different representations are orthogonal. Hence,

$$n_1 = \begin{cases} 0 & \text{if} \quad \mathbf{\Gamma}_A \neq \mathbf{\Gamma}_B \\ 1 & \text{if} \quad \mathbf{\Gamma}_A = \mathbf{\Gamma}_B \end{cases}$$

That is, not only is (3-77) valid, but A_1 can occur only once.

If we apply this to triple integrals of the form (3-48), we have the extremely important result that the transition moment will vanish and there will be no infrared line unless the Kronecker product of the representations for which ψ_n and ψ_p are a basis contains a representation belonging to x (or to y or z in three dimensions, since $\mathbf{p} = e\mathbf{r}$).

In practice, this means that the representations to which x, y, or z belong are infrared active; all others (in the harmonic approximation) are inactive. Returning to $CHCl_3$, for which

$$\Gamma_{\text{vib}} = 3A_1 + 3E \tag{2-201}$$

we see that this agrees with Table 2-14, since z belongs to A_1 and (x, y) belongs to E. A second example is the XY_3 molecule belonging to the group D_{3h} of Table 2-17. Although there is a normal mode belonging to A_1, it is *not* infrared active, since the first power of x, y, or z is not a basis for this representation. A final example is the group C_{2v} of Table 2-13. By (2-199), H_2O has only infrared modes A_1 and B_2, whereas some molecules with this symmetry may have other combinations involving A_1, B_1, and B_2.

Returning to $CHCl_3$ and to the rules which we determined in the previous section, we may summarize all the results we have obtained regarding selection rules in the manner of Table 3-3.

It is worth noting that the structure of $CHCl_3$ shown in Fig. 2-27 is essentially an assumed one. That is, it seems reasonable that the three Cl atoms will be symmetrically disposed with respect to the CH-axis. If the normal mode amplitudes are worked out, it is found that ν_1 is primarily associated with CH stretching and ν_4 with the bending of this axis. If we study the mole-

TABLE 3-3 ***Summary of $CHCl_3$ Infrared Frequencies***

Representation	Infrared activity
A_1	allowed
A_2	forbidden
E	allowed
$A_1 \times A_1$	allowed
$A_1 \times A_2$	forbidden
$A_1 \times E$	allowed
$A_2 \times A_2$	allowed
$A_2 \times E$	allowed
$E \times E$	allowed
A_1^n	allowed
A_2^n (n even)	allowed
A_2^n (n odd)	forbidden
E^n	allowed

cule $CDCl_3$, in which the hydrogen in chloroform is replaced by deuterium (D), with twice the mass, it is found that ν_1 and ν_4 are considerably lower. On the other hand, frequencies associated with CCl stretching are virtually unchanged. Figure 3-5(b) illustrates this.

3-6 The Raman effect

Returning to the induced dipole moment which led to Eq. (3-28), let us consider the effect of a sinusoidal wave on this dipole. Let the wave be

$$\mathscr{E} = \mathscr{E}_m \cos \omega t$$

so that the induced moment is

$$p = \alpha \mathscr{E}_m \cos \omega t \tag{3-80}$$

provided that α is a constant in the presence of this field. For a monatomic molecule this is probably true, but a diatomic (or multiatom) molecule will execute normal vibrations under the influence of the field, and α itself will vary sinusoidally. Thus, we write α as

$$\alpha = \alpha_0 + (\Delta\alpha_0) \cos \omega_v t \tag{3-81}$$

where the cosine term causes α to vary about its equilibrium value α_0 with a vibrational frequency ω_v. Putting (3-81) in (3-80) gives

$$\begin{aligned} p &= [\alpha_0 + (\Delta\alpha_0) \cos \omega_v t]\mathscr{E}_m \cos \omega t \\ &= \alpha_0 \mathscr{E}_m \cos \omega t + \frac{\Delta\alpha_0}{2}[\cos (\omega + \omega_v)t + \cos (\omega - \omega_v)t] \end{aligned} \tag{3-82}$$

The scattered radiation then has three components; the original frequency ω corresponds to *Rayleigh scattering* and the components with frequencies $(\omega \pm \omega_v)$ represent *Raman scattering*.

Rayleigh showed theoretically that the intensity of light scattered out of a beam is proportional to $1/\lambda^4$ and hence that the predominantly blue color of the sky is caused by the greater scattering of components of sunlight at shorter wavelengths by atmospheric molecules. This type of scattering is what we generally observe when light is scattered by a liquid or gas. We can give a simple dimensional argument, due to Duffey,[5] for this λ^{-4} dependence. Let the incident field $\mathscr{E} = \mathscr{E}_m \cos \omega t$ be scattered by an atom of volume V. It seems reasonable to assume that the maximum amplitude A of the scat-

[5] G. H. Duffey, *Physical Chemistry*, McGraw-Hill Book Company, New York, 1962.

tered wave is proportional to both V and to $\mathscr{E}_m$. Regarding the scattering atom as a point-source, the scattered wave will be spherical. Then its intensity (in watts/m^2) depends on $1/r^2$, where r is the radial distance from the atom. Further, A^2 has the same behavior, and we may express this as

$$A = \frac{k\mathscr{E}_m V}{r}$$

where k is a proportionality constant with the dimensions of length^{-2}. Since only wavelength remains to be considered, it follows that

$$k = \frac{C}{\lambda^2}$$

where C is a dimensionless constant. We thus obtain

$$A = \frac{C\mathscr{E}_m V}{r\lambda^2}$$

The intensity of the scattered wave, as measured by A^2, will consequently depend on λ^{-4}.

The Raman effect is more difficult to observe. The usual experimental arrangement for liquids is shown in Fig. 3-6. One line of mercury passes through the liquid filter and strikes the sample, which is in a cell with a blackened end to prevent direct reflection of the exciting light into the spectrograph. This is because the Raman lines are extremely faint and can be seen only when the plate is heavily overexposed with respect to the Rayleigh-scattered exciting line. In fact, the lower frequency $(\omega - \omega_v)$, called the *Stokes line*, is usually the only one observed; the high frequency line, the *anti-Stokes* line, is generally too weak. Further, under high resolution,

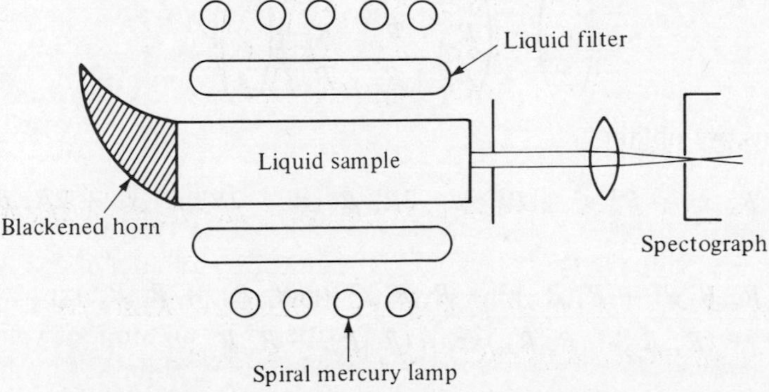

3-6 Experimental arrangement for Raman measurements on a liquid

the Raman and Rayleigh lines are actually bands, that is, groups of closely spaced lines. This is because the scattering is not due solely to the purely vibrational modes; energy can be absorbed and emitted by rotational modes as well. However, we shall consider the Raman effect to be vibrational in the discussion which follows.

The selection rules for the Raman effect are based on the symmetry properties of the polarizability tensor $\boldsymbol{\alpha}$ of Eq. (3-29). Under a symmetry operation R, $\boldsymbol{\alpha}$ is transformed into a new tensor α' which by (2-56) is

$$\boldsymbol{\alpha}' = \mathbf{R}^{-1}\boldsymbol{\alpha}\mathbf{R} = \tilde{\mathbf{R}}\boldsymbol{\alpha}\mathbf{R} \tag{3-83}$$

since $\mathbf{R}$ is orthogonal.

PROBLEM 3-8

Show that (3-83) in component form may be written

$$\alpha'_{xx} = R^2_{xx} + R^2_{xy}\alpha_{yy} + R^2_{xz}\alpha_{zz} + 2R_{xx}R_{xy}\alpha_{xy} + 2R_{xx}R_{xz}\alpha_{xz} + 2R_{xy}R_{xz}\alpha_{yz}$$

$$\begin{aligned}\alpha'_{xy} = {} & R_{xx}R_{yz}\alpha_{xx} + R_{xy}R_{yy}\alpha_{yy} + R_{xz}R_{yz}\alpha_{zz} + (R_{xx}R_{yy} + R_{xy}R_{yx})\alpha_{xy} \\ & + (R_{xx}R_{yz} + R_{xz}R_{yx})\alpha_{xz} + (R_{xy}R_{yz} + R_{xz}R_{yy})\alpha_{yz} \\ & \text{etc.}\end{aligned}$$

where we are using x, y, z as the indices on both $\mathbf{R}$ and $\boldsymbol{\alpha}$.

Now let $\mathbf{r}'$ be a vector generated from $\mathbf{r}$ by the same operation R. Then

$$\mathbf{r}' = \mathbf{R}\mathbf{r}$$

or

$$\begin{pmatrix} x' \\ y' \\ z' \end{pmatrix} = \begin{pmatrix} R_{xx} & R_{xy} & R_{xz} \\ R_{yx} & R_{yy} & R_{yz} \\ R_{zx} & R_{zy} & R_{zz} \end{pmatrix} \begin{pmatrix} x \\ y \\ z \end{pmatrix}$$

From this, we obtain

$$x'^2 = R^2_{xx}x^2 + R^2_{xy}y^2 + R^2_{xz}z^2 + 2R_{xx}R_{xy}xy + 2R_{xx}R_{xz}xz + 2R_{xy}R_{xz}yz$$

and

$$\begin{aligned}x'y' = {} & R_{xx}R_{yz}x^2 + R_{xy}R_{yy}y^2 + R_{xz}R_{yz}z^2 + (R_{xx}R_{yy} + R_{xy}R_{yx})xy \\ & + (R_{xx}R_{yz} + R_{xz}R_{yx})xz + (R_{xy}R_{yz} + R_{xz}R_{yy})yz\end{aligned}$$

Thus, the components $\alpha_{xx}, \alpha_{xy}, \ldots, \alpha_{zz}$ of $\boldsymbol{\alpha}$ have exactly the same transformation properties as the quadratic functions $x^2, xy, \ldots, z^2$. This is analogous to the situation for **p**, whose components transform like the linear functions x, y, and z. We can then tell immediately from the character tables which modes are Raman active. For example, for C_{3v} of Table 2-14, we see that quadratic functions go with A_1 and E, the same modes that are infrared active. Further, the Raman selection rule is of the same form as the infrared rule, and is applied in the same way. A more rigorous treatment of Raman selection rules will be found in Eyring, Walter, and Kimball.[1]

A Raman spectrum for carbon tetrachloride CH_4, as given by Tobias,[6] is shown in Fig. 3-7. As stated above, the high frequency shifts are much weaker than those on the low frequency side. The exciting frequency is the blue line of the mercury arc at 4360 A or 22,938 cm^{-1}. This causes the intense center line, which is Rayleigh scattered. Table 3-4 is the character table with basis functions for the tetrahedral group T_d, to which CH_4 belongs.

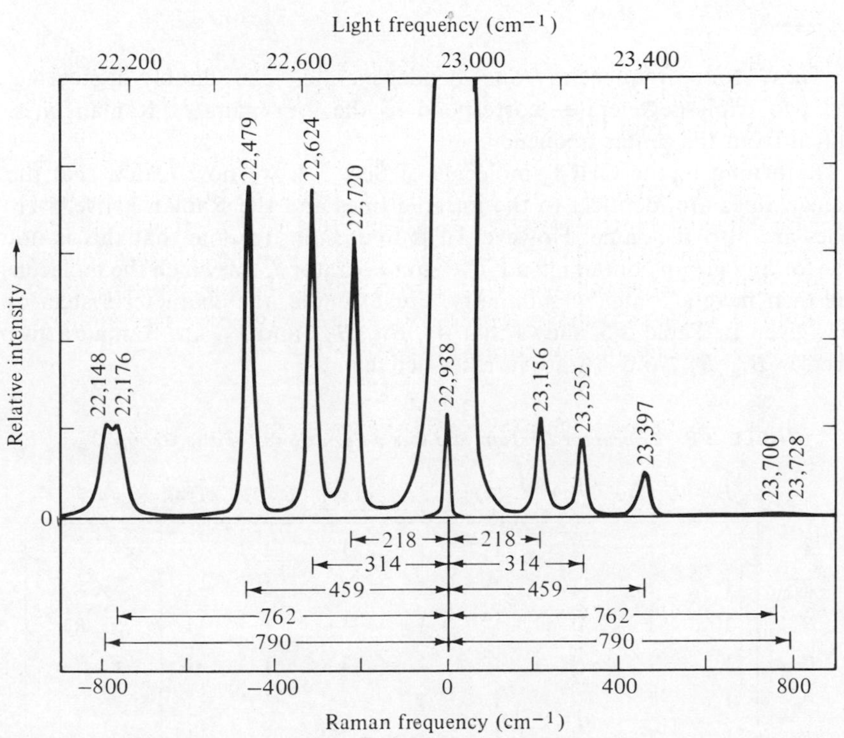

3-7 Raman spectrum of methane (CH_4)

[6]R. S. Tobias, *J. Chem. Educ.*, **44**, 2 (1967).

TABLE 3-4 ***Character System and Basis Functions for the Group T_d***

	E	$8C_3$	$3C_2$	$6S_4$	$6\sigma_d$	
A_1	1	1	1	1	1	$x^2 + y^2 + z^2$
A_2	1	1	1	−1	−1	
E	2	−1	2	0	0	$(2z^2 - x^2 - y^2, x^2 - y^2)$
F_1	3	0	−1	1	−1	(R_x, R_y, R_z)
F_2	3	0	−1	−1	1	$(x, y, z), (xy, yz, xz)$

PROBLEM 3-9

Show that the vibrational modes for CH_4 may be expressed as

$$\Gamma_{\text{vib}} = A_1 + E + 2F_2$$

These four frequencies—one nondegenerate, one doubly-degenerate, and two triply-degenerate—correspond to the four pairs of Raman lines, shifted from the center frequency.

Returning to the $CHCl_3$ molecule of Sec. 3-4, we now realize that the Raman lines are identical to the infrared lines and the Raman active overtones are also the same. However, it is interesting to note that this is not true for any group containing an inversion operator J, for which the molecule will then have a center of symmetry. For example, the character system of D_{2h}, given in Table 3-5, shows that A_g, B_{1g}, B_{2g}, and B_{3g} are Raman active whereas B_{1u}, B_{2u}, and B_{3u} are infrared active.

TABLE 3-5 ***Character System and Basis Functions for the Group D_{2h}***

	E	$C_2(z)$	$C_2(y)$	$C_2(x)$	J	$\sigma(xy)$	$\sigma(xz)$	$\sigma(yz)$	
A_g	1	1	1	1	1	1	1	1	x^2, y^2, z^2
B_{1g}	1	1	−1	−1	1	1	−1	−1	xy, R_z
B_{2g}	1	−1	1	−1	1	−1	1	−1	xz, R_y
B_{3g}	1	−1	−1	1	1	−1	−1	1	yz, R_x
A_u	1	1	1	1	−1	−1	−1	−1	
B_{1u}	1	1	−1	−1	−1	−1	1	1	z
B_{2u}	1	−1	1	−1	−1	1	−1	1	y
B_{3u}	1	−1	−1	1	−1	1	1	−1	x

Our treatment of the Raman effect has emphasized the applications of group theory, with very little attention being paid to its physical interpretation. A modern approach to the theory of the Raman effect, with emphasis on vibrations in solids, has been given by Loudon.[7] For further details on applications to molecules, the classic treatises of Herzberg[8] and Wilson, Decius, and Cross[9] should be consulted. Specific details for typical molecules belonging to the groups T_d, D_{4h}, O_h, D_{3h}, C_{3v}, and C_{2v} are listed in the book by Ferraro and Ziomek.[10]

[7] R. Loudon, *Radar Res. Estab. J.*, No. 51, (April, 1964).

[8] G. Herzberg, *Molecular Structure and Molecular Spectra*, II: *Infrared and Raman Structure of Polyatomic Molecules*, D. Van Nostrand, Co., Inc., Princeton, N. J., 1945.

[9] E. B. Wilson, Jr., J. C. Decius, and P. C. Cross, *Molecular Vibrations*, McGraw-Hill Book Company, New York, 1955.

[10] J. R. Ferraro and J. S. Ziomek, *Introductory Group Theory*, Plenum Press, New York, 1969.

CHAPTER FOUR MOLECULAR BINDING

4-1 *Introduction to the structure of molecules*

Near the end of the previous chapter, we pointed out that one of the goals of infrared and Raman experiments was to study the shape of molecules. It seems very logical, for example, to assume that the three Cl atoms in the $CHCl_3$ molecule of Fig. 2-27 are symmetrically placed with respect to one of the other two atoms. It even seems reasonable to put the heavier carbon atom near the center of the structure. But these notions are really just guesses and we would like a more systematic approach to this question, and a quantitative one, if possible. We shall show in this chapter that the answer we are seeking may be obtained by combining group theory with the Schroedinger equation. Some of the molecules we shall consider are water (H_2O), methane (CH_4), and singly-ionized hydrogen (H_2^+).

It should also be mentioned that we shall not consider what might be expected to be one of the simplest symmetrical systems: ozone (O_3). Intuition would lead us to expect that this molecule should be an equilateral triangle, since there is no obvious reason for any oxygen atom to be different from any other one. Hence, the symmetry should be that of the group C_{3v}, with normal modes as computed in Chapter 2. It will turn out, however, that this guess is incorrect; the ozone molecule is not quite an equilateral triangle; instead, its structure is similar to H_2O (Fig. 2-22). The central oxygen atom at the apex of the "vee" is bound to the two atoms at the ends of the legs but these atoms are *not* bound to one another. We shall consider H_2O in some detail in this chapter, presenting the arguments about why it assumes C_{2v} symmetry. Some of this analysis should also carry over to O_3, although the latter has a more complicated electronic structure.

4-2 The Schroedinger equation for the hydrogen atom

Extending Eq. (3-56) to a three-dimensional rectangular coordinate system, we obtain

$$-\frac{\hbar^2}{2m}\left(\frac{\partial^2}{\partial x^2}+\frac{\partial^2}{\partial y^2}+\frac{\partial^2}{\partial z^2}\right)\psi + V\psi = E\psi \tag{4-1}$$

or

$$-\frac{\hbar^2}{2m}\nabla^2\psi + V\psi = E\psi \tag{4-2}$$

where the del or Laplacian operator is

$$\nabla^2 = \frac{\partial^2}{\partial x^2}+\frac{\partial}{\partial y^2}+\frac{\partial^2}{\partial z^2}$$

and ψ is a function of x, y, and z.

Let us apply Eq. (4-2) to the hydrogen atom, and for this problem it is convenient to use spherical coordinates. We shall regard the position of the proton as fixed at the origin, as shown in Fig. 4-1, and the electron as being at a point whose spherical coordinates are (r, θ, ϕ). The force between the particles, expressing Coulomb's law in MKS units, is

$$F = -\frac{e^2}{4\pi\epsilon_0 r^2}$$

and the potential energy $V(r)$, which depends on r, becomes

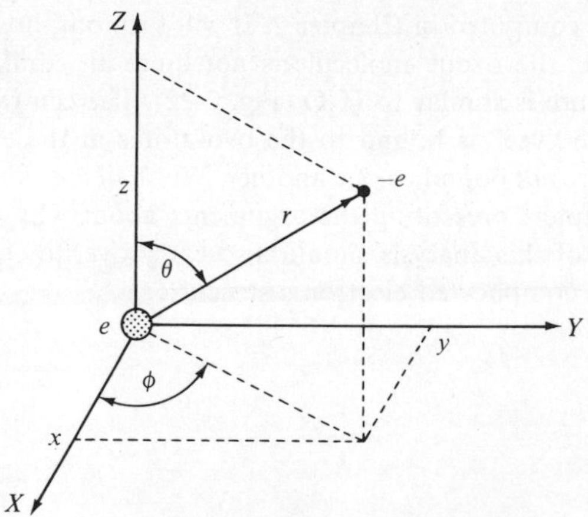

4-1 The hydrogen atom

$$V(r) = -\int F\,dr = -\frac{e^2}{4\pi\epsilon_0 r} \tag{4-3}$$

The Schroedinger equation in spherical coordinates is then

$$\frac{1}{r^2}\frac{\partial}{\partial r}\left(r^2\frac{\partial\psi}{\partial r}\right) + \frac{1}{r^2\sin^2\theta}\frac{\partial^2\psi}{\partial\phi^2} + \frac{1}{r^2\sin\theta}\frac{\partial}{\partial\theta}\left(\sin\theta\frac{\partial\psi}{\partial\theta}\right) + \frac{2m}{\hbar^2}[E - V(r)]\psi = 0 \tag{4-4}$$

This partial differential equation can be separated by letting

$$\psi(r, \theta, \phi) = R(r)\Theta(\theta)\Phi(\phi) \tag{4-5}$$

where $R(r)$ is a function of r alone, with similar definitions for $\Theta(\theta)$ and $\Phi(\phi)$. Computing the appropriate derivatives and substituting into Eq. (4-4) gives

$$\frac{1}{r^2R}\frac{d}{dr}\left(r^2\frac{dR}{dr}\right) + \frac{1}{(r^2\sin^2\theta)}\frac{d^2\Phi}{d\phi^2} + \frac{1}{(r^2\sin\theta)}\frac{d}{d\theta}\left(\sin\theta\frac{d\Theta}{d\theta}\right) + \frac{2m}{\hbar^2}[E - V(r)] = 0$$

By the use of a separation constant, which is written as m_l^2 for convenience, we obtain the two equations

$$\frac{d^2\Phi}{d\phi^2} = -m_l^2\Phi \tag{4-6}$$

and

$$\frac{1}{R}\frac{d}{dr}\left(r^2\frac{dR}{dr}\right) - \frac{m_l^2}{\sin^2\theta} + \frac{1}{\Theta\sin\theta}\frac{d}{d\theta}\left(\sin\theta\frac{d\Theta}{d\theta}\right) + \frac{2m}{\hbar^2}[E - V(r)]r^2 = 0 \tag{4-7}$$

Equation (4-7) can be further separated by the use of another constant, denoted as $l(l + 1)$, to give

$$\frac{1}{\sin\theta}\frac{d}{d\theta}\left(\sin\theta\frac{d\Theta}{d\theta}\right) - \frac{m_l^2}{\sin^2\theta}\Theta + l(l + 1)\Theta = 0 \tag{4-8}$$

$$\frac{1}{r^2}\frac{d}{dr}\left(r^2\frac{dR}{dr}\right) - \frac{l(l + 1)}{r^2}R + \frac{2m}{\hbar^2}[E - V(r)]R = 0 \tag{4-9}$$

The Schroedinger equation (4-4) has thus been converted into the three ordinary differential equations (4-6), (4-8), and (4-9); we recognize Eq. (4-6) as the classical harmonic motion equation with solutions

$$\Phi_m = e^{\pm im_l\phi} \tag{4-10}$$

where we have chosen the complex exponential form of Eq. (2-91). Note that Φ_m is labelled with a subscript corresponding to the integer m_l.

PROBLEM 4-1

Show that the solution given by (4-10) is multivalued at $\phi = 0$ or 2π unless m_l is an integer.

But if Φ is multivalued, then so is ψ of Eq. (4-5) and therefore the probability density ψ^2 as defined in Sec. 3-2 is ambiguous. Hence, we restrict the quantum number m_l to the values

$$m_l = 0, \pm 1, \pm 2, \ldots$$

Note that we now have complex functions and therefore the definition ψ^2 for the probability density is modified to $\psi^*\psi = |\psi|^2$, where ψ^* is the complex conjugate of ψ, obtained by replacing i with $-i$.

Equation (4-8) is known as the *associated Legendre equation* and has finite solutions only when l is an integer and when m_l is also an integer such that $|m_l| \leq l$. To prove these statements involves going into the details of obtaining the solutions. Instead we shall use the same approach as was taken for the Hermite equation. Let us label the associated Legendre functions $\Theta_{l,m}(\theta)$ with the subscripts l and m corresponding to the values of l and m_l in the equation. Then the first few solutions can be verified by differentiation to be

$$\Theta_{0,0} = 1$$

$$\Theta_{1,0} = \cos\theta, \qquad \Theta_{1,\pm 1} = \sin\theta \tag{4-11}$$

$$\Theta_{2,0} = (3\cos^2\theta - 1), \qquad \Theta_{2,\pm 1} = \sin\theta\cos\theta, \qquad \Theta_{2,\pm 2} = \sin^2\theta$$

PROBLEM 4-2

Show that the subscripts on the $\Theta_{l,m}$ of Eq. (4-11) are correct.

Equation (4-9) can be solved by substituting

$$R(r) = e^{-\alpha r} r^l L(r) \tag{4-12}$$

where $\alpha^2 = -2mE/\hbar^2$ and $L(r)$ is a new function to be determined. The equation which $L(r)$ obeys is known as the *associated Laguerre equation* and has finite solutions only if

$$n^2 = \frac{-me^4}{8h^2 E\epsilon_0^2} \tag{4-13}$$

where $n = 1, 2, 3, \ldots$, and if

$$l = 0, 1, 2, \ldots, n - 1$$

This problem thus involves the three quantum numbers n, l, and m_l, but the energy levels depend only on n, which is called the *principal quantum number*. These values are then

$$E_n = -\frac{me^4}{8n^2h^2\epsilon_0^2}, \quad (n = 1, 2, 3, \cdots) \tag{4-14}$$

and this result is identical to the expression obtained from the older Bohr theory. The first few associated Laguerre functions $L_{n,l}(r)$ are

$$\begin{gathered} L_{1,0} = 1 \\ L_{2,0} = (1 - \alpha r), \qquad L_{2,1} = 1 \\ L_{3,0} = (3 - 6\alpha r + 2\alpha^2 r^2), \qquad L_{3,1} = (2 - \alpha r), \qquad L_{3,2} = 1 \end{gathered} \tag{4-15}$$

where we have again labelled the $L_{n,l}$ with subscripts.

PROBLEM 4-3

(a) Show that the function $\psi = R\Theta\Phi$ corresponding to the lowest value of E_n is

$$\psi = e^{-\alpha r}$$

(b) Show that $\alpha = 1/r_1$, where r_1 is the radius of the *first Bohr orbit*.

PROBLEM 4-4

By substituting $R_{3,0} = e^{-\alpha r} L_{3,0}$ into Eq. (4-9), show that we verify Eq. (4-13) for $n = 3$.

Combining the individual solutions of Eqs. (4-10), (4-11), (4-15) we obtain the expressions for the wave functions as given in Table 4-1.

TABLE 4-1 ***Wave Functions, Quantum Numbers and Energy Levels for the Hydrogen Atom***

n	l	m_l	E_n	ψ_{nlm}
1	0	0	-13.6 eV	$e^{-\alpha r}$
2	0	0	-3.4	$(1-\alpha r)e^{-\alpha r}$
2	1	0	-3.4	$re^{-\alpha r}\cos\theta$
2	1	± 1	-3.4	$re^{-\alpha r}\sin\theta\, e^{\pm i\phi}$
3	0	0	-1.5	$(3-6\alpha r+2\alpha^2 r^2)e^{-\alpha r}$
3	1	0	-1.5	$(2-\alpha r)re^{-\alpha r}\cos\theta$
3	1	± 1	-1.5	$(2-\alpha r)re^{-\alpha r}\sin\theta\, e^{\pm i\phi}$
3	2	0	-1.5	$r^2 e^{-\alpha r}(3\cos^2\theta-1)$
3	2	± 1	-1.5	$r^2 e^{-\alpha r}\sin\theta\cos\theta\, e^{\pm i\phi}$
3	2	± 2	-1.5	$r^2 e^{-\alpha r}\sin^2\theta\, e^{\pm 2i\phi}$

PROBLEM 4-5

Verify that all possible combinations of quantum numbers for $n = 3$ are shown in Table 4-1.

In Fig. 4-2, we have plotted the probability density $\psi^*_{nlm}\psi_{nlm}$ for a number of the functions listed in the table. The following points should be kept in mind when using this figure:

1. The probability densities are all symmetric about the z-axis—that is, they are independent of ϕ—since $e^{-i\phi}e^{i\phi} = 1$.
2. The densities decay as we get farther from the nucleus in accordance with $e^{-\alpha r}$.
3. For the higher values of n, the functions $R_{n,l}$ have *nodes* (points where the function vanishes), so the densities show corresponding blank regions.

4-3 Multielectron atoms

We were able to solve the Schroedinger equation in spherical coordinates for a system consisting of a moving electron and a stationary nucleus. When we try to apply this equation to helium, an atom with two electrons, the potential energy term V involves the force of attraction between each electron and the nucleus and also the repulsive force between the two electrons, as we shall show explicitly in a later section of this chapter. It then becomes impossible to separate the Schroedinger equation into three ordinary differential equations, which we were able to do for hydrogen, and it is necessary to use approximate methods. For three-electron atoms and beyond, the situation is even more complicated.

The approximation that we shall use is to assume initially that the interaction between individual electrons can be neglected and that each electron moves in an electric field which is the original field of the nucleus modified by the presence of all the other electrons surrounding the nucleus and it is further assumed that the resulting potential energy depends only on r, the distance from the nucleus, and not on the angles θ and ϕ. Such a potential energy function $V(r)$ is said to be *spherically symmetric*. With this approximation, the Schroedinger equation is still separable, and, in particular, we obtain Eq. (4-9), but with a more complicated form of $V(r)$.

PROBLEM 4-6

Show that (4-4) is separable for any $V(r)$ which is a function only of r.

Although we can no longer obtain an exact solution of Eq. (4-9), we would expect that any solution would involve two quantum numbers like n and l of the hydrogen atom, and that the complete solution of the Schroedinger equation will then involve three quantum numbers. Unlike Eq. (4-9), however, we can guess that the total energy will depend on both n and l, instead of n alone.

Calling these quantum numbers n, l, and m_l, as before, we can specify the *state* of each electron in the atom by giving the associated wave function $\psi_{n,l,m}$ and energy $E_{n,l}$. Since both $\psi_{n,l,m}$ and $E_{n,l}$ are known when n, l, and m_l are known, the state of an electron is also specified by giving its quantum numbers. It was initially thought that electrons in atoms could occupy states governed only by the amount of energy available in the system to which they belonged. Pauli showed that this assignment to states is determined by the *exclusion principle*, which says that in any atom there can be only one electron in a given quantum state. However, we must include the specification of *spin* as part of the designation of a state, so that only two electrons, one of each kind of spin, can have the same three quantum numbers n, l, and m_l.

It is convenient in discussing the possible atoms forming the periodic table to use a notation developed by spectroscopists. The values of the quantum number l are denoted by letters, as follows:

$$\begin{matrix} 0 & 1 & 2 & 3 & 4 & \cdots \\ s & p & d & f & g & \cdots \end{matrix}$$

TABLE 4-2 ***Electronic Configurations of Elements 1 through 54***

Atomic number	Element	Electron configuration							
		1s	2s	2p	3s	3p	3d	4s	4p
1	H	1							
2	He	2							
3	Li	2	1						
4	Be	2	2						
5	B	2	2	1					
6	C	2	2	2					
7	N	2	2	3					
8	O	2	2	4					
9	F	2	2	5					
10	Ne	2	2	6					
11	Na	2	2	6	1				
12	Mg	2	2	6	2				
13	Al	2	2	6	2	1			
14	Si	2	2	6	2	2			
15	P	2	2	6	2	3			
16	S	2	2	6	2	4			
17	Cl	2	2	6	2	5			
18	Ar	2	2	6	2	6			
19	K	2	2	6	2	6		1	
20	Ca	2	2	6	2	6		2	
21	Sc	2	2	6	2	6	1	2	
22	Ti	2	2	6	2	6	2	2	
23	V	2	2	6	2	6	3	2	
24	Cr	2	2	6	2	6	5	1	
25	Mn	2	2	6	2	6	5	2	
26	Fe	2	2	6	2	6	6	2	
27	Co	2	2	6	2	6	7	2	

Atomic number	Element	Electron configuration										
		1s	2s	2p	3s	3p	3d	4s	4p	4d	5s	5p
28	Ni	2	2	6	2	6	8	2				
29	Cu	2	2	6	2	6	10	1				
30	Zn	2	2	6	2	6	10	2				
31	Ga	2	2	6	2	6	10	2	1			
32	Ge	2	2	6	2	6	10	2	2			
33	As	2	2	6	2	6	10	2	3			
34	Se	2	2	6	2	6	10	2	4			
35	Br	2	2	6	2	6	10	2	5			
36	Kr	2	2	6	2	6	10	2	6			
37	Rb	2	2	6	2	6	10	2	6		1	
38	Sr	2	2	6	2	6	10	2	6		2	
39	Y	2	2	6	2	6	10	2	6	1	2	
40	Zr	2	2	6	2	6	10	2	6	2	2	
41	Nb	2	2	6	2	6	10	2	6	4	1	
42	Mo	2	2	6	2	6	10	2	6	5	1	
43	Te	2	2	6	2	6	10	2	6	6	1	
44	Ru	2	2	6	2	6	10	2	6	7	1	
45	Rh	2	2	6	2	6	10	2	6	8	1	
46	Pd	2	2	6	2	6	10	2	6	10		
47	Ag	2	2	6	2	6	10	2	6	10	1	
48	Cd	2	2	6	2	6	10	2	6	10	2	
49	In	2	2	6	2	2	10	2	6	10	2	1
50	Sn	2	2	6	2	2	10	2	6	10	2	2
51	Sb	2	2	6	2	2	10	2	6	10	2	3
52	Te	2	2	6	2	6	10	2	6	10	2	4
53	I	2	2	6	2	6	10	2	6	10	2	5
54	Xe	2	2	6	2	6	10	2	6	10	2	6

where the symbols come from the historical names sharp, principal, diffuse, and fundamental for spectral series. The principal quantum number n is indicated by an integer in front of the letter which denotes l and the number of electrons with the same values of n and l is indicated by an exponent. To be explicit, hydrogen, the first element, has one electron. Its state of lowest energy, called the *ground state*, is denoted as $1s$, since $n = 1$ and $l = 0$. Helium, element number 2, can have two electrons in its ground state, one with spin up and the other with spin down, so that the symbol is $1s^2$. The next element, lithium, has the ground state $1s^2 2s$ and beryllium is $1s^2 2s^2$. Boron, with five electrons, is $1s^2 2s^2 2p$, since the fifth electron cannot be an s electron, carbon is $1s^2 2s^2 2p^2$ and nitrogen is $1s^2 2s^2 2p^3$. We may proceed in this way until we reach neon, element 10, which has the arrangement $1s^2 2s^2 2p^6$. As Table 4-1 shows, it is possible to have two electrons for $n = 1$ and eight for $n = 2$. Table 4-2 summarizes these possibilities.

When we add p electrons, however, an additional feature enters which we should consider next. The transition from element 3 to element 4 can be accomplished only be adding an electron whose spin is opposite to that of the $2s$ electron already present in lithium. But Table 4-1 indicates that we may have three kinds of p electrons; those for which m_l is $+1$, 0, or -1. Thus, many arrangements of spin are possible. For example, nitrogen is $1s^2 2s^2 2p^3$. The three p electrons could all have spin up; then they each have one of the three possible m_l values. This distribution of spins turns out to be the ground state. On the other hand, two electrons with opposite spin could have the same value of m_l, a third electron could have a different value, and there will be one of the three values of m_l not corresponding to any electron. If the electronic arrangement $1s^2 2s^2 2p^3$ is called a *configuration*, then we see that the configuration of nitrogen permits a number of states, only one of which is the ground state.

The electrons belonging to a given value of n form *shells* and when an atom has the maximum number of electrons for each n value, the shell is *closed*. In addition, in the shells for larger values of n (and hence l), we find *subshells* corresponding to the maximum l value and these can also be closed. Table 4-2 lists the elements with atomic numbers 1 through 54 (this group is of primary interest in solid state physics) and shows their electron configuration as determined by spectroscopic methods.

This table enables us to locate atoms whose outermost shells are similar. For example, the inert gases shown are helium, with a filled shell corresponding to $n = 1$; neon, with a filled $n = 2$ shell; argon, with a filled $n = 3$ shell; krypton, with a filled $4p$ subshell; and xenon, with a filled $5p$ subshell. These four elements illustrate how the exclusion principle can account for the repetition of chemical properties in the periodic table. In a later chapter we shall consider the periodic series shown in Table 4-3; carbon (as diamond), silicon, germanium, and tin (in the gray modification) all have the same

TABLE 4-3 ***Ground State Configuration of Semiconductors Which Crystallize in the Diamond Structure***

Element	Atomic number	Ground state
Carbon	6	$1s^22s^22p^2$
Silicon	14	$1s^22s^22p^63s^23p^2$
Germanium	32	$1s^22s^22p^63p^23p^63d^{10}4s^24p^2$
Tin	50	$1s^22s^22p^63s^23p^63d^{10}4s^24p^64d^{10}5s^25p^2$

$n = 1, l = 0, m_l = 0$

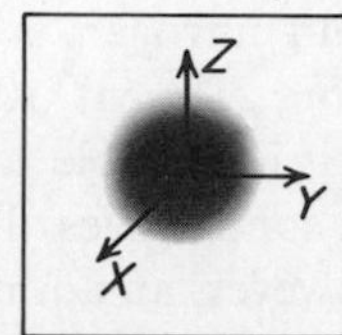

$n = 2, l = 0, m_l = 0$

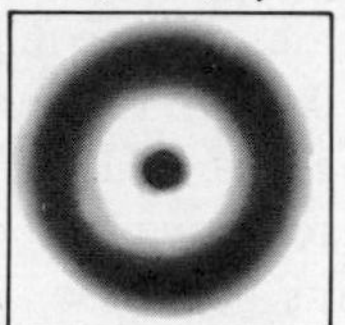

$n = 2, l = 1, m_l = 0$

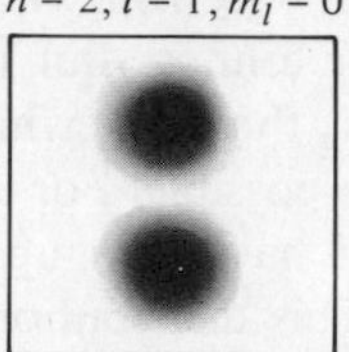

$n = 2, l = 1, m_l = \pm 1$

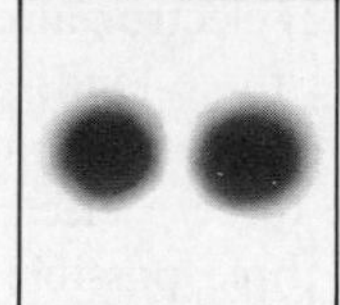

$n = 3, l = 0, m_l = 0$

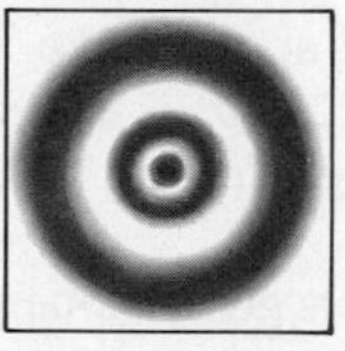

$n = 3, l = 1, m_l = 0$

$n = 3, l = 1, m_l = \pm 1$

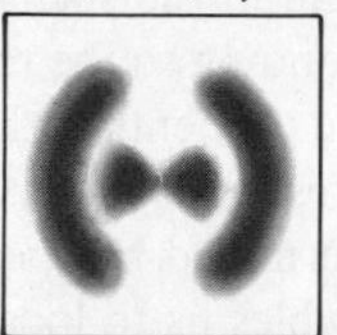

$n = 3, l = 2, m_l = 0$

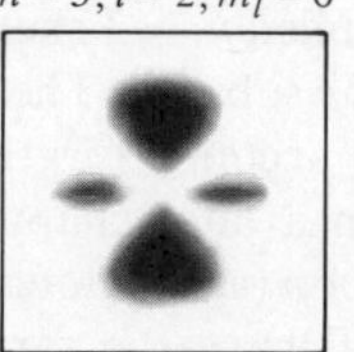

$n = 3, l = 2, m_l = \pm 1$

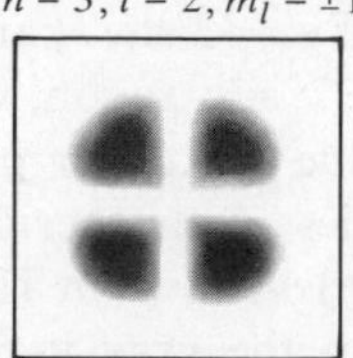

$n = 3, l = 2, m_l = \pm 2$

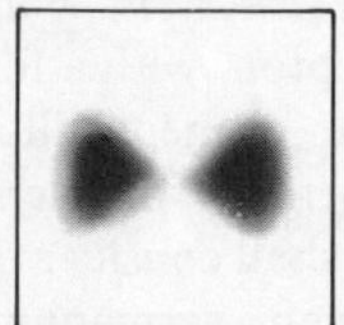

4-2 Probability densities for the hydrogen atom

crystal structure. In addition, they have a similar electronic configuration. That is, they all have four electrons in their last, unfilled shell. These electrons are called the *valence electrons* and are responsible for the binding of the atoms in the crystal. These four elements form crystals exhibiting *covalent bonding*. They also illustrate another interesting property of atoms, namely, that no outer shell possesses more than eight electrons. Hence, they form crystals in which each atom is joined to four neighboring atoms by a covalent bond representing two shared electrons and the atoms all effectively have a closed shell of the maximum size. We shall return to this concept when we consider the water molecule.

Referring back to Table 4-2, we see that it also indicates that since every atom can have two s electrons and six p electrons at the most, the limitation of eight electrons in the outer shell or subshell means that d, f, and higher electrons go into inner shells. A reason for this can be given in terms of the shapes of the probability distributions of Fig. 4-2. We note that only the s orbits (that is, those for which $l = 0$) are spherically symmetric and this means that on the average they represent a higher probability of finding the electron close to the origin. In simple language, we are implying that s electrons are more tightly bound, so that a $4s$ electron, for example, would have a lower energy than a $3p$ electron. Using the principle that s and p states are generally filled before higher ones, we can almost predict the complete sequence of filling, which is $1s$, $2s$, $2p$, $3s$, $3p$, $4s$, $3d$, $4p$, $5s$, $4d$, $5p$, $6s$, $4f$, $5d$, $6p$, $7s$, $6d$.

4-4 Atomic orbitals

It was indicated in Problem 4-6 that the Schroedinger equation is separable for any spherically symmetric $V(r)$. The solutions to the θ-equation and the ϕ-equation for any atom are thus identical to (4-10) and (4-11). These solutions, written as

$$Y_{l,m}(\phi, \theta) = \Theta_{l,m}(\theta)\Phi_m(\phi) \tag{4-16}$$

are known as *spherical harmonics*. They represent the angular part of the complete (but approximate) solution of the form

$$\phi_{n,l,m}(r, \phi, \theta) = R_{n,l}(r)Y_{l,m}(\phi, \theta) \tag{4-17}$$

where the radical functions $R_{n,l}$ are no longer the associated Laguerre functions

$$R_{n,l}(r) = e^{-\alpha r}r^l L_{n,l}(r) \tag{4-12}$$

but something more complicated. In fact, they must be determined numerically or in some convenient approximate fashion, as has been done by Slater[1] and Zener.[2] This work has been summarized by Eyring, Walter, and Kimball.[3]

We might guess that the solution to the Schroedinger equation for a multielectron atom would involve a function of r which depends on r to some power and also contains an exponentially decaying term. That is, the Slater-Zener functions should have the form

$$\psi(r, \theta, \phi) = Nf(r)\, X_{l,m}(\theta, \phi) \tag{4-18}$$

where N is a normalization constant and where $f(r)$ is given by the expression

$$f(r) = r^{(n^*-1)} e^{-(Z-s)r/n^* r_1} \tag{4-19}$$

The quantity Z in this expression is the atomic number, r_1 is the first Bohr orbit (Problem 4-3), n^* is an *effective quantum number*, and s is a *screening constant*. The quantum number n^* differs somewhat from n at higher values (that is, it is not always an integer) and the term $(Z - s)$ expresses the effective nuclear charge due to screening by the inner electrons. The functions ψ given by (4-18) are called *atomic orbitals*, since the magnitude squared indicates approximately the shape and extent of an orbit for any of the outer or valence electrons.

The radial portions $f(r)$ of the Slater-Zener orbitals indicate that the probability density $|\psi^2|$ falls off approximately exponentially. This is unsatisfactory because it means that these functions do not have the nodes indicated by Fig. 4-2 and these are known to exist for multielectron atoms and for molecules; they are an inherent part of the solutions to the Schroedinger equation. In addition, exponential solutions (with or without nodes) are "fuzzy"; the criterion for determining when $|\psi^2|$ is roughly equal to zero is rather arbitrary.

Fortunately, we can overlook these difficulties by realizing that the symmetry of an atomic orbital depends only on the spherical harmonic or angular portion $Y_{l,m}(\phi, \theta)$; the radial part $f(r)$ merely specifies the decay properties. Further, these angular functions are the same for any atom when our assumption that V is a function of r, only, holds—this being known as the *central-field approximation*. We shall therefore ignore the dependence on r from now on and just the angular parts of the atomic orbitals will be used in conjunction with symmetry arguments.

A more convenient form of these angular functions for use with group theory may be obtained for we know that a linear combination of solutions

[1] J. C. Slater, *Phys. Rev.*, **36**, 57 (1930).

[2] C. Zener, *Phys. Rev.*, **36**, 51 (1930).

[3] H. Eyring, J. Walter, and G. E. Kimball, *Quantum Chemistry*, John Wiley & Sons, Inc., New York, 1944.

TABLE 4-4 ***The Angular Parts of the Hydrogen Atom Wave Functions***

l	m_l	Spherical harmonic
0	0	$s = 1/2\sqrt{\pi}$
1	1	$p_{+1} = (\sqrt{3}/2\sqrt{2\pi}) \sin\theta\, e^{i\phi}$
1	0	$p_0 = (\sqrt{3}/2\sqrt{\pi}) \cos\theta$
1	−1	$p_{-1} = (\sqrt{3}/2\sqrt{2\pi}) \sin\theta\, e^{-i\phi}$
2	2	$d_{+2} = (\sqrt{15}/4\sqrt{2\pi}) \sin^2\theta\, e^{i2\phi}$
2	1	$d_{+1} = (\sqrt{15}/2\sqrt{2\pi}) \sin\theta \cos\theta\, e^{i\phi}$
2	0	$d_0 = (\sqrt{5}/4\sqrt{\pi})(3\cos^2\theta - 1)$
2	−1	$d_{-1} = (\sqrt{15}/2\sqrt{2\pi}) \sin\theta \cos\theta\, e^{-i\phi}$
2	−2	$d_{-2} = (\sqrt{15}/4\sqrt{2\pi}) \sin^2\theta\, e^{-i2\phi}$

of a differential equation is also a solution. Let us therefore extract from Table 4-1 the angular parts $\Theta_{l,m}(\theta)\Phi_m(\phi) = Y_{m,l}(\theta, \phi)$, normalize these functions, and list them in Table 4-4. Note that the symbols used for these first nine spherical harmonics indicate the corresponding value of l and m_l. We then form the nine linear combinations of Table 4-5, including an additional normalizing factor of $1/\sqrt{2}$ for the combinations involving the sum or difference of two spherical harmonics. These *angular orbitals* are assigned the symbols shown in the table and they are plotted to scale in Fig. 4-3. The

TABLE 4-5 ***Angular Atomic Orbitals***

Symbol	Angular orbital
s	$1/2\sqrt{\pi}$
p_x	$\frac{1}{2}(p_{+1} + p_{-1}) = (\sqrt{3}/2\sqrt{\pi}) \sin\theta \cos\phi$
p_y	$\frac{1}{2}[-i(p_{+1} - p_{-1})] = (\sqrt{3}/2\sqrt{\pi}) \sin\theta \sin\phi$
p_z	$p_0 = (\sqrt{3}/2\sqrt{\pi}) \cos\theta$
$d_{x^2-y^2}$	$\frac{1}{2}(d_{+2} + d_{-2}) = (\sqrt{15}/4\sqrt{\pi}) \sin^2\theta \cos 2\phi$ $= (\sqrt{15}/4\sqrt{\pi}) \sin^2\theta (\cos^2\phi - \sin^2\phi)$
d_{xy}	$\frac{1}{2}[-i(d_{+2} - d_{-2})] = (\sqrt{15}/4\sqrt{\pi}) \sin^2\theta \sin 2\phi$ $= (\sqrt{15}/2\sqrt{\pi}) \sin^2\theta \cos\phi \sin\phi$
d_{xz}	$\frac{1}{2}(d_{+1} + d_{-1}) = (\sqrt{15}/2\sqrt{\pi}) \sin\theta \cos\theta \cos\phi$
d_{yz}	$\frac{1}{2}[-i(d_{+1} - d_{-1})] = (\sqrt{15}/2\sqrt{\pi}) \sin\theta \cos\theta \sin\phi$
d_{z^2}	$d_0 = (\sqrt{5}/4\sqrt{\pi})(3\cos^2\theta - 1)$

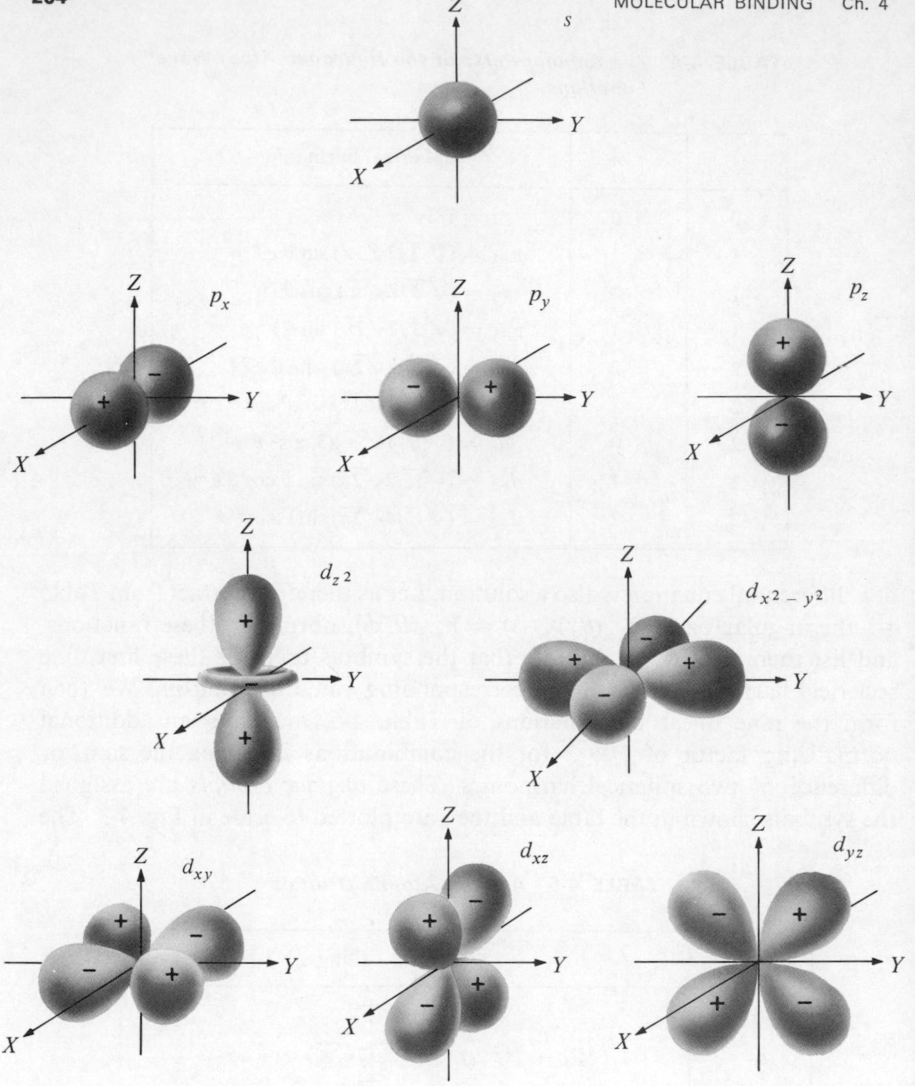

4-3 The *s*, *p*, and *d* atomic orbitals

subscripts on the symbols indicate the association of a function with a rectangular coordinate. For example, p_x is symmetrical about the *x*-axis, with a similar relation for p_y and p_z. We also see this from the defining equations for spherical coordinates, which are

$$\begin{aligned} x &= r \sin\theta \cos\phi \\ y &= r \sin\theta \sin\phi \\ z &= r \cos\theta \end{aligned} \tag{4-20}$$

Under a symmetry operation, r is invariant, so we can say that $p_x = (\sqrt{3}/2\sqrt{\pi}) \sin\theta \cos\phi$ varies as x, and so on. We also note that d_{xy}, d_{yz}, d_{zx}, and $d_{x^2-y^2}$ have the form indicated by their respective subscripts. For d_{z^2}, we use the relation

$$3\cos^2\theta - 1 = \frac{3z^2}{r^2} - 1$$

to establish the identification. In general, the s, p, and d orbitals are sufficient for our needs in this book. Expressions for f orbitals and polar plots will be found in Gray.[4]

Let us again emphasize that the angular orbitals of Fig. 4-3 ignore the exponential decay with r and are normalized only with respect to θ and ϕ. For example, for the s orbital of Table 4-5, we may verify that

$$\int_0^{\pi}\int_0^{2\pi}\left(\frac{1}{2\sqrt{\pi}}\right)^2 \sin\theta\, d\theta\, d\phi = 1$$

where

$$dS = \sin\theta\, d\theta\, d\phi$$

is an element of area on a unit sphere. On the other hand, the integral

$$\int_0^{\infty}\int_0^{\pi}\int_0^{2\pi}\left(\frac{1}{2\sqrt{\pi}}\right)^2 r^2 \sin\theta\, d\theta\, d\phi\, dr$$

is improper. Complete normalization of this state requires dealing with the function

$$\psi_{100} = N_{100}e^{-r/a_1}$$

as specified by Table 4-1 and Problem 4-3.

PROBLEM 4-7

Show that

$$N_{100} = \frac{1}{\pi^{1/2}a_1^{3/2}}$$

Figure 4-2 shows the decay with r for the orbitals s, $p_z = p_0$, and $d_z^2 = d_0$; these are the three functions which correspond to the following quantum number combinations:

[4] H. B. Gray, *Electrons and Chemical Bonding*, W. A. Benjamin, New York, 1964.

n	l	m_l
1	0	0
2	1	0
3	2	0

In the introduction to this chapter, we mentioned the shape and symmetry properties of molecules. It should now be apparent that the directional properties of the angular orbitals will have some bearing on this problem. These considerations were treated quantitatively by Pauling[5] and by Mulliken.[6] The four valence electrons of Table 4-3 are in fact one pair of s electrons and one pair of p electrons in each atom. That is, two of the electrons have one energy and two have another energy. It has been suggested that a more symmetrical possibility is for all four electrons to have the same energy. The way this is achieved is to combine an s and a p orbital to form an *sp hybrid*. That is, we assume that an electron can have a wave function of the form

$$\psi = as + bp_i \tag{4-21}$$

where p_i is one of the three orbitals p_x, p_y, or p_z. Choosing p_z, for example, this becomes

$$\psi = \frac{a + b\sqrt{3}\cos\theta}{2\sqrt{\pi}} \tag{4-22}$$

The coefficients in (4-21) obey the relation

$$a^2 + b^2 = 1 \tag{4-23}$$

if ψ is to be normalized.

PROBLEM 4-8

Prove (4-23).

Define an angle α such that

$$a = \cos\alpha, \qquad b = \sin\alpha \tag{4-24}$$

[5] L. Pauling, *The Nature of the Chemical Bond*, 3rd ed., Cornell University Press, Ithaca, N.Y., 1960.

[6] R. S. Mulliken, *Phys. Rev.*, **43**, 279 (1933).

Then (4-24) is another form of (4-23), and (4-22) becomes

$$\psi = \frac{\cos \alpha + \sqrt{3} \sin \alpha \cos \theta}{2\sqrt{\pi}} \qquad \textbf{(4-25)}$$

This function will have a maximum with respect to θ when $\theta = 0°$ or

$$\psi = \frac{\cos \alpha + \sqrt{3} \sin \alpha}{2\sqrt{\pi}}$$

and with respect to α when

$$\frac{\partial \psi}{\partial \alpha} = -\sin \alpha + \sqrt{3} \cos \alpha = 0 \quad \text{or} \quad \alpha = 60°$$

so that $$\sin \alpha = \frac{\sqrt{3}}{2} = b, \qquad \cos \alpha = \frac{1}{2} = a$$

Hence $$\psi = \frac{1}{2} s + \frac{\sqrt{3}}{2} p_z \qquad \textbf{(4-26)}$$

is an *sp* hybrid orbital.

An orbital of this kind has directional properties which are more pronounced than those of p_z alone, shown at the end of the second row of Fig. 4-3. Combining the p_z function with the *s* function and using the coefficients of (4-22), we obtain the *sp* hybrid illustrated in Fig. 4-4. This figure uses the symbolism of Sebera,[7] in which a vertical line represents a positive function and a horizontal line a negative function. When two regions of the same sign combine,

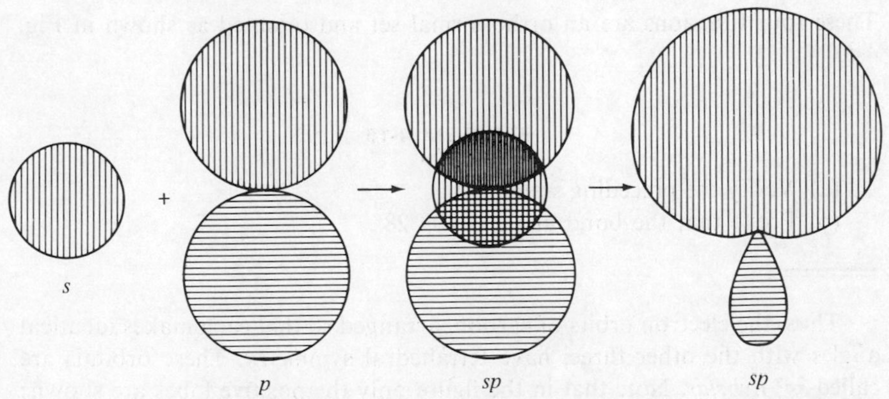

4-4 The construction of an *sp* hybrid orbital

[7] D. K. Sebera, *Electronic Structure and Chemical Bonding*, Blaisdell Publishing Co., Waltham, Mass., 1964.

the resulting amplitude increases; two regions of opposite sign cancel. The *sp* hybrid consequently has an elongated lobe along the positive z-axis and a shortened one in the other direction, with a node as shown.

PROBLEM 4-9

Verify the shape of the *sp* hybrid both graphically and analytically.

Our discussion so far has been restricted to a single combination of an s orbital and a p orbital. That is, we have taken one of the two s functions belonging to the four valence electrons of carbon and combined it with one of the two p functions. If we imagine an electron raised or *promoted* to a p orbit, then we can generate four hybrids which are identical except for orientation, so that the functions s, p_x, p_y, and p_z may be combined to give the four functions

$$\begin{aligned}\psi_1 &= \frac{s + p_x + p_y + p_z}{2}\\ \psi_2 &= \frac{s + p_x - p_y - p_z}{2}\\ \psi_3 &= \frac{s - p_x + p_y - p_z}{2}\\ \psi_4 &= \frac{s - p_x - p_y + p_z}{2}\end{aligned} \tag{4-27}$$

These four functions are an orthonormal set and oriented as shown in Fig. 4-5.

PROBLEM 4-10

(a) Verify the preceding statement.
(b) Prove that the bond angle is 109°28′.

Thus, the electron orbits of carbon, arranged so that each makes identical angles with the other three, have tetrahedral symmetry. These orbitals are called *sp^3 hybrids*. Note that in the figure only the positive lobes are shown; the negative lobes play no role, since they are so small.

Hybrid orbitals conform to the physical symmetry of the molecule or crystal in which they are located. For example, molecules of the form XY_3

can be planar, so that there are three equivalent XY bonds. We can visualize these sp^2 *hybrids* as combinations of s, p_x, and p_y, written as

$$\begin{aligned} \psi_1 &= a_1 s + b_1 p_x + c_1 p_y \\ \psi_2 &= a_2 s + b_2 p_x + c_2 p_y \\ \psi_3 &= a_3 s + b_3 p_x + c_3 p_y \end{aligned} \tag{4-28}$$

If we arbitrarily place ψ_1 along the x-axis, then $c_1 = 0$. Since ψ_2 and ψ_3 are equivalent, we also have $a_2 = a_3$. The orthogonality requirement means that we have three relations like

$$a_2 a_3 + b_2 b_3 + c_2 c_3 = 0$$

and normalization gives

$$a_i^2 + b_i^2 + c_i^2 = 1, \quad (i = 1, 2, 3)$$

These relations reduce (4-28) to

$$\begin{aligned} \psi_1 &= \frac{1}{\sqrt{3}} s + \frac{\sqrt{2}}{\sqrt{3}} p_x \\ \psi_2 &= \frac{1}{\sqrt{3}} s - \frac{1}{\sqrt{6}} p_x + \frac{1}{\sqrt{2}} p_y \\ \psi_3 &= \frac{1}{\sqrt{3}} s - \frac{1}{\sqrt{6}} p_x - \frac{1}{\sqrt{2}} p_y \end{aligned} \tag{4-29}$$

The relative magnitude of the lobes in these hybrid orbitals is measured by a quantity called the *Pauling strength*, defined as the maximum amplitude of a given function. For example, the function s has a magnitude $1/2\sqrt{\pi}$ and p_z has a magnitude $\sqrt{3}/2\sqrt{\pi}$. It is customary to remove the factor $1/2\sqrt{\pi}$ which is common to all functions of Table 4-4. Then the Pauling strength of the sp hybrid of (4-25) is

$$\psi_{\max} = \frac{1}{2} + \frac{\sqrt{3}}{2}\sqrt{3} = 2$$

For the sp^3 tetrahedral orbital, we must choose θ and ϕ to correspond to a specific direction in Fig. 4-5. For ψ_1, we have

$$\begin{aligned} \phi &= 45° \\ \theta &= \arctan(\sqrt{2}) \end{aligned} \tag{4-30}$$

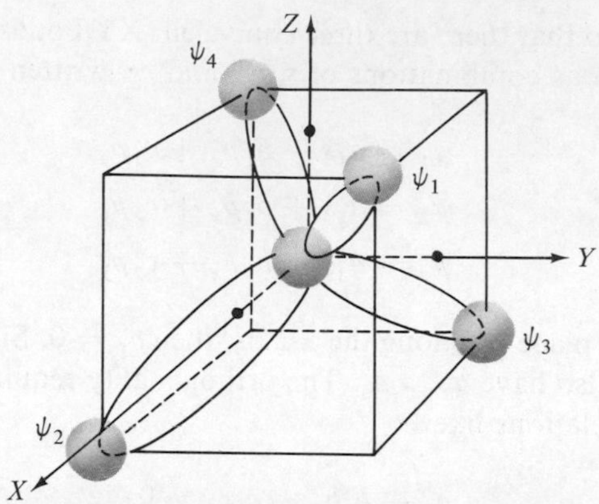

4-5 Four hybrid orbitals arranged in a tetrahedral configuration

PROBLEM 4-11

Prove (4-30).

Using $\sin\phi = \cos\phi = \sqrt{2}/2$ and $\sin\theta = \sqrt{2}/\sqrt{3}$, $\cos\theta = 1/\sqrt{3}$ gives

$$\psi_{\max} = 1$$

for the Pauling strength.

When carbon atoms with tetrahedral orbitals combine to form a crystal which has the well-known *diamond structure* which we shall consider later in detail, we intuitively imagine each of the four valence electrons forming a covalent bond by joining with one electron from each of the neighboring atoms. To again use the simple approach of Sebera, consider the two sp^3 hybrids of Fig. 4-6(a). They lie along the internuclear axis of two C atoms and overlap to a certain extent. In the region of overlap, the combined amplitude is enhanced. This, in turn, lowers the energy of the electrons, for there is now a higher probability that they will be closer to the nuclei. Such an orbital arrangement is called a *bonding orbital* and is what we have previously called

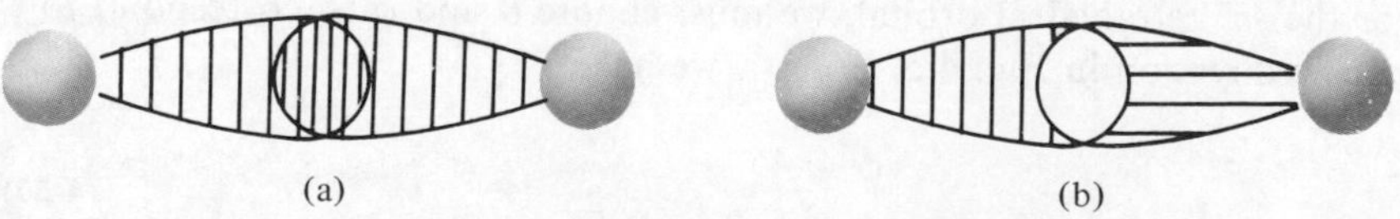

4-6 (a) A bonding orbital; (b) an anti-bonding orbital

a covalent bond. In contrast to this, a positive p orbital overlapping a negative one, as shown in Fig. 4-6(b), gives a region of zero probability along the internuclear line; that is, a node is created. Since the electrons are effectively excluded from this region, their resulting energy is higher and the bond is not stable. We then have an *anti-bonding orbital.* In the diamond lattice, this does not occur, for the long lobes of the sp^3 hybrids are all positive in sign. Thus, tetrahedral bonding is very stable.

4-5 Group theory and hybrid bonds

Some of the concepts of the previous section can be extended by the use of group theory. Consider a planar XY_3 molecule as an example and assume the bonding orbitals are formed from a mixture of s and p functions. Although the bonds lie in the plane of the molecule (Fig. 4-7), the symmetry

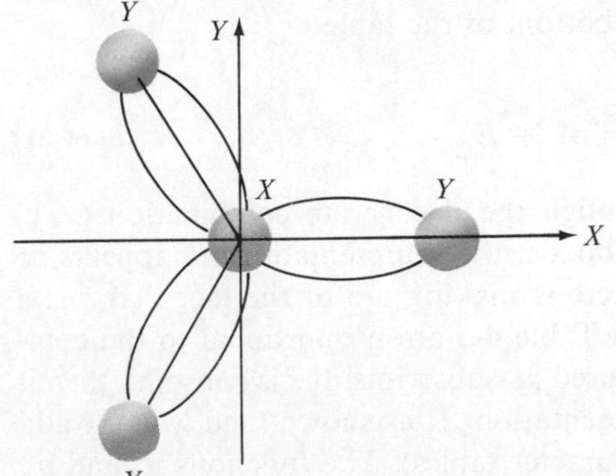

4-7 The sp^2 hybrids for a planar XY_3 molecule. (*Note:* Do not confuse the molecular labels X and Y with the axis labels *X* and *Y*)

elements are those of the group D_{3h} of Chapter 1, since the p functions lie along each of the three axes. The character system is given in Table 4-6. The reducible representation has a character system which we obtain by using the three orbitals of Fig. 4-7 as a basis. That is, they are invariant under E so that

$$\chi(E) = 3$$

All three are shifted by C_3, giving

$$\chi(C_3) = 0$$

and the same is true for S_3. For the horizontal reflection

$$\chi(\sigma_h) = 3$$

TABLE 4-6 ***Character System and Basis Functions for the Group D_{3h}***

	E	$2C_3$	$3C_2$	σ_h	$2S_3$	$3\sigma_v$	
A_1'	1	1	1	1	1	1	s, x^2+y^2, z^2
A_2'	1	1	−1	1	1	−1	R_z
E'	2	−1	0	2	−1	0	$(x, y), (x^2-y^2, xy)$
A_1''	1	1	1	−1	−1	−1	
A_2''	1	1	−1	−1	−1	1	z
E''	2	−1	0	−2	1	0	$(xz, yz), (R_x, R_y)$
Γ_σ	3	0	1	3	0	1	
Γ_π	6	0	−2	0	0	0	

and so on, obtaining the reducible representation Γ_σ (the subscript σ will be explained shortly) shown at the bottom of the table.

Empirically, we see that

$$\Gamma_\sigma = A_1' + E' \tag{4-31}$$

and this is puzzling, because, although the degenerate combination (x, y) of E' seems reasonable for planar bonds, the representation A_1' appears to play no role. What we are doing here is making use of the fact that s is a constant and the other functions in Table 4-5 are proportional to the combinations $x, y, z, x^2 - y^2, xy$, etc., used as subscripts. If s is constant, then it belongs to the fully symmetric representation A_1' as shown (and we may add it to the first line of all previous character tables). The functions p_x and p_y, however, transform as a pair, just like (x, y) and p_z, of course, belong to A_2''. Thus, we have shown that the hybrid wave function consistent with D_{3h} symmetry is

$$\psi_\sigma = as + bp_x + cp_y \tag{4-32}$$

However, this is not the only possible combination, for we also see that d_z^2 belongs to A_1'. Since d_{z^2} lies along the z-axis, it is perfectly symmetrical with respect to the xy-plane and is therefore an acceptable substitute for s. This gives the *pd hybrid*

$$\psi_\sigma = ad_{z^2} + bp_x + cp_y \tag{4-33}$$

In the same way, we are permitted to use the d combination belonging to E', obtaining a hybrid

$$\psi_\sigma = as + bd_{xy} + cd_{x^2-y^2} \tag{4-34}$$

and a *d function* $\psi_\sigma = ad_{z^2} + bd_{xy} + cd_{x^2-y^2}$ **(4-35)**

For the tetrahedral bonding of the CH_4 molecule or the diamond lattice, we use the four hybrid orbitals of Fig. 4-5 as a basis.

PROBLEM 4-12

Show that the characters of this reducible representation are

	E	$8C_3$	$3C_2$	$6S_4$	$6\sigma_d$
Γ_{tetra}	4	1	0	0	2

The symmetry group for CH_4 is the cubic group T_d discussed in Sec. 1-4. Its character system, as given in Table 4-7, shows that

$$\Gamma_{\text{tetra}} = A_1 + F_2 \tag{4-36}$$

TABLE 4-7 ***Character System and Basis Functions for the Group T_d***

	E	$8C_3$	$3C_2$	$6S_4$	$6\sigma_d$	
A_1	1	1	1	1	1	s
A_2	1	1	1	−1	−1	
E	2	−1	2	0	0	$(d_{z^2}, d_{x^2-y^2})$
F_1	3	0	−1	1	−1	(R_x, R_y, R_z)
F_2	3	0	−1	−1	1	(d_{xy}, d_{yz}, d_{zx})

Hence, the hybrid orbitals are identical sp^3 hybrids oriented as shown in the figure. Further, they may also be sd^3 hybrids (where d explicitly refers to xy, yz, or zx and not to the other two). Although either sp^3 or sd^3 hybrids are acceptable from a symmetry point of view, Cotton[8] points out that in carbon the $3d$ energy levels, the lowest d levels available, are about 10 eV above the $2p$ levels (where 1 eV is 1.6×10^{-19} joules or about 23 kcal/mol). This is a tremendous amount of energy in a molecule, so that carbon bonds are certainly sp^3 hybrids.

An example that we can study in more depth is the water molecule, which was shown in Chapter 2 to have C_{2v} symmetry. The configuration of oxygen,

[8] F. A. Cotton, *Chemical Applications of Group Theory*, Interscience, New York, 1963.

from Table 4-2, is $1s^2 2s^2 2p^4$, and we shall therefore assume that only s and p functions participate in the binding; we shall denote them as s_O, p_{xO}, p_{yO}, and p_{zO}. Each hydrogen has a single $1s$ electron, denoted s_1 and s_2, to correspond to Fig. 4-8; the effect of the symmetry operations on these five functions is listed in Table 4-8 and the projections are given in Table 4-9.

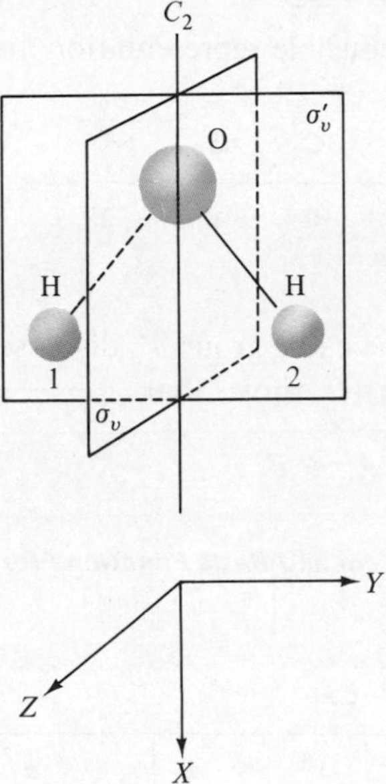

4-8 The H_2O molecule

(Note that the coordinates in Fig. 4-8 have been altered from those in Fig. 2-20.)

The reducible representation for these six functions as a basis has the character system

	E	C_2	σ_v	σ_v'
Γ_{H_2O}	6	0	2	4

as may be easily seen from Table 4-8. This set of characters is equivalent to the sum

$$\Gamma_{H_2O} = 3A_1 + 0A_2 + B_1 + 2B_2$$

which is obtained from Table 2-13 and confirmed by Table 4-9.

The interpretation of these results has been given by Slater[9] and by

[9] J. C. Slater, *Quantum Theory of Molecules and Solids I, Electronic Structure of Molecules*, McGraw-Hill Book Company, New York, 1963.

TABLE 4-8 ***Effect of Symmetry Operations for the Group C_{2v} on the Basis Functions of the Water Molecule***

	E	C_2	σ_v	σ'_v
s_O	s_O	s_O	s_O	s_O
p_{xO}	p_{xO}	p_{xO}	p_{xO}	p_{xO}
p_{yO}	p_{yO}	$-p_{yO}$	$-p_{yO}$	p_{yO}
p_{zO}	p_{zO}	$-p_{zO}$	p_{zO}	$-p_{zO}$
s_1	s_1	s_2	s_2	s_1
s_2	s_2	s_1	s_1	s_2

TABLE 4-9 ***Projected Functions for the Water Molecule***

Representation	Basis function	Projected function
A_1	s_O	$4s_O$
A_2		0
B_1		0
B_2		0
A_1	p_{xO}	$4p_{xO}$
A_2		0
B_1		0
B_2		0
A_1	p_{yO}	0
A_2		0
B_1		0
B_2		$4p_{yO}$
A_1	p_{zO}	0
A_2		0
B_1		$4p_{zO}$
B_2		0
A_1	s_1	$2(s_1 + s_2)$
A_2		0
B_1		0
B_2		0
A_1	s_2	0
A_2		0
B_1		0
B_2		$2(s_1 - s_2)$

Ballhausen and Gray.[10] The functions s_O and p_{xO} belonging to A_1 form an sp hybrid along the z-axis, equidistant from the two hydrogen atoms and creating a bonding orbital with the combination $(s_1 + s_2)$ [Fig. 4-9(a)]. On the other hand, the orbital p_y forms a bond with the combination $(s_1 - s_2)$ in the manner indicated in Fig. 4-9(b). There is then a p_z orbital left in Table

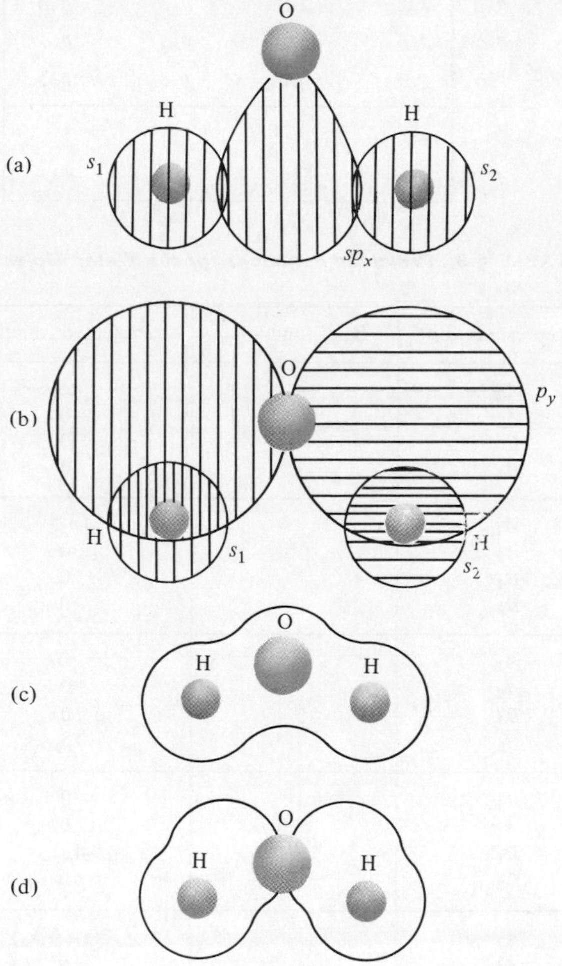

4-9 Orbitals in the water molecule: (a) a bonding orbital composed of an sp hybrid overlapping two s functions; (b) a bonding orbital composed of a pure p function overlapping two s functions; (c) part (a) shown as a single orbital; (d) part (b) shown as a single orbital.

[10]C. J. Ballhausen and H. B. Gray, *Molecular Orbital Theory*, W. A. Benjamin, New York, 1965.

4-9, and this is seen to take no part in the bonding, so that it is called a *nonbonding orbital*. Before considering the significance of these facts, let us go back to the 1916 theory of G.N. Lewis, as described by Linnett.[11] Groups of two or eight electrons in molecules and crystals appear to have very high stability. Methane (CH_4) particularly demonstrates this, since chemists often write its formula as

$$\begin{array}{c} \mathrm{H} \\ \mathrm{H} : \mathrm{C} : \mathrm{H} \\ \mathrm{H} \end{array}$$

where we imagine a pair of electrons shared in each covalent bond, and each H atom has effectively two electrons while the C atom has eight. This molecule has the stable tetrahedral structure we have already discussed. In the diamond lattice, every C atom has eight effective electrons and again we have a stable structure. We also note that when we include spin in this theory, we need merely change the formula to

$$\begin{array}{c} \mathrm{H} \\ \times\,\cdot \\ \mathrm{H} \overset{\cdot}{\times} \mathrm{C} \overset{\times}{\cdot} \mathrm{H} \\ \cdot\,\times \\ \mathrm{H} \end{array}$$

where the dot denotes spin up and the cross is for spin down.

For water, the Lewis formula is written as

$$\begin{array}{c} \times\,\cdot \\ \overset{\cdot}{\times} \mathrm{O} \overset{\times}{\cdot} \mathrm{H} \\ \cdot\,\times \\ \mathrm{H} \end{array}$$

and the atoms still have closed shells of two and eight valence electrons, respectively. In this case, however, there are two pairs of bonding orbitals and two *lone pairs* which play no part in the bonding. These are nonbonding orbitals like p_z just above. This formula for H_2O leads to two physical pictures: either we can regard the bonding orbitals as composed of pure p_x functions and p_y functions, obtaining a 90° bond angle, or we can think of the two bonding pairs and two lone pairs as a tetrahedral set, so that the bond angle is 109°28′. Since the true angle is in between, but close to the larger value, it is possible that H_2O has a modified tetrahedral structure, the bond angle being decreased from 109.5° to 104.5° because of the bonding effect of the electron cloud surrounding the H nuclei. This is the explanation favored by Ballhausen and Gray.[10] Further information on the application of the lone pair idea will be found in a descriptive article of Fowles.[12]

[11] J. W. Linnett, *American Scientist*, **52**, 459 (1964).

[12] G. W. H. Fowles, *J. Chem. Educ.*, **34**, 187 (1957).

On the other hand, Slater[9] works directly with the symmetry properties of C_{2v}, so that Fig. 4-9 serves as the beginning of his electronic structure for H_2O. The situation of Fig. 4-9(a) may be combined into a single orbital composed of contributions from s_O, p_{xO}, and $(s_1 + s_2)$, as shown in Fig. 4-9(c). This takes care of one covalent bond with two electrons. The second covalent bond is derived from Fig. 4-9(b), as shown in Fig. 4-9(d). Then there is a lone pair in the nonbonding orbital p_z and a second lone pair in the $s_O p_{xO}$ hybrid belonging to A_1; this orbital is nonbonding, because the other hybrid of this type has already formed the bond. In this type of structure, we cannot explicitly assign oxygen electrons to particular bonds, since the six electrons are divided up among a bonding hybrid, a nonbonding hybrid, a bonding p_y orbital, and a nonbonding p_z orbital. We therefore say that the structure is *delocalized.*

4-6 *Molecular orbitals and the LCAO method*

We have seen that the Schroedinger equation is exactly soluble for only a single atom: the Laplacian operator of an electron moving around a nucleus leads to a separable equation in spherical coordinates. The helium atom, with two electrons, does not permit this. To see why, let us label the coordinates of the two electrons (with the nucleus assumed fixed) as r_1, θ_1, ϕ_1 and r_2, θ_2, ϕ_2, respectively. The potential energy V_{12} of each electron due to the charge on the other is

$$V_{12} = \frac{e^2}{4\pi\epsilon_0 r_{12}} \tag{4-37}$$

where r_{12} is the interelectron distance (Fig. 4-10). If Eq. (4-2) is extended to cover this situation, it becomes

$$-\frac{\hbar^2}{2m}(\nabla_1^2 + \nabla_2^2)\psi - \frac{Ze^2}{4\pi\epsilon_0 r_1}\psi - \frac{Ze^2}{4\pi\epsilon_0 r_2}\psi + \frac{e^2}{4\pi\epsilon_0 r_{12}}\psi = E\psi \tag{4-38}$$

where $Z = 2$.

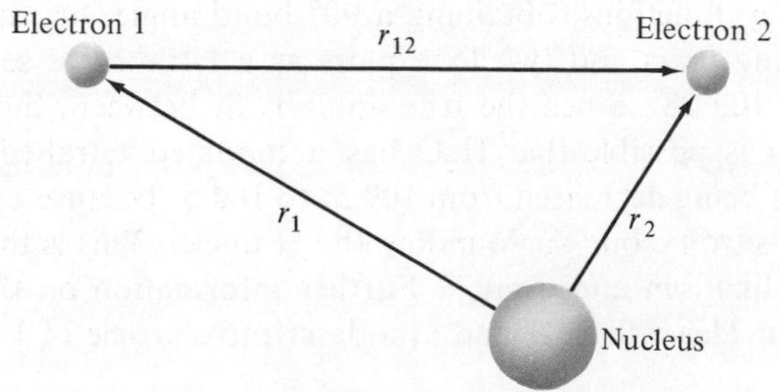

4-10 Coordinates for the helium atom

PROBLEM 4-13

Show that a solution of (4-38) is

$$\psi(r_1, \theta_1, \phi_1, r_2, \theta_2, \phi_2) = R_1(r_1)\Theta_1(\theta_1)\Phi_1(\phi_1)R_2(r_2)\Theta_2(\theta_2)\Phi_2(\phi_2)$$

provided $V_{12} = 0$.

Unfortunately, V_{12} does not vanish; in fact, when the electrons are close together it will be a very large quantity and cannot be neglected.

On the other hand, a three-particle system for which we can find a solution to the wave equation is the single ionized hydrogen molecule H_2^+ of Fig. 4-11(a). Assuming the internuclear distance R is a constant, the Schroedinger equation can be separated for this set of particles in *confocal ellipsoidal coordinates*. This system is generated as follows: we place the two nuclei at positions $\pm R/2$ on the z-axis of a rectangular system. Next, we define two coordinates by

$$\mu = \frac{r_a + r_b}{R}, \qquad \nu = \frac{r_a - r_b}{R} \tag{4-39}$$

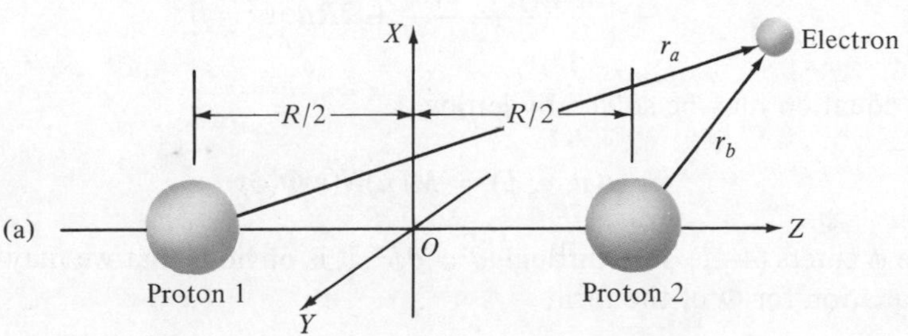

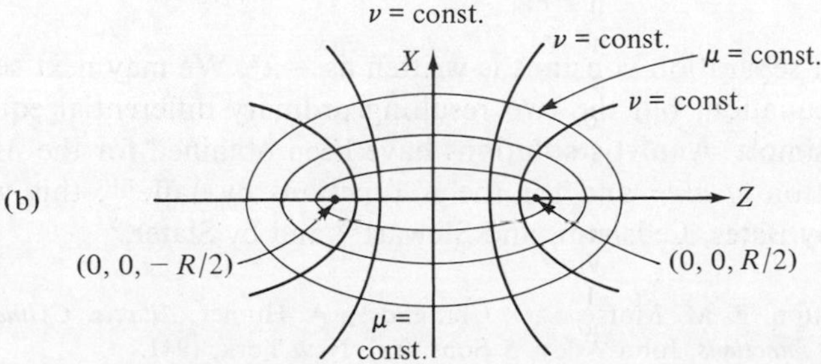

4-11 (a) Coordinates for the singly ionized hydrogen molecule H_2^+; (b) confocal ellipsoidal coordinates

and a third one ϕ which is identical to ϕ in the spherical system. Then the surfaces for which μ is constant are ellipsoids of revolution with foci at $\pm R/2$ and the surfaces of constant ν are hyperboloids of revolution with the same foci; the two sets of surfaces are orthogonal, as shown in Fig. 4-11(b). The third set of orthogonal surfaces are the planes of constant ϕ. The Schroedinger equation for the single electron is

$$-\frac{\hbar^2}{2m}\nabla^2\psi - \frac{e^2}{4\pi\epsilon_0 r_a}\psi - \frac{e^2}{4\pi\epsilon_0 r_b}\psi = E\psi \tag{4-40}$$

The Laplacian in ellipsoidal coordinates, computed by a standard method explained in Appendix III of Eyring, Walter, and Kimball,[3] is

$$\nabla^2 = \frac{4}{R^2(\mu^2 - \nu^2)}\left[\frac{\partial}{\partial\mu}\left\{(\mu^2 - 1)\frac{\partial}{\partial\nu}\right\} + \frac{\partial}{\partial\nu}\left\{(1 - \nu^2)\frac{\partial}{\partial\nu}\right\} + \frac{\mu^2 - \nu^2}{(\mu^2 - 1)(1 - \nu^2)}\frac{\partial^2}{\partial\phi^2}\right]$$

The Schroedinger equation then becomes

$$\frac{\partial}{\partial\mu}\left[(\mu^2 - 1)\frac{\partial\psi}{\partial\mu}\right] + \frac{\partial}{\partial\nu}\left[(1 - \nu^2)\frac{\partial\psi}{\partial\nu}\right] + \left[\frac{1}{\mu^2 - 1} + \frac{1}{1 - \nu^2}\right]\frac{\partial^2\psi}{\partial\phi^2} + \left[\frac{R^2E(\mu^2 - \nu^2)}{4} + 2R\mu\right]\psi = 0 \tag{4-41}$$

This equation may be solved by letting

$$\psi(\mu, \nu, \phi) = M(\mu)N(\nu)\Phi(\phi)$$

Since ϕ enters (4-41) only through $\partial^2\psi/\partial\phi^2$, it is obvious that we may obtain an equation for Φ of the form

$$\frac{d^2\Phi}{d\phi^2} = -\lambda^2\Phi \tag{4-42}$$

where the first separation constant is written as $-\lambda^2$. We may next separate the μ and ν equation, but the two resulting ordinary differential equations are far from simple. Analytic solutions have been obtained for the M functions by Stratton *et al.*[13] and for the N functions by Jaffe[14]; this work is summarized by Bates, Ledsham, and Stewart[15] and by Slater.[9]

[13] J. A. Stratton, P. M. Morse, L. J. Chu and R. A. Hutner, *Elliptic, Cylinder, and Spheroidal Wave Functions*, John Wiley, & Sons, Inc., New York, 1941.

[14] G. Jaffe, *Z. Physik*, **87**, 535 (1934).

[15] D. R. Bates, K. Ledsham, and A. L. Stewart, *Trans. Roy. Soc.*, **246**, 215 (1953).

The quantum number λ in (4-42) will have values just like m_l in (4-6); that is, $\lambda = 0, \pm 1, \pm 2, \ldots$. The complete solutions $\psi(\mu, \nu, \phi)$ are called *molecular orbitals* and they are designated like the corresponding atomic orbitals which would result if the two protons were united at the origin. In addition, each value $|\lambda|$ of the magnitude of λ corresponds to a different state, unlike the atomic case, where the values

$$m_l = l, l-1, \ldots, 0, \ldots, -l+1, -l$$

all belong to the same energy; that is, they are degenerate (in the absence of a magnetic field). In analogy with $s, p, d, \ldots$, the values $|\lambda| = 0, 1, 2, \ldots$ are designated $\sigma, \pi, \delta, \ldots$, and the possible molecular orbitals are then $1s\sigma$, $2s\sigma$, $2p\pi$, $3s\sigma$, $3p\sigma$, $3p\pi$, $3d\sigma$, $3d\pi$, $3d\delta, \ldots$.

To appreciate the significance of these symbols, we may consider in more detail the nature of the nodal surfaces, as discussed by Partington.[16] For the hydrogen atom of Fig. 4-2, we can see that the number of nodal surfaces is related to the quantum numbers in a specific way. Let

n_ϕ = number of nodal meridian planes through the z-axis
n_θ = number of nodal cones with apex at the origin
n_r = number of nodal spheres with center at origin

Then the figure shows that

$$n_\phi = |m_l|, \qquad n_\theta = l - |m_l|, \qquad n_r = n - l - 1 \tag{4-43}$$

PROBLEM 4-14

Verify (4-43) for Fig. 4-2.

We further note that the total number n_n of nodes is

$$n_n = n_r + n_\theta + n_\phi = n - 1 \tag{4-44}$$

In the case of H_2^+, we then specify the quantum numbers in terms of the number of nodes. That is, the principal quantum number n is $(n_n + 1)$, and so on. The nodal surfaces are indicated in Fig. 4-12 in two ways: the upper figure has the z-axis in the plane of the paper; for the lower figure, it is normal. The dotted rectangles are nodal planes. Starting with the state $1s\sigma$, the value $n = 1$ indicates that there are no nodal surfaces. The symmetry

[16] J. R. Partington, *An Advanced Treatise on Physical Chemistry*, Vol. 5, Longmans, Green & Co., Inc., London, 1954.

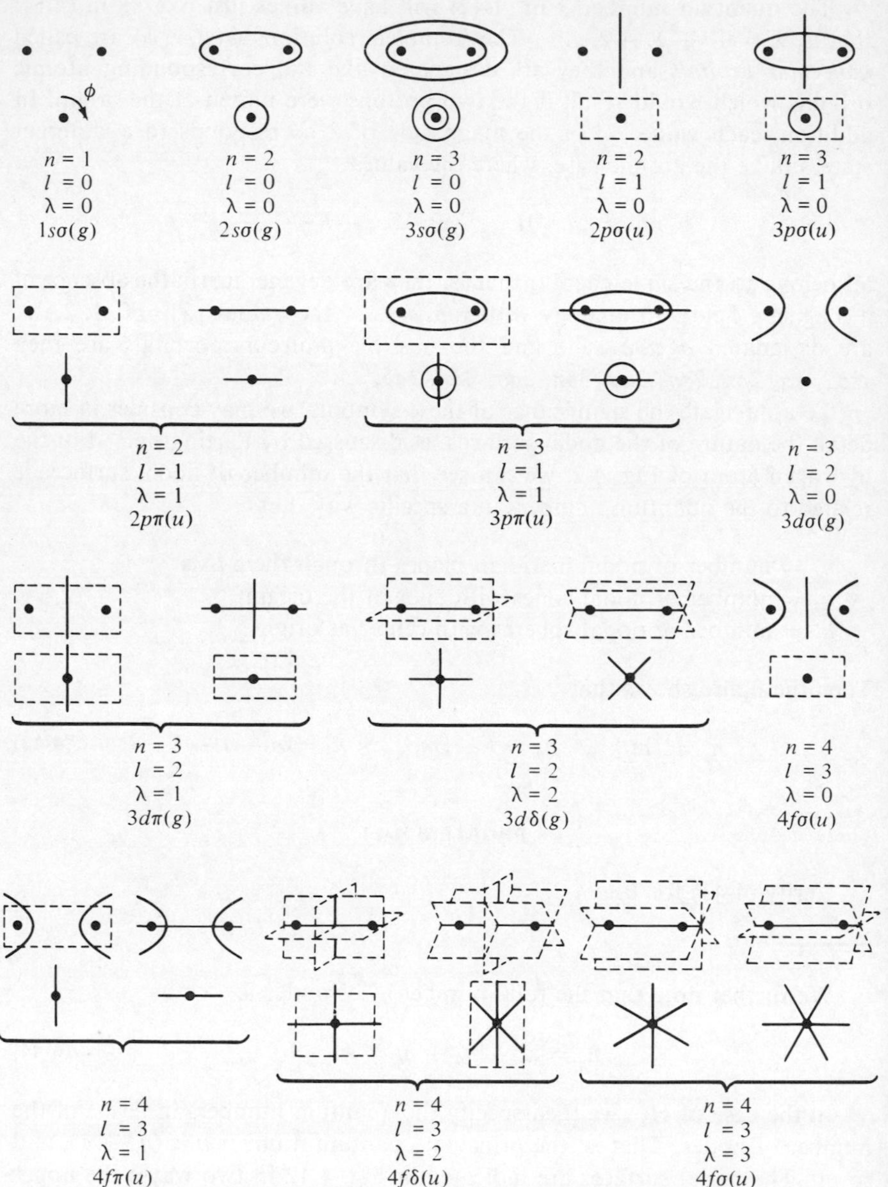

4-12 Nodal surfaces of the molecular orbitals obtained for H_2^+

with respect to the origin (g for even and u for odd) follows the convention introduced in connection with Table 2-13. The next figure, for a $2s\sigma$ electron, corresponds to $n = 2$ or $n_r = 1$, and there is one nodal surface. By (4-43), the azimuthal quantum number l may be expressed as

$$l = n_\theta + n_\phi \tag{4-45}$$

and it expresses the total number of nodal spheres and planes. We would thus regard l for H_2^+ as the sum of the number of hyperboloids and planes. Since $l = 0$ for the $2s\sigma$ electron, the nodal surface is an ellipsoid. Similarly, for the $3s\sigma$ electron, there are two ellipsoids.

For the $2p\sigma$ electron, with one nodal surface, the combination $l = 1$, $\lambda = 0$ means there is no nodal plane, so the surface is a hyperboloid which, however, coincides with the limiting plane bisecting the proton axis. Carrying on in this way, a $2p\pi$ electron has one nodal surface, and, since $\lambda = 1$, it is a plane through the z-axis. Here there are two possibilities, corresponding to the solutions $\sin \phi$ or $\cos \phi$, and the two orientations are shown. The same thing happens, for example, in the $3d\delta$ case, with solutions $\sin 2\phi$ and $\cos 2\phi$.

PROBLEM 4-15

Interpret the $4f\delta$ solution.

As we might expect from the treatment of the hydrogen atom, the solutions involve certain combinations of integers, and these lead to a determination of the allowed values of E. When E is calculated for various values of R, the curves of Fig. 4-13 are obtained. The abscissa used is the ratio of R to a_1, the radius of the first Bohr orbit ($a_1 = 0.5$ A). The ordinate is the ratio of E to E_1, the ionization energy of hydrogen, which is the energy required to remove an electron in the first Bohr orbit from the atom ($-E_1 = 13.6$ eV $= 1$ *rydberg*). We can predict the nature of these curves roughly by realizing that at the left-hand end the energies along the vertical axis are those corresponding to a single electron moving around a charge of magnitude $+2e$; this is equivalent to singly-ionized helium He^+. The energy levels, by (4-14), would be

$$E_n = -\frac{mZ^2e^4}{8n^2h^2\epsilon_0^2} = -13.6\frac{Z^2}{n^2}$$

or

$$\frac{E_n}{E_1} = -\frac{Z^2}{n^2} = -4, -\frac{4}{4}, -\frac{4}{9}, -\frac{4}{16}, \dots \tag{4-46}$$

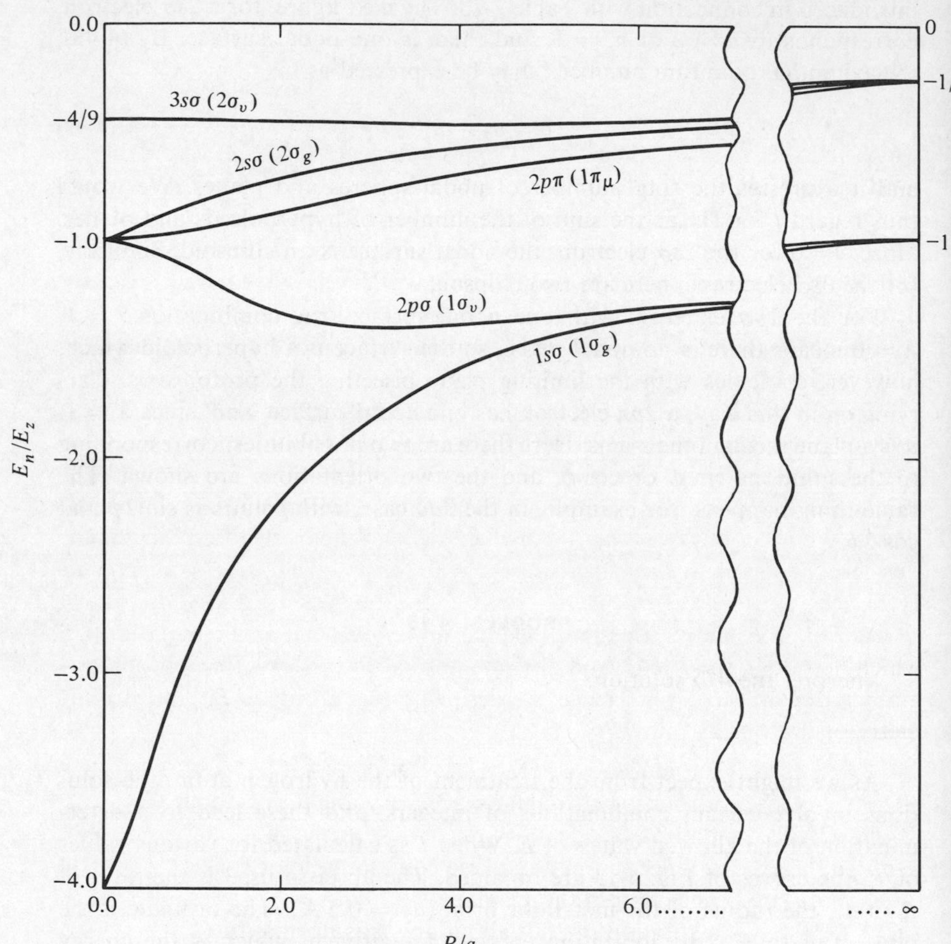

4-13 Transition of energy levels from He^+ (left-hand edge) to H (right-hand edge)

Although we have not proved that Z enters (4-14) as Z^2, it is easy to see why it does. The potential energy of an electron of charge $-e$ in the field of a nucleus of charge Ze will be

$$V = -\frac{Ze^2}{4\pi\epsilon_0 r}$$

which reduces to (4-3) when $Z = 1$. Since e appears in (4-13) as e^4, then Z must appear as Z^2.

On the right-hand side of the graph, the energy levels correspond to those of a single electron moving around a single proton with the other proton at infinity. This is simply the hydrogen atom, with levels as we have previously determined them. Further, the levels on the two sides must be joined in a continuous fashion, for we can start with He^+ and move one proton to infinity in a continuous fashion.

To fully understand the behavior of these curves, let us start at the lowest level on the right, the $1s$ level of hydrogen, which splits into two levels as R/a_1 approaches zero. The lower one is the $1s$ level of He^+, and it is labelled $1s\sigma$, since $\lambda = 0$ for this solution; an alternate notation is shown in parentheses. We note, however that the hydrogen $1s$ level can also join the helium $2p\sigma$ level. To explain this surprising fact, we must consider the symmetry of H_2^+.

As Fig. 4-11 shows, this molecule has an *infinite-fold* axis of rotation C_∞ which lies along the z-axis, and any line normal to this axis through the origin is a dihedral axis C_2. Hence, we designate this group as $D_{\infty h}$, with symmetry operations as follows:

E,	the identity
C_ϕ,	a rotation through an arbitrary angle $\pm\phi$ about C_∞
$C_\phi^2, C_\phi^3, \ldots,$	powers of C_ϕ
$\sigma_v^{(a)}, \sigma_v^{(b)}, \ldots,$	a reflection in any plane containing the C_∞-axis
σ_h,	reflection in the xy-plane
$C_2^{(a)}, C_2^{(b)}, \ldots,$	a rotation of 180° about any axis normal to C_∞
J,	inversion through O

In working out the Cayley table, we also need $S_\phi = \sigma_h C_\phi$, but σ_h drops out, since J has the same effect. There are an infinite number of representations, and the character system is given in Table 4-10. The representations are indicated in two ways; the new notation is based on the use of Greek capital letters to correspond to the lower case Greek letters used for λ of (4-42). The superscripts are added as follows:

Σ^+, symmetric with respect to a plane of symmetry through the molecular axis

Σ^-, antisymmetric with respect to a plane of symmetry through the molecular axis

In order to identify representations with specific molecular orbitals, we

TABLE 4-10 *Character System and Basis Functions for the Molecular Point Group $D_{\infty h}$*

		E	$2C_\phi$	$\cdots$	$\sigma_v^{(a)}$	$\cdots$	J	$2JC_\phi$	$\cdots$	$C_2{}^{(a)}$	$\cdots$	
Σ_g^+	A_{1g}	1	1	$\cdots$	1	$\cdots$	1	1	$\cdots$	1	$\cdots$	$s, x^2 + y^2, z^2$
Σ_u^+	A_{1u}	1	1	$\cdots$	1	$\cdots$	−1	−1	$\cdots$	−1	$\cdots$	z
Σ_g^-	A_{2g}	1	1	$\cdots$	−1	$\cdots$	1	1	$\cdots$	−1	$\cdots$	R_z
Σ_u^-	A_{2u}	1	1	$\cdots$	−1	$\cdots$	−1	−1	$\cdots$	1	$\cdots$	
Π_g	E_{1g}	2	$2\cos\phi$	$\cdots$	0	$\cdots$	2	$2\cos\phi$	$\cdots$	0	$\cdots$	$(xz, yz), (R_x, R_y)$
Π_u	E_{1u}	2	$2\cos\phi$	$\cdots$	0	$\cdots$	−2	$-2\cos\phi$	$\cdots$	0	$\cdots$	(x, y)
Δ_g	E_{2g}	2	$2\cos 2\phi$	$\cdots$	0	$\cdots$	2	$2\cos 2\phi$	$\cdots$	0	$\cdots$	$(x^2 - y^2, xy)$
Δ_u	E_{2u}	2	$2\cos 2\phi$	$\cdots$	0	$\cdots$	−2	$-2\cos 2\phi$	$\cdots$	0	$\cdots$	
$\vdots$	$\vdots$	$\vdots$	$\vdots$	$\vdots$	$\vdots$	$\vdots$	$\vdots$	$\vdots$	$\vdots$	$\vdots$	$\vdots$	

need analytic expressions for these functions and these we do not have. However, we can obtain approximate expressions by assuming the molecular orbitals are equivalent to *linear combinations of atomic orbitals,* abbreviated as the *LCAO method.* If the electron were in a $1s$ orbit around proton 1, we would expect the solution ψ to be very close to the orbital $1s$, since the other proton would have a very small effect. A similar argument holds for proton 2. It therefore seems reasonable to assume (and we shall demonstrate this in Chapter 5) that the molecular orbital is a linear combination of s_1 and s_2, where the subscripts refer to the protons. This symmetry is like that of the H_2O molecule with the oxygen removed, so that the linear combinations are $s_1 + s_2$ which goes with A_{1g} or Σ_g^+ of Table 4-10 and $s_1 - s_2$, which goes with A_{1u} or Σ_u^+. The first representation is symmetric and the second one causes $s_1 - s_2$ to reverse sign under an inversion or a two-fold rotation.

Next, we consider what happens to $s_1 + s_2$ as the protons come closer together. The two functions s_1 and s_2 overlap heavily as shown in Fig. 4-14(a) and eventually reduce to a single one with spherical symmetry. But under the

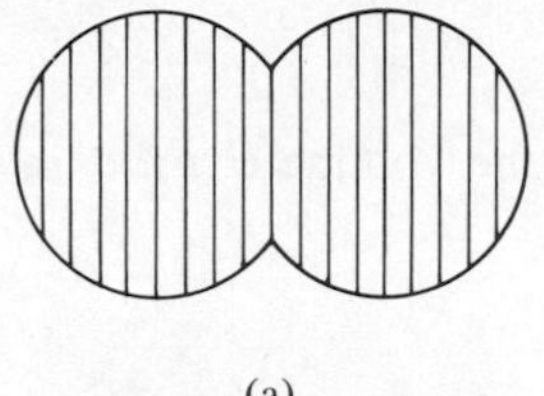

(a)

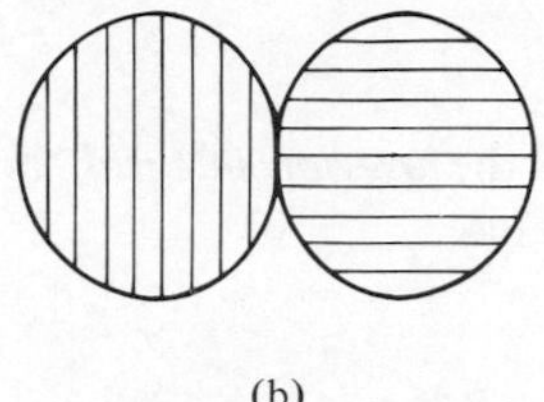

(b)

4-14 (a) The sum of two s functions; (b) the difference of two s functions

same conditions the antisymmetric function $s_1 - s_2$ develops the node shown in Fig. 4-14(b) and this is equivalent to a p function. Thus, an electron in a $1s\sigma$ molecular orbital in H_2^+ which is approximately equal to, and has the same symmetry as $(s_1 + s_2)$ or $(s_1 - s_2)$, will go to either a $1s$ or a $2p$ state in He^+. On the other hand, it can only go to a $1s$ state in H. The remaining curves of Fig. 4-13 are generated in the same way; not all possibilities are shown, however. We note that the molecular orbitals $2s\sigma$, $2p\sigma$, and so on belong to Σ representations in Table 4-10, while the function $2p\pi$ belongs to a Π representation, etc.

4-7 The nature of bonding orbitals

Let us consider in more detail the LCAO method as applied to the singly-ionized hydrogen molecule. We write the approximate molecular orbital solution ψ as the linear combination

$$\psi = N(\psi_1 + \psi_2) \tag{4-47}$$

where ψ_1, ψ_2 are arbitrary atomic orbitals and N is the normalization constant. But we have already seen that an antisymmetric function is also an acceptable solution, so that we have two combinations. We write these as

$$\psi_+ = N_+(\psi_1 + \psi_2) \tag{4-48}$$

and

$$\psi_- = N_-(\psi_1 - \psi_2) \tag{4-49}$$

Squaring (4-48) gives

$$\psi_+^2 = N_+^2(\psi_1^2 + 2\psi_1\psi_2 + \psi_2^2) \tag{4-50}$$

PROBLEM 4-16

Prove that

$$N_+ = (2 + 2S)^{-1/2} \tag{4-51}$$

where

$$S = \int \psi_1\psi_2 \, dV_{r\theta\phi} \tag{4-52}$$

is called the *overlap integral*, and $dV_{r\theta\phi}$ is the volume element in spherical coordinates.

The quantity S in (4-52) must obey the inequality

$$0 < S < 1 \tag{4-53}$$

for it is a measure of the overlap of ψ_1 centered on proton 1 with ψ_2 centered on proton 2. When the two protons are very far apart, the overlap is negligible and S almost vanishes; if they are coincident, $\psi_1 = \psi_2$ and $S = 1$. In a similar way, we find that

$$N_- = (2 - 2S)^{-1/2} \tag{4-54}$$

Now consider a point halfway between the two protons, where $\psi_1 = \psi_2$. Then

$$\psi_+^2 = \frac{(2\psi_1)^2}{(2 + 2S)} = \frac{2\psi_1^2}{1 + S} > \psi_1^2 \tag{4-55}$$

This result indicates that the probability of finding an electron halfway between the two protons is greater than finding it at the same distance from either proton alone—that is, we have a bonding orbital. On the other hand, we may show in the same way that

$$\psi_-^2 < \psi_1^2 \tag{4-56}$$

and this linear combination is thus anti-bonding.

We have already seen that if ψ_1 is chosen as an s atomic orbital, then

$$\psi_+ = N_+(s_1 + s_2), \qquad \psi_- = N_-(s_1 - s_2) \tag{4-57}$$

so that the symmetric combination is bonding and the antisymmetric one is anti-bonding. This in fact is the meaning of the symbols in parentheses for the two lowest curves of Fig. 4-13. The $1\sigma_g$ state of lower energy is symmetric and the higher $1\sigma_u$ state is antisymmetric.

Bonds of the $1s\sigma$(or $1\sigma_g$) type in which the overlap is heavy or the probability is raised along the line joining the protons are called *σ bonds*. We may visualize the formation of the σ bonds or of a σ anti-bonding state as shown in Fig. 4-15(a). However, it is also possible to obtain σ bonding and anti-bonding states from p orbitals: in particular, p_z orbitals if the z-axis is the internuclear axis. This situation is shown in Fig. 4-15(b). (The shape of the orbitals is obtained intuitively.) The upper combination increases the prob-

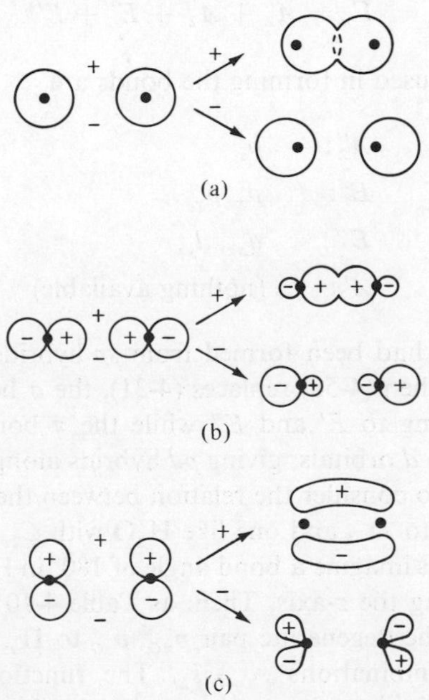

4-15 (a) Formation of a σ bonding (upper) or a σ antibonding (lower) state from s orbitals; (b) formation of a σ bonding or a σ antibonding state from p orbitals; (c) formation of a π bonding or a π antibonding state from p orbitals

ability of finding the electron between the protons and the lower one reduces it; these are the combinations $(p_{z1} + p_{z2})$ generated by the projection operator of the representation A_{1g} or Σ_g^+ of Table 4-10 and $(p_{z1} - p_{z2})$ belonging to A_{1u} or Σ_u^+.

Next, let us ask how p_x or p_y orbitals contribute to the bonding. Schematically, we obtain Fig. 4-15(c), which shows an increase or a decrease in ψ^2 in the normal direction. The upper portion of this sketch shows what is known as a *π bond*; the terms σ bond or π bond correspond to the value of λ in the united atom. We may continue in this fashion, obtaining $\delta, \ldots,$ bonds, but σ and π bonds are generally sufficient for the applications of group theory. We note that π bonds are degenerate pairs of functions belonging to the representation Π_u.

Returning to Table 4-6, we now see that the reducible representation Γ_σ specifies the symmetry of the planar XY_3 molecule when σ bonds only are considered. If π bonds are also possible, then we have a 6×6 reducible representation. Under C_3, all orbitals are shifted so that the character is 0; under C_2, the two orbitals on the C_2 axis reverse sign giving $\chi(C_2) = -2$, etc. The decomposition is then

$$\Gamma_\pi = A_2' + A_2'' + E' + E'' \tag{4-58}$$

so that the orbitals used in forming the bonds are

A_2'':	p_z
E':	p_x, p_y
E'':	d_{xz}, d_{yz}
A_2':	(nothing available)

By (4-31), σ bonds had been formed from *sp* hybrids combining s with p_x and p_y functions. When (4-58) replaces (4-31), the σ bonds are formed from the orbitals belonging to E' and E'' while the π bonds are formed by p_z orbitals and the two d orbitals, giving *pd* hybrids along the z-direction.

It is interesting to consider the relation between the properties of a linear molecule belonging to $D_{\infty h}$ and one like H_2O with C_{2v} symmetry. Following Hochstrasser[17], let us imagine a bond angle of 180° in H_2O; that is, it assumes the form HOH along the z-axis. Then, as Table 4-10 shows, s_O belongs to Σ_g, p_{zO} to Σ_u, and the degenerate pair p_{xO}, p_{yO} to Π_u. For the H atoms, we have the linear combinations $s_1 \pm s_2$. The function $s_1 + s_2$, belonging to Σ_g, combines with s_O to form a bonding orbital, and $s_1 - s_2$, belonging to Σ_u, forms an orbital with $s_1 - s_2$, whereas the pair (p_x, p_y) is nonbonding.

[17] R. N. Hochstrasser, *Molecular Aspects of Symmetry*, W. A. Benjamin, New York, 1966.

Next, let us imagine the bond to be 90° and, further, that the C_2-axis is along the y-direction rather than the z-direction as indicated in Table 4-9 (the x-direction would also be acceptable). Then the combination $s_1 + s_2$ still goes with A_1 but $s_1 - s_2$ now goes with B_1, s_O and p_{yO} go with A_1, p_{zO} goes with B_1, and p_{xO} with B_2. Then the sp_y hybrid belonging to A_1 forms a bond with $s_1 + s_2$ and there is also a lone pair. The p_z orbital bonds with $s_1 - s_2$ and the p_x orbital is nonbonding.

Thus, we realize that as the bond angle is changed continuously from 90° to 180°, there should be a transition or *correlation* between the states corresponding to the two limiting values. We can indicate the correlations in the manner of Table 4-11. If we can estimate the relative values of the energies

TABLE 4-11 ***Correlation of Orbitals for the H_2O Molecule***

D_h			
Representation	O	H	Type
Σ_g	s_O	$s_1 + s_2$	bonding
Σ_u	p_z	$s_1 - s_2$	bonding
Π_u	(p_x, p_y)		nonbonding

C_{2v}			
Representation	O	H	Type
A_1	$\left.\begin{matrix} s \\ p_y \end{matrix}\right\}$	$s_1 + s_2$	$\left\{\begin{matrix} \text{bonding} \\ \text{nonbonding} \end{matrix}\right.$
B_1	p_z	$s_1 - s_2$	bonding
B_2	p_x		nonbonding

at 90° and at 180°, we can draw curves like those of Fig. 4-13, but qualitative in nature. This gives what is known as a *correlation diagram*. Our discussion of Fig. 4-13 indicates that $\sigma_1 + \sigma_2$ corresponds to a lower energy for the hydrogens than $s_1 - s_2$ (which for H_2O is bonding) and both should be lower than the nonbonding state. This is indicated on the right-hand side of Fig. 4-16.

The transition from Π_u to B_2 is the easier one to explain; this is a nonbonding orbital for any angle and the energy should be constant. The other orbital belonging to Π_u, namely p_y, is nonbonding at 180° but gradually overlaps with $s_1 + s_2$ and mixes with s_O as the molecule is bent. Hence,

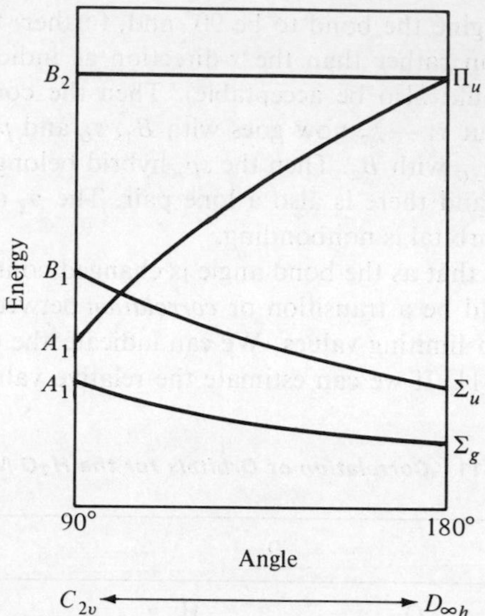

4-16 Correlation diagram for the transition from a 90° to a 180° interbond angle in the H_2O molecule

this portion of the correlation diagram shows a sharp drop in energy. On the other hand, the Σ_u to B_1 transition shows a rise because the s_1 and s_2 clouds, which are of opposite sign, come closer together.

PROBLEM 4-17

What is the energy rise in going from Σ_g to A_1?

This treatment of molecular orbital theory has been highly qualitative, emphasizing the uses of group theory in predicting the nature of electronic bonds. For more extensive treatments of the physical aspect of the subject, we refer the reader to sources such as Streitweiser,[18] who discusses in addition the application of group theory to the LCAO method (which we shall consider in Chapter 5) and to Parr,[19] who (along with Ballhausen and Gray[10]) reprints a number of original papers in this field.

[18] A. Streitweiser, *Molecular Orbital Theory*, John Wiley & Sons, Inc., New York, 1961.

[19] R. G. Parr, *The Quantum Theory of Molecular Electronic Structure*, W. A. Benjamin, New York, 1963.

CHAPTER FIVE

ELECTRONS IN CRYSTALS

5-1 Bragg-von Laue scattering

Consider a plane-wave X-ray striking a crystal lattice at an angle θ to the atomic planes (Fig. 5-1). Let us assume, as Bragg did, that the angle of reflection is the same as the incident angle; that is, the planes act as mirrors. If

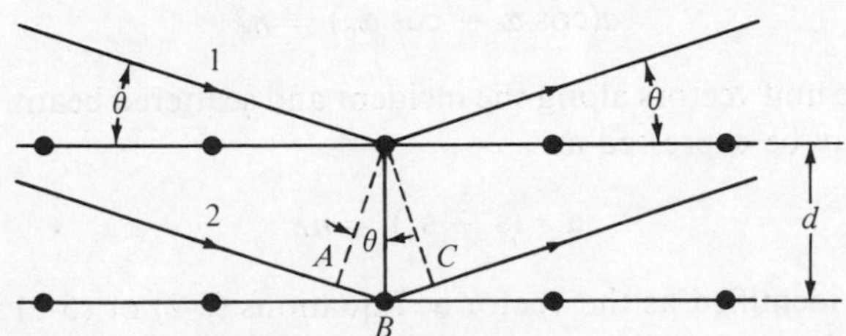

5-1 Derivation of the Bragg law

two rays, labelled 1 and 2, are reflected as shown, then they will reinforce one another in the reflected direction only if they are perfectly in phase; that is, the additional distance *ABC* travelled by ray 2 must be an integer n times the wavelength λ of the beam. Since

$$AB = BC = d \sin \theta$$

where d is the spacing between two adjacent parallel planes, then

$$n\lambda = 2d \sin \theta \tag{5-1}$$

which is the *Bragg Law* for X-ray diffraction.

Although this relation has been experimentally verified, its validity does not depend on the restrictive (and dubious) assumption of mirror-like reflection. It has been shown by von Laue that a more general treatment is possible. Consider the beam of Fig. 5-2 making an angle α_0 to a line of atoms

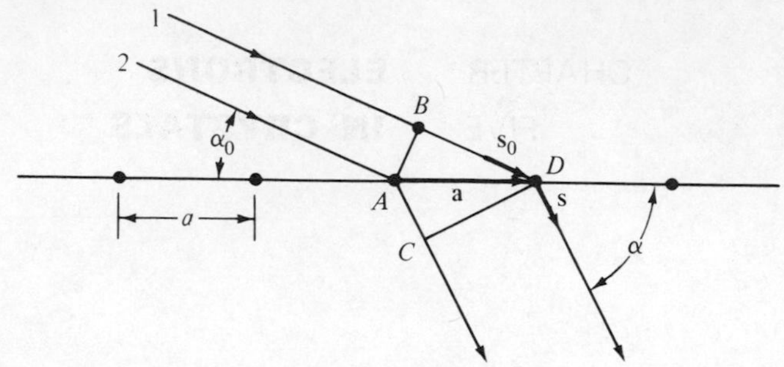

5-2 Derivation of the Bragg-Laue equations

with spacing a. Let the diffracted beam make an angle α to this line. Then the condition for reinforcement of rays 1 and 2 is that the difference in the distances travelled by these two rays as the result of scattering by the atoms is $n\lambda$ (just as for the Bragg law). This difference is determined by the quantity $(\overline{AC} - \overline{BD})$, so that

$$a(\cos \alpha - \cos \alpha_0) = n\lambda \tag{5-2}$$

If $\mathbf{s}_0$ and $\mathbf{s}$ are unit vectors along the incident and scattered beam, respectively, then (5-2) may be expressed as

$$\mathbf{a} \cdot (\mathbf{s} - \mathbf{s}_0) = n\lambda \tag{5-3}$$

where $\overrightarrow{AD}$ is identified as the vector $\mathbf{a}$. Equations (5-2) or (5-3) indicate that the scattered beam may lie anywhere on a cone whose axis is along the line of atoms and whose apex half-angle is α. If the scattering occurs from two intersecting lines of atoms, then reinforcement can occur only where the cones meet; that is, along some sort of line. Similarly, in three dimensions the lines will intersect at points, determined by three equations of the form of (5-3). These will be

$$\begin{aligned} \mathbf{a}_1 \cdot (\mathbf{s} - \mathbf{s}_0) &= n_1\lambda \\ \mathbf{a}_2 \cdot (\mathbf{s} - \mathbf{s}_0) &= n_2\lambda \\ \mathbf{a}_3 \cdot (\mathbf{s} - \mathbf{s}_0) &= n_3\lambda \end{aligned} \tag{5-4}$$

where $\mathbf{a}_1$, $\mathbf{a}_2$, $\mathbf{a}_3$ are vectors joining neighboring atoms along three different lattice directions and n_1, n_2, n_3 are integers. Equations (5-4) are the *Laue equations.*

We may show that these equations are equivalent to the Bragg law by introducing the *Miller indices*. For algebraic convenience, consider a simple cubic lattice, for which

$$a_1 = a_2 = a_3 = a \tag{5-5}$$

The Miller indices h, k, l of an atomic plane are defined as the three smallest integers related to the intercepts m_1a, m_2a, m_3a of this plane along the axis of the cube by the relations

$$h:k:l = \frac{1}{m_1}:\frac{1}{m_2}:\frac{1}{m_3} \tag{5-6}$$

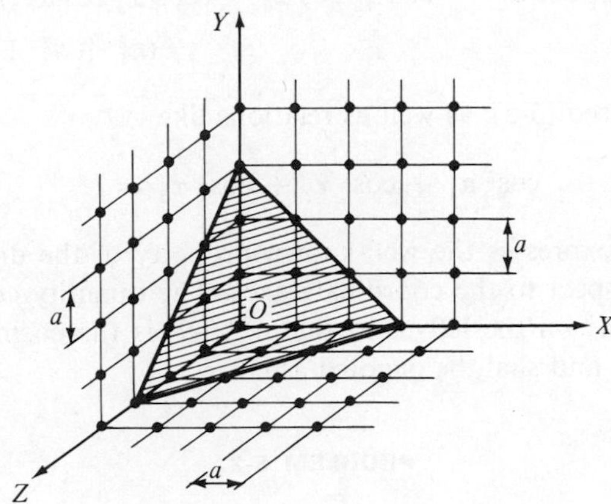

5-3 A (111)-plane

For example, Fig. 5-3 shows intercepts all equal to $3a$, so that $h:k:l = \frac{1}{3}:\frac{1}{3}:\frac{1}{3} = 1:1:1$. This plane is called the (111)-plane, and so are all the planes parallel to it. If the intercepts are $2a$, $1a$, and ∞ (i.e., parallel to the z-axis), then we have a (120)-plane.

PROBLEM 5-1

Consider a simple cubic crystal with the axis of a coordinate system $OXYZ$ along the edges of the cube. Let d be the perpendicular distance from the atom at the origin to the nearest (hkl)-plane. Show that

$$d = \frac{a}{\sqrt{h^2+k^2+l^2}} \tag{5-7}$$

where a is the lattice constant.

Writing (5-4) in scalar form gives

$$
\begin{aligned}
a_1(\cos\alpha_1 - \cos\alpha_{01}) &= n_1\lambda \\
a_2(\cos\alpha_2 - \cos\alpha_{02}) &= n_2\lambda \\
a_3(\cos\alpha_3 - \cos\alpha_{03}) &= n_3\lambda
\end{aligned} \tag{5-8}
$$

Then squaring and adding these three equations, we obtain

$$
\begin{aligned}
2a^2(1 - \cos\alpha_1\cos\alpha_{01} - \cos\alpha_2\cos\alpha_{02} - \cos\alpha_3\cos\alpha_{03}) \\
= (n_1^2 + n_2^2 + n_3^2)\lambda^2
\end{aligned} \tag{5-9}
$$

where we have used (5-5), as well as relations like

$$\cos^2\alpha_1 + \cos^2\alpha_2 + \cos^2\alpha_3 = 1 \tag{5-10}$$

Equation (5-10) expresses the well-known property of the direction cosines of a line with respect to the coordinate axes. The quantity composed of all the negative terms on the left-hand side of (5-9) is the cosine of the angle between $\mathbf{s}_0$ and $\mathbf{s}$ and shall be denoted as $\cos\phi$.

PROBLEM 5-2

Use vector algebra to prove that

$$\cos\phi = \cos\alpha_1\cos\alpha_{01} + \cos\alpha_2\cos\alpha_{02} + \cos\alpha_3\cos\alpha_{03} \tag{5-11}$$

Using Eq. (5-11) in (5-9) gives

$$2a^2(1 - \cos\phi) = (n_1^2 + n_2^2 + n_3^2)\lambda^2 \tag{5-12}$$

The term in parentheses on the left is converted with the identity

$$\sin^2\frac{\phi}{2} = \frac{1 - \cos\phi}{2}$$

Since the angle between $\mathbf{s}_0$ and $\mathbf{s}$ is by definition identical to 2θ of the Bragg law (Fig. 5-4), we see that (5-12) becomes

$$2\sin\theta = \frac{\lambda}{a}\sqrt{n_1^2 + n_2^2 + n_3^2} \tag{5-13}$$

There are some sets of values of the integers n_1, n_2, n_3 in (5-13) which may be written as

$$n_1 = nh, \qquad n_2 = nk, \qquad n_3 = nl \tag{5-14}$$

where n is an arbitrary integer and h, k, l are the Miller indices of an atomic

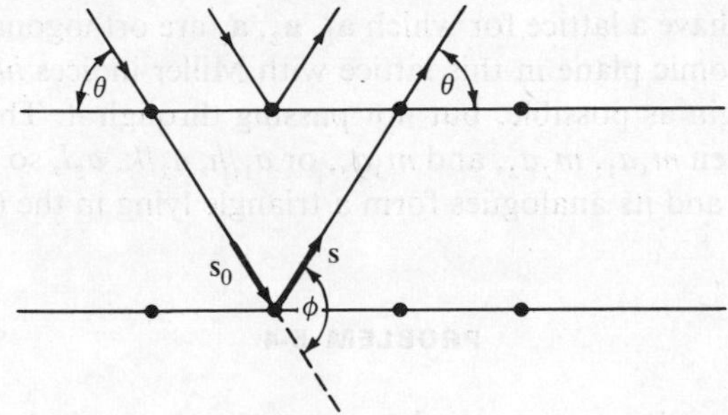

5-4 Connection between Bragg and Bragg-Laue relations

plane. Using (5-14) and (5-7) in (5-13) gives

$$2d_{hkl}\sin\theta = n\lambda \tag{5-15}$$

In this result, the subscripts on d_{hkl} indicate that we are specifying the distance between two adjacent (hkl)-planes. We have thus shown that the Laue and Bragg relations are equivalent.

Let us now introduce the concept of the *reciprocal lattice*. The general crystal lattice can be specified by three noncoplanar vectors $\mathbf{a}_1$, $\mathbf{a}_2$, $\mathbf{a}_3$. We then define three new vectors $\mathbf{b}_1$, $\mathbf{b}_2$, $\mathbf{b}_3$ by the relations

$$\mathbf{a}_i \cdot \mathbf{b}_j = \delta_{ij}, \quad (i, j = 1, 2, 3) \tag{5-16}$$

where the Kronecker delta δ_{ij} is defined as

$$\delta_{ij} = \begin{cases} 1 & \text{if} \quad i = j \\ 0 & \text{if} \quad i \neq j \end{cases}$$

PROBLEM 5-3

(a) Show that $\mathbf{a}_1$ is perpendicular to the plane of $\mathbf{b}_2$ and $\mathbf{b}_3$, and that there are five other relations of this type.

(b) Prove that

$$\mathbf{b}_1 = \frac{\mathbf{a}_2 \times \mathbf{a}_3}{[\mathbf{a}_1 \mathbf{a}_2 \mathbf{a}_3]} \tag{5-17}$$

where the denominator of (5-17) is the scalar triple product or box product.

(c) Show that the reciprocal lattice of a simple cubic (SC) structure is SC, the reciprocal of a body-centered cubic (BCC) is face-centered cubic (FCC), and the reciprocal of a FCC crystal is BCC.

Suppose we have a lattice for which $\mathbf{a}_1$, $\mathbf{a}_2$, $\mathbf{a}_3$ are orthogonal but unequal. Construct an atomic plane in this lattice with Miller indices *hkl* and lying as close to the origin as possible, but not passing through it. The intercepts of this plane are then m_1a_1, m_2a_2, and m_3a_3, or a_1/h, a_2/k, a_3/l, so that the vector $[(\mathbf{a}_1/h) - \mathbf{a}_2/k)]$ and its analogues form a triangle lying in the (*hkl*)-plane.

PROBLEM 5-4

(a) Show that the vector $\mathbf{r}_b$, whose components in *reciprocal space* are **h**, **k**, and **l**, is perpendicular to the (*hkl*)-plane in direct space.

(b) Show that

$$d_{hkl} = \frac{1}{r_b} \tag{5-18}$$

by generalizing (5-6) to an orthogonal lattice with unequal spacing on the three axes.

Wilson has shown that the three Laue equations may be combined into a single equation in reciprocal space.

PROBLEM 5-5

Show that Eqs. (5-4) are equivalent to

$$\frac{\mathbf{s} - \mathbf{s}_0}{\lambda} = (n_1\mathbf{b}_1 + n_2\mathbf{b}_2 + n_3\mathbf{b}_3) \tag{5-19}$$

Since n_1, n_2, n_3 in (5-19) are integers, the quantity on the right must be a vector $\mathbf{r}_R$ in reciprocal space joining two lattice points. It is designated as $\overrightarrow{OS}$ in Fig. 5-5. Equation (5-19) may then be written as

$$\frac{\mathbf{s} - \mathbf{s}_0}{\lambda} = \mathbf{r}_R \tag{5-20}$$

Geometrically, this indicates that constructing a vector $-s_0/\lambda$ from O to a point R (which will not in general be a reciprocal lattice point) permits the direction $\mathbf{s}$ of the diffracted beam to be determined by simply joining R to S, since the vector $\overrightarrow{RS}$ must be identical to $\mathbf{s}/\lambda$. Further, we note that R lies on the perpendicular bisector $\overline{RT}$ of the vector $\overrightarrow{OS}$ or $\mathbf{r}_R$. Hence, we have the

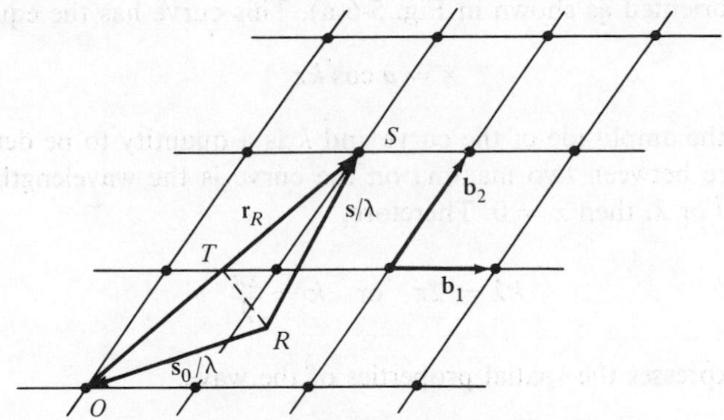

5-5 An Ewald construction

very important conclusion that the Bragg-Laue law may be expressed geometrically in the following way: take any vector joining two points of the reciprocal lattice and construct its perpendicular bisector. If an incident beam in the direct lattice has the proper direction $\mathbf{s}_0$ and wavelength λ so that the vector $-\mathbf{s}_0/\lambda$ joins one reciprocal lattice point to the perpendicular bisector, then the beam will satisfy the Bragg law and its direction of scattering is automatically determined by connecting the perpendicular bisector to the other reciprocal lattice point.

It is customary in solid state and semiconductor theory to change the definition (5-16) slightly and work in *k-space*. We define three vectors $\mathbf{k}_1$, $\mathbf{k}_2$, $\mathbf{k}_3$ by

$$\mathbf{a}_i \cdot \mathbf{k}_j = 2\pi\delta_{ij} \tag{5-21}$$

so that k-space is just like reciprocal space, but the distances are altered by a factor of 2π. The Wilson form (5-20) of the Bragg law is unchanged, since we need merely multiply by 2π to obtain

$$\frac{2\pi\mathbf{s}}{\lambda} - \frac{2\pi\mathbf{s}_0}{\lambda} = 2\pi\mathbf{r}_R \tag{5-22}$$

The quantity $2\pi/\lambda$ appearing in both terms on the left is known in the theory of wave motion as the *propagation constant* or *wave number* k, or

$$k = \frac{2\pi}{\lambda} \tag{5-23}$$

That is, k is proportional to the number of waves per unit length. To see its significance, consider a long piece of wire bent into the shape of a cosine curve and oriented as shown in Fig. 5-6(a). This curve has the equation

$$x = a \cos kz \tag{5-24}$$

where a is the amplitude of the curve and k is a quantity to be determined. The distance between two maxima on the curve is the wavelength, so that when $z = 0$ or λ, then $x = 0$. Therefore,

$$k\lambda = 2\pi \quad \text{or} \quad k = \frac{2\pi}{\lambda}$$

so that k expresses the spatial properties of the wave.

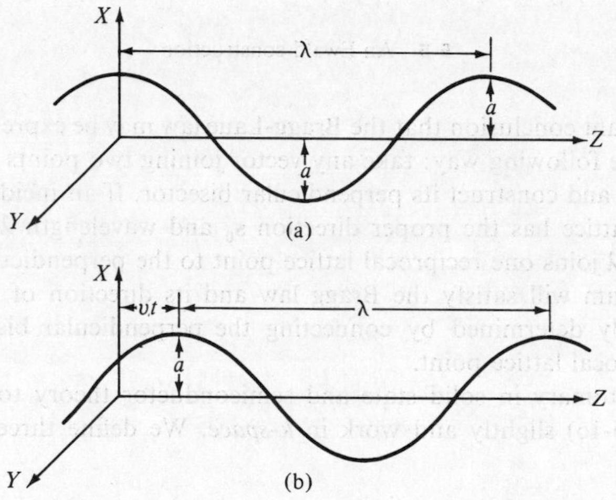

5-6 (a) A cosine wave; (b) a cosine wave in motion

Now let the rigid curve move to the right with velocity v [Fig. 5-6(b)]. At a time t, the entire curve will have moved a distance vt, and Eq. (5-24) is modified to become

$$x = a \cos \left[\frac{2\pi}{\lambda}(z - vt) \right] \tag{5-25}$$

PROBLEM 5-6

(a) Verify that (5-25) describes a cosine wave moving to the right with velocity v.

(b) Show that (5-25) satisfies the partial differential equation

$$v^2 \frac{\partial^2 x}{\partial z^2} = \frac{\partial^2 x}{\partial t^2} \tag{5-26}$$

(c) Show that (5-25) may be written as

$$x = a \cos (kx - \omega t)$$

where $\omega = vk$ is the angular frequency of the wave.

(d) Show that (5-26) is also satisfied by

$$x = ae^{\pm i(kx - \omega t)}$$

In three dimensions, we specify both the direction and the reciprocal wavelength by the *propagation vector* **k**. For the scattered X-ray beam of amplitude **A**, this quantity is defined as

$$\mathbf{k} = \frac{2\pi \mathbf{s}}{\lambda} \tag{5-27}$$

and appears in the solution

$$\mathbf{A} = \mathbf{A}_{\max} e^{i(\mathbf{k} \cdot \mathbf{r} - \omega t)} \tag{5-28}$$

to the three-dimensional wave equation

$$\nabla^2 \mathbf{A} = \frac{1}{v^2} \frac{\partial^2 \mathbf{A}}{\partial t^2} \tag{5-29}$$

where

$$\mathbf{r} = x\mathbf{i} + y\mathbf{j} + z\mathbf{k}$$

PROBLEM 5-7

Show that (5-28) is a solution of (5-29).

Returning to (5-22), we may use (5-23) to obtain

$$k\mathbf{s} - k\mathbf{s}_0 = 2\pi \mathbf{r}_R \tag{5-30}$$

Using the fact that **s** and $\mathbf{s}_0$ are unit vectors and defining a new vector $\mathbf{r}_K$ by

$$\mathbf{r}_K = 2\pi \mathbf{r}_R \tag{5-31}$$

then (5-30) becomes

$$\mathbf{k} - \mathbf{k}_0 = \mathbf{r}_K \tag{5-32}$$

This equation may be put in a form which returns us to the geometric interpretation given in connection with Fig. 5-5.

PROBLEM 5-8

(a) Show that (5-32) is equivalent to

$$2\mathbf{k}_0 \cdot \mathbf{r}_K + \mathbf{r}_K^2 = 0 \tag{5-33}$$

(b) Show that Eq. (5-33) defines a plane generated by a variable vector $\mathbf{k}$ which is the perpendicular bisector of a constant vector $\mathbf{r}_K$.

This problem restates in k-space the result we know to be valid for reciprocal space. That is, $\mathbf{r}_K$ is a vector joining any two lattice points in k-space, as indicated by Fig. 5-7. Any vector $\mathbf{k}_0$ joining O to the plane QR bisecting

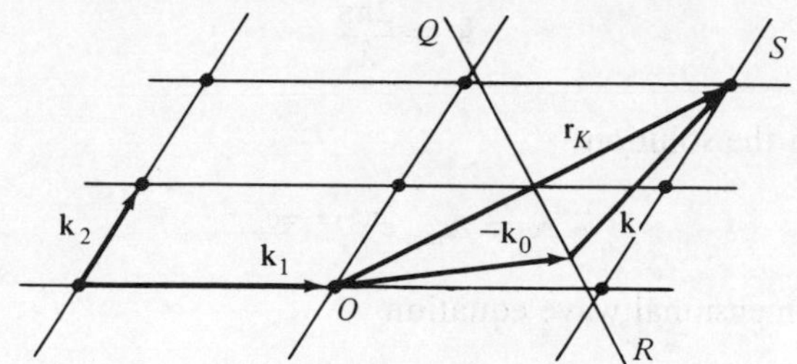

5-7 The Ewald construction in k-space

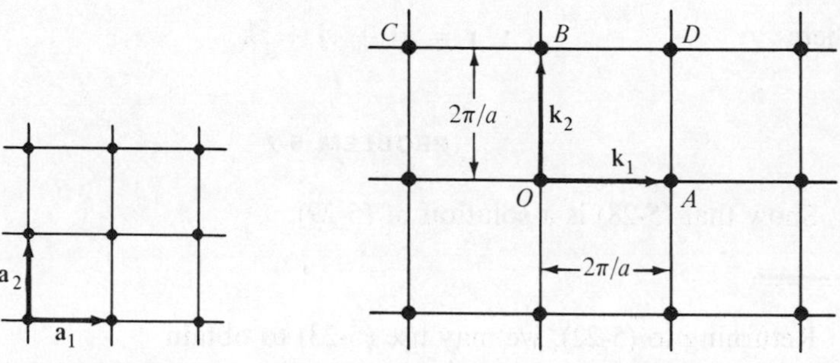

5-8 The square lattice in direct space

5-9 The k-space reciprocal lattice of the previous figure

$\overrightarrow{OS}$, and perpendicular to it, will satisfy the Bragg-Laue law. The vector $\mathbf{k}_0$ has its initial point fixed, but the endpoint may fall anywhere on this plane.

Since these perpendicularly bisecting planes can be constructed for all vectors like $\overrightarrow{OS}$, we shall restrict ourselves to vectors which join a lattice point O with its neighbors. The smallest region enclosed by all such intersecting planes is known as the *first Brillouin zone*; the next smallest region

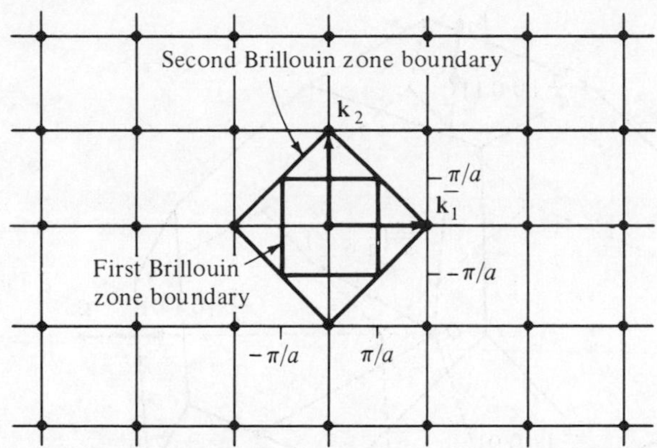

5-10 The first two Brillouin zones of the square lattice

would be the second Brillouin zone, and so forth. Consider as an example the square lattice of Fig. 5-8. The *k*-space vectors, by (5-21), are determined from the equations

$$\begin{aligned} \mathbf{a}_1 \cdot \mathbf{k}_1 &= 2\pi, \qquad & \mathbf{a}_2 \cdot \mathbf{k}_2 &= 2\pi \\ \mathbf{a}_1 \cdot \mathbf{k}_2 &= 0, \qquad & \mathbf{a}_2 \cdot \mathbf{k}_1 &= 0 \end{aligned} \tag{5-34}$$

These equations tell us that $\mathbf{k}_1$, for example, is collinear with $\mathbf{a}_1$ and has a magnitude $\mathbf{k}_1 = 2\pi/a$, where $a = a_1 = a_2$. The reciprocal lattice then looks like the direct lattice; only the size is changed (Fig. 5-10). Since the point O has four nearest neighbors (points such as A and B) and four next-nearest neighbors (such as C and D), the boundaries of the first and second Brillouin zones have the form indicated in Fig. 5-10.

For a three-dimensional FCC crystal, the first Brillouin zone looks like a cube with the corners cut off, as shown in Fig. 5-11. As before, if the propagation vector $\mathbf{k}_0 = 2\pi\mathbf{s}_0/\lambda$ of an X-ray beam connects O with any point on the surface of the zone, the Bragg law will be obeyed.

PROBLEM 5-9

Verify Fig. 5-11. The figures in brackets give the coordinates of the points shown.

PROBLEM 5-10

A hexagonal lattice has a unit cell which appears as shown in Fig. 5-12. The quantities a and c are called *lattice parameters*.

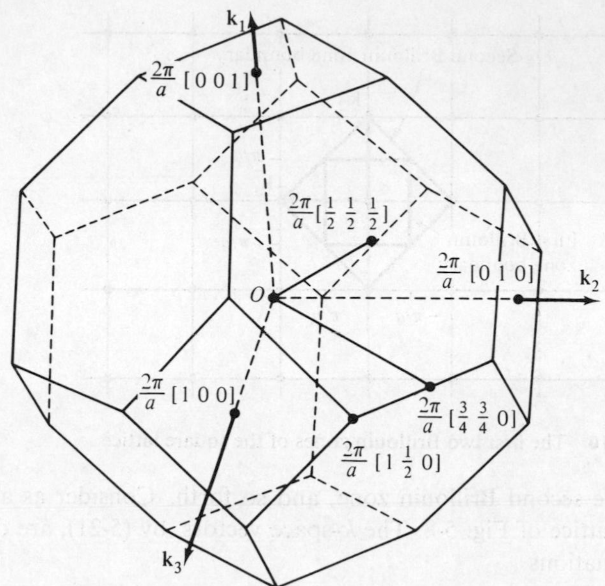

5-11 The first Brillouin zone for the face-centered cubic lattice

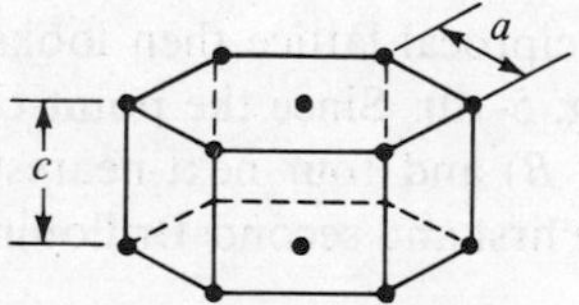

5-12 A hexagonal lattice

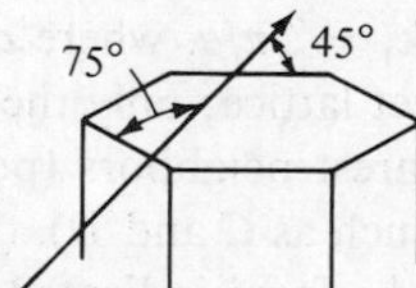

5.13 The incident wave for a hexagonal lattice

(a) By choosing an appropriate set of noncoplanar vectors to specify the geometry of this lattice, find the shape and dimensions of the Brillouin zone in terms of *a* and *c*.

(b) An X-ray beam oriented as shown in Fig. 5-13 is scattered by a crystal with dimensions given by

$$a = 2.0 \times 10^{-10}\ \text{m}, \qquad c = 3.0 \times 10^{-10}\ \text{m}$$

Show on a *k-space diagram* that the Bragg law is satisfied for a wavelength of about 3.3×10^{-10} m, and find the scattering direction from the diagram.

5-2 *Energy bands in solids*

A hydrogen atom is regarded as an electron of negative charge $-e$ circulating around a proton of positive charge e. The force between them, by Coulomb's law, is proportional to $1/r^2$, where r is their separation. The potential energy of the electron may then be expressed as

$$V = -\int F\,dx = C\int \frac{dr}{r^2}$$

where $-C$ represents all the constants in Coulomb's law. Integrating, we see that

$$V = -\frac{C}{r}$$

so that the potential energy for the electron in the hydrogen atom may be pictured as shown in Fig. 5-14.

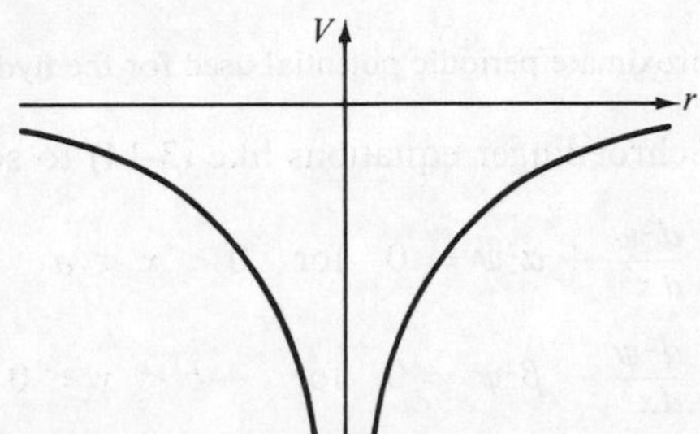

5-14 The potential energy for an electron in the hydrogen atom

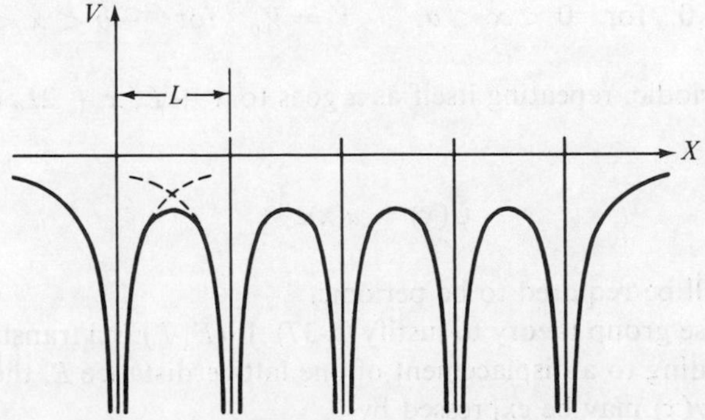

5-15 A one-dimensional hydrogen crystal

If it were possible to make a one-dimensional hydrogen crystal, the potential energy as a function of distance would be expected to appear as shown in Fig. 5-15, where the separation L between atoms is the lattice constant. For mathematical convenience, we first replace the potential energy curve of Fig. 5-15 with an idealized curve of rectangular shape and then shift its position with respect to both the x-axis and the V-axis, as shown in Fig. 5-16.

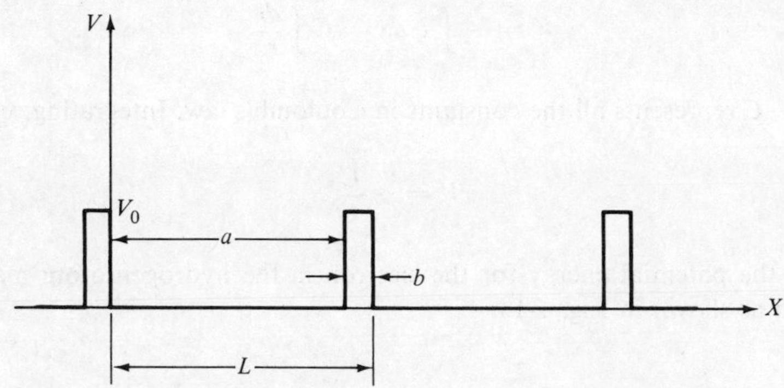

5-16 The approximate periodic potential used for the hydrogen crystal

We now have two Schroedinger equations like (3-14) to solve; namely,

$$\frac{d^2\psi}{dx^2} + \alpha^2\psi = 0 \quad \text{for} \quad 0 < x < a$$

$$\frac{d^2\psi}{dx^2} - \beta^2\psi = 0 \quad \text{for} \quad -b < x < 0 \tag{5-35}$$

where
$$\alpha^2 = \frac{2mE}{\hbar^2}, \qquad \beta^2 = \frac{2m(V_0 - E)}{\hbar^2} \tag{5-36}$$

since $\quad V = 0 \quad \text{for} \quad 0 < x < a, \qquad V = V_0 \quad \text{for} \quad -b < x < a$

Since V is periodic, repeating itself as x goes to $x + L$, $x + 2L$, etc., we will try a solution

$$\psi(x) = u(x)e^{ikx} \tag{5-37}$$

where $u(x)$ will be required to be periodic.

We may use group theory to justify (5-37). If $(E|\,T)$ is a translation operator corresponding to a displacement of one lattice distance L, then the effect of $(E|\,T)$ on $\psi(x)$ may be expressed by

$$(E|\,T)\psi(x) = \psi(x + L)$$

From this, we may write

$$(E|T)^2\psi(x) = \psi(x + 2L)$$

and so on, so that for an arbitrary integer N

$$(E|T)^N\psi(x) = \psi(x + NL) \tag{5-38}$$

If we have a hydrogen crystal with N atoms and of total length NL, then, as in connection with Fig. 1-18, we may require atom number $(N + 1)$ to be

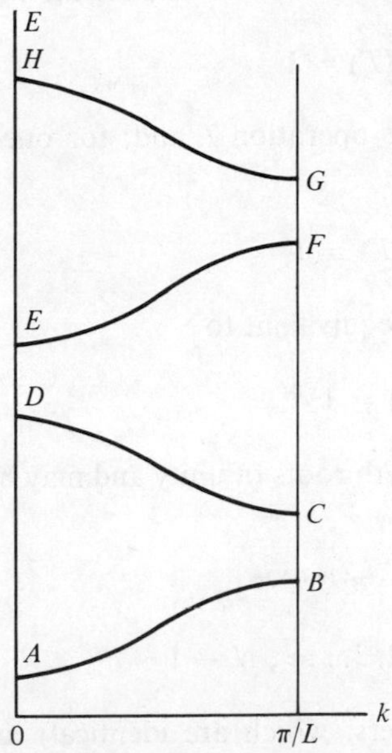

5-17 The reduced zone E vs. k curves

coincident with atom number 1. This was called a *cyclic* or *Born-von Karman boundary condition* in Chapter 1. It may be expressed as

$$\psi(x + NL) = \psi(x) \tag{5-39}$$

which is equivalent to

$$(E|T)^N = (E|0) \tag{5-40}$$

We recall now that a translation group is Abelian; that is,

$$T^mT^n = T^nT^m \tag{5-41}$$

for every pair of integers m and n, where we shorten $(E|T)$ to T. From this, we see that every translation $T, T^2, T^3, \ldots, T^{N-1}, T^N = E$ is in a class by itself, for

$$(T^m)^{-1}T^nT^m = (T^m)^{-1}T^{n+m} = (T^m)^{-1}T^{m+n} = T^n$$

Thus, the translation group has N classes and (as we saw at the beginning of Chapter 2) must therefore have N representations. Let $D(T)$ denote the determinants of these representations, where the symbol D comes from the German word *Darstellung* for representation. These are all one-dimensional, for by (2-66)

$$\sum_{k=1}^{N} d_k^2 = N \tag{5-42}$$

and $d_k = 1$ is the only way of satisfying (5-42). The effect of making the group finite and of order N, as indicated by (5-40), is to impose the condition

$$D^N(T) = \mathbf{I} \tag{5-43}$$

on the matrix representation for the operation T and, for one-dimensional matrices, the condition

$$\chi^N(T) = 1 \tag{5-44}$$

on the characters. Further, (5-44) is equivalent to

$$\chi(T) = 1^{1/N} \tag{5-45}$$

The quantities on the right are the Nth roots of unity and may have complex values, so that for T^m

$$\chi(T^m) = e^{(2\pi il/N)m} \tag{5-46}$$

where
$$m, l = 0, 1, 2, 3, \ldots, N-1 \tag{5-47}$$

The representations (or the characters, which are identical) are shown in Table 5-1. The rows correspond to $m = 0, 1, \ldots, N-1$ and the columns to $l = 0, 1, \ldots, N-1$.

PROBLEM 5-11

According to (2-153), any two rows (or columns) of Table 5-1 should have a scalar product which vanishes, and the scalar product of any row (or column) with itself should be equal to the order of the group. Show that (2-153) is *not* obeyed, but that it would be if one of the two representations involved in the scalar product were replaced by its complex conjugate. (This seems reasonable, since the elements of Table 5-1 are complex quantities.)

We have thus shown that the character $\chi^{(l)}(T^m)$ for the lth representation and the operation T^m is

$$\chi^{(l)}(T^m) = e^{2\pi iml/N} \tag{5-48}$$

To connect these results with (5-37), let us evaluate $\psi(x)$ at $x + mL$, where m is the integer defined by (5-47). Then

$$\psi(x + mL) = u(x + mL)e^{ik(x+mL)} \tag{5-49}$$

TABLE 5-1 ***Representations (or Characters) for a One-Dimensional Cyclic Translation Group***

	E	T	T^2	$\cdots$	T^{N-1}
Γ_0	1	1	1	$\cdots$	1
Γ_1	1	$e^{2\pi i/N}$	$e^{2(2\pi i/N)}$	$\cdots$	$e^{(N-1)(2\pi i/N)}$
Γ_2	1	$e^{2(2\pi i/N)}$	$e^{2\{2(2\pi i/N)\}}$	$\cdots$	$e^{(N-1)\{2(2\pi i/N)\}}$
$\vdots$	$\vdots$	$\vdots$	$\vdots$	$\cdots$	$\vdots$
Γ_{N-1}	1	$e^{(N-1)(2\pi i/N)}$	$e^{(N-1)\{2(2\pi i/N)\}}$	$\cdots$	$e^{(N-1)^2(2\pi i/N)}$

The function $\psi(x + lL)$ will be identical to $\psi(x)$—that is, $\psi(x)$ will have the lattice periodicity—provided

$$u(x + mL) = u(x) \tag{5-50}$$

and

$$e^{ik(x+mL)} = e^{ikx} \tag{5-51}$$

We satisfy (5-50) by simply requiring $u(x)$ to have the same periodicity as $\psi(x)$ or $V(x)$. To make (5-51) valid, we see that the relation

$$e^{ikmL} = 1 \tag{5-52}$$

must be true. But (5-52) and (5-46) may be combined to give

$$ikmL = 2\pi ilmN$$

so that (5-52) holds if

$$k = \frac{2\pi l}{NL} \tag{5-53}$$

Using (5-23), we may interpret this to mean that when a crystal of length NL

is bent round on itself, it will sustain a wave which just fits this circular structure. This, of course, is what we need for constructive interference. Equation (5-37) is known as the *Bloch theorem* in solid state theory and as *Floquet's theorem* in mathematical physics.

To apply (5-37), we substitute it into (5-35), obtaining

$$\begin{aligned} \frac{d^2u}{dx^2} + 2ik\frac{du}{dx} + (\alpha^2 - k^2)u = 0, \quad (0 < x < a) \\ \frac{d^2u}{dx^2} + 2ik\frac{du}{dx} - (\beta^2 + k^2)u = 0, \quad (-b < x < 0) \end{aligned} \tag{5-54}$$

The solutions to these equations can be verified to be

$$\begin{aligned} u_1 &= Ae^{i(\alpha-k)x} + Be^{-i(\alpha+k)x} \\ u_2 &= Ce^{(\beta-ik)x} + De^{-(\beta+ik)x} \end{aligned} \tag{5-55}$$

We impose the boundary conditions

$$u_1(0) = u_2(0), \qquad \frac{du_1(0)}{dx} = \frac{du_2(0)}{dx} \tag{5-56}$$

Since $u(k)$ is required to be periodic, we have, in addition

$$u_1(a) = u_2(-b), \qquad \frac{du_1(a)}{dx} = \frac{du_2(-b)}{dx} \tag{5-57}$$

Inserting Eq. (5-55) into Eqs. (5-56) and (5-57) gives the following four homogeneous equations for A, B, C, and D

$$\begin{aligned} &A + B = C + D \\ &i(\alpha - k)A - i(\alpha + k)B = (\beta - ik)C - (\beta + ik)D \\ &e^{i(\alpha-k)a}A + e^{-i(\alpha+k)a}B = e^{-(\beta-ik)b}C + e^{(\beta+ik)b}D \\ &i(\alpha - k)e^{i(\alpha-k)a}A - i(\alpha + k)e^{-i(\alpha+k)a}B \\ &\qquad = (\beta - ik)e^{-(\beta-ik)b}C - (\beta + ik)e^{(\beta+ik)b}D \end{aligned} \tag{5-58}$$

and expanding the determinant of the coefficients after equating it to zero, gives

$$\frac{\beta^2 - \alpha^2}{2\alpha\beta}\sinh\beta b\sin\alpha a + \cosh\beta b\cos\alpha a = \cos k(a + b) \tag{5-59}$$

All quantities involved in Eq. (5-59) except E (through α and β) and k are already determined. Since $\cos kL$ on the right-hand side can only assume values from -1 to 1, it follows that a value for E exists only when the left-hand side, which we denote by $L(E)$, lies in the same range. Figure 5-18(a) is a plot of the left-hand side of Eq. (5-59) as a function of E with V_0 chosen to have the numerical value

$$\frac{mV_0a^2}{\hbar^2} = 72$$

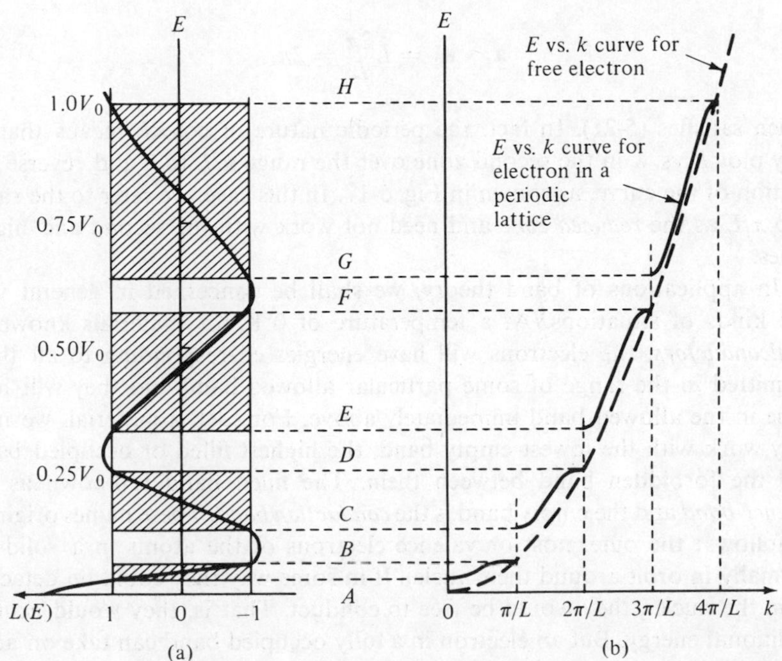

5-18 (a) Allowed and forbidden bands for a one-dimensional crystal; (b) E vs. k curve for the periodic crystal and for a free electron

In addition, we shall let $b/a = \frac{1}{24}$. The shaded areas of the graph show the ranges of E for which values exist. These ranges are *allowed energy bands*, and the unshaded regions are *forbidden energy bands*. According to the figure, the allowed bands have the approximate limits

$$0.03V_0 \leqq E \leqq 0.06V_0$$
$$0.13V_0 \leqq E \leqq 0.25V_0$$
$$0.33V_0 \leqq E \leqq 0.57V_0$$
$$0.64V_0 \leqq E \leqq 0.99V_0$$

We can obtain another important way of representing energy bands by realizing that (5-59) can be solved for E as a function of k. Since $\cos k(a + b) = \cos kL$ has the value $+1$ when $k = 0$ and -1 when $k = \pi/L$, then each value of k in the range 0 to π/L corresponds to a distinct value of E, as shown in Fig. 5-18(b). (We shall consider the dashed curve in this figure in connection with Problem 5-15.)

We also realize that the range 0 to π/L corresponds to the first Brillouin zone for a one-dimensional lattice of lattice constant L, for

$$\mathbf{a}_1 \cdot \mathbf{k}_1 = L\frac{2\pi}{L} = 2\pi$$

which satisfies (5-21). In fact, the periodic nature of $\cos kL$ means that we may plot E vs. k in the second zone over the range π/L to 0 and reverse this portion of the curve, as shown in Fig. 5-17. In this case, we refer to the range 0 to π/L as the *reduced zone* and need not work with the second and higher zones.

In applications of band theory, we shall be concerned in general with two kinds of situations. At a temperature of 0°K for materials known as *semiconductors*, the electrons will have energies corresponding to all those permitted in the range of some particular allowed band and they will have none in the allowed band immediately above. For such a material, we need only work with the lowest empty band, the highest filled or occupied band, and the forbidden band between them. The filled band is known as the *valence band* and the empty band is the *conduction band*. These names originate as follows: the outermost or valence electrons of the atoms in a solid are normally in orbit around their nuclei. If in some way they could be detached from the nuclei, they would be free to conduct. That is, they would acquire additional energy. But an electron in a fully occupied band can take on additional energy only by going up to the next higher band, since there are no available energies in a full band. Hence, the empty band corresponds to free or conduction electrons and the filled band to the bound or valence electrons.

Materials for which some allowed band contains only a portion of all the electrons that can occupy it are *conductors* and the highest occupied energy within this band is the *Fermi energy* or *Fermi level*. The Fermi energy in a semiconductor generally lies in the forbidden band. The reasons for this are discussed in many texts on semiconductor physics.

5-3 Brillouin zones in three dimensions

We have used two types of energy band diagrams in this chapter; one is like Fig. 5-18(a), which simply shows allowed and forbidden bands, and the other is like Fig. 5-18(b), which not only shows the bands but also the way in which

E depends on k. We would next like to inquire about the nature of the band structure and the E vs. k curves for a more realistic situation. That is, what happens in a three-dimensional periodic lattice? An exact answer to this question is impossible but an approximate one may be obtained by a method of Armstrong.[1] The three-dimensional Schroedinger equation, which is

$$-\frac{\hbar^2}{2m}\nabla^2\psi + V\psi = E\psi \tag{4-2}$$

where
$$\nabla^2 = \frac{\partial^2}{\partial x^2} + \frac{\partial^2}{\partial y^2} + \frac{\partial^2}{\partial z^2} \tag{5-60}$$

may be converted into a different form by using a Taylor expansion. A function $f(x, y, z)$ at the points $(x \pm a), y, z$ is expressed as

$$f(x \pm a, y, z) = f(x, y, z) \pm a\frac{\partial}{\partial x}f(x, y, z) + \frac{a^2}{2}\frac{\partial^2}{\partial x^2}f(x, y, z) \pm \cdots \tag{5-61}$$

Adding the expression for $f(x + a, y, z)$ to that for $f(x - a, y, z)$ gives

$$\frac{\partial^2}{\partial x^2}f(x, y, z) = \frac{f(x + a, y, z) + f(x - a, y, z) - 2f(x, y, z)}{a^2} \tag{5-62}$$

When we apply this result to terms such as $\partial^2\psi/\partial x^2$ in the Schroedinger equation, we must bear in mind that the Taylor expansion is valid to second order, so that we are assuming that higher derivatives are small. We have seen for the harmonic oscillator, for example, that ψ can oscillate quite rapidly within a short distance of the force center (Fig. 3-1). Hence, for the approximation of (5-62) to be valid, the electron must be very close to the attractive center, and when we apply this to a crystal, the electron must be close to the atoms. This is what is known as the *tight-binding approximation.* Our present treatment of the tight-binding approximation is extremely intuitive; a more elaborate discussion will be given in Sec. 5-9.

For simplicity, we shall consider a simple cubic crystal with primitive vectors $\mathbf{a}_1 = \mathbf{a}_2 = \mathbf{a}_3$, where $a_1 = a_2 = a_3 = a$. Let the potential at the lattice points be alternatively V_1 and V_2. To solve the wave equation, we shall use the three-dimensional form of (5-37), which has the form

$$\psi(\mathbf{r}, \mathbf{k}) = u(\mathbf{r})e^{i\mathbf{k}\cdot\mathbf{r}} \tag{5-63}$$

We recognize this as correct, since it reduces to (5-37) when $\mathbf{r} = x\mathbf{i}$ and $\mathbf{k} = k\mathbf{i}$. Further, let $u = u_1$ when $V = V_1$ and $u = u_2$ when $V = V_2$ When the potential is V_1, the Schroedinger equation becomes

[1] H. L. Armstrong, *Amer. J. Phys.*, **35**, 328 (1967).

$$-\frac{\hbar^2}{2m}\left(\frac{\partial^2}{\partial x^2}+\frac{\partial^2}{\partial y^2}+\frac{\partial^2}{\partial z^2}\right)\psi + V_1\psi = E\psi \tag{5-64}$$

But by (5-62)

$$\frac{\partial^2\psi}{\partial x^2} = \frac{\psi(x+a, y, z) + \psi(x-a, y, z) - 2\psi(x, y, z)}{a^2}$$

where
$$\psi(x, y, z) = u_1 e^{i\mathbf{k}\cdot\mathbf{r}}$$
and
$$\psi(x \pm a, y, z) = u_2 e^{i\mathbf{k}\cdot(\mathbf{r}\pm\mathbf{a}_1)} = u_2 e^{i\mathbf{k}\cdot\mathbf{r}} e^{\pm i\mathbf{k}\cdot\mathbf{a}_1}$$

with corresponding expressions for the other derivatives. Substituting these six relations back into (5-64) gives

$$-\frac{\hbar^2}{2ma^2}[e^{i\mathbf{k}\cdot\mathbf{a}_1} + e^{-i\mathbf{k}\cdot\mathbf{a}_1} + e^{i\mathbf{k}\cdot\mathbf{a}_2} + e^{-i\mathbf{k}\cdot\mathbf{a}_2} + e^{i\mathbf{k}\cdot\mathbf{a}_3} + e^{-i\mathbf{k}\cdot\mathbf{a}_3}]u_2 e^{i\mathbf{k}\cdot\mathbf{r}}$$
$$- 6u_1 e^{i\mathbf{k}\cdot\mathbf{r}} + V_1 u_1 e^{i\mathbf{k}\cdot\mathbf{r}} = E u_1 e^{i\mathbf{k}\cdot\mathbf{r}}$$

or

$$\left[\frac{2ma^2}{\hbar^2}(E - V_1) - 6\right]u_1 + 2[\cos(\mathbf{k}\cdot\mathbf{a}_1) + \cos(\mathbf{k}\cdot\mathbf{a}_2) + \cos(\mathbf{k}\cdot\mathbf{a}_3)]u_2 = 0 \tag{5-65}$$

When $V = V_2$, a similar calculation gives

$$2[\cos(\mathbf{k}\cdot\mathbf{a}_1) + \cos(\mathbf{k}\cdot\mathbf{a}_2) + \cos(\mathbf{k}\cdot\mathbf{a}_3)]v_1 + \left[\frac{2ma^2}{\hbar^2}(E - V_2) - 6\right]u_2 = 0 \tag{5-66}$$

For these two equations to have consistent solutions, the determinant of the coefficients in square brackets must vanish, giving

$$E = \frac{1}{2}(V_1 + V_2) + \frac{3\hbar^2}{ma^2} \pm \left[\frac{1}{4}(V_1 - V_2)^2 + \frac{\hbar^4}{m^2a^4}\{\cos(\mathbf{k}\cdot\mathbf{a}_1) + \cos(\mathbf{k}\cdot\mathbf{a}_2) + \cos(\mathbf{k}\cdot\mathbf{a}_3)\}^2\right]^{1/2} \tag{5-67}$$

Equation (5-67) may be interpreted graphically by considering its two-dimensional form. For a given value of E, V_1, and V_2, let us determine how the variable $\mathbf{k}\cdot\mathbf{a}_1$, or k_1a, depends on the variable $\mathbf{k}\cdot\mathbf{a}_2 = k_2a$. Setting the term in k_3a equal to zero, and rearranging (5-67), we obtain

$$\cos(k_1a) + \cos(k_2a) = E_r \tag{5-68}$$

where E_r is the constant that is obtained by consolidating all terms in E, V_1, V_2, m, $\hbar$, and a. Since E_r is obtained from E essentially by a shift in scale, it is called a *relative* or *reduced energy* (see Sec. 5-9). Equation (5-68) is the trace in the (k_1a, k_2a)-plane of the *constant energy surfaces*. A plot of (k_1a) vs. (k_2a) for values of C from 2.0 to -2.0 is shown in Fig. 5-19. Near the origin, these surfaces are circles with center at the origin, and at the upper right-hand corner, they are circles centered on the corner. We may picture them in three dimensions as shown in Fig. 5-20, and as we expect, there is a sphere at each corner. The interior surfaces are quite complicated.

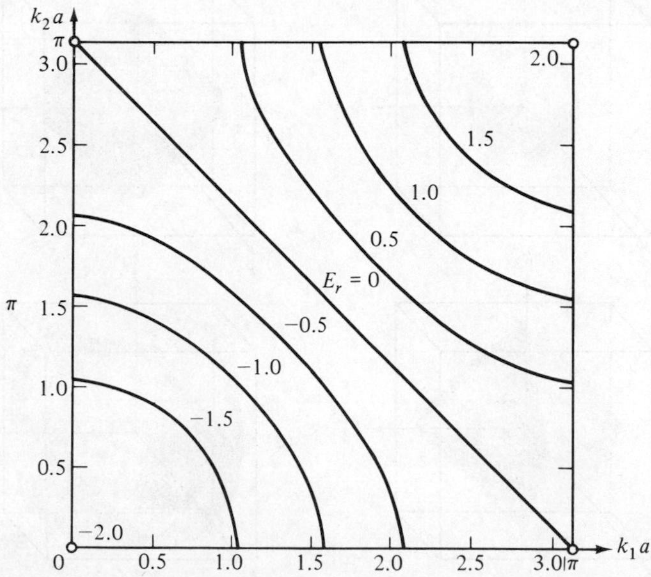

5-19 Constant energy curves for the tight-binding approximation

PROBLEM 5-12

Let $V_1 = -6.0$ eV, $V_2 = -4.0$ eV, and $a = 3.0$ A. Find the E vs. k curves for the tight-binding approximation. (*Hint:* Let $k_2 = k_3 = 0$.)

5-4 Free electron energy bands

If we assume that an electron moving in a crystal sees a constant potential energy to a first approximation, then we may take this constant value as zero without loss of generality. This is called the *free electron* or *empty lattice* approximation. The solution of the wave equation under these conditions is a simple matter and it would appear at first glance that the approximation would be so unrealistic as to be useless. It turns out, however, that the information obtained in this way is a great help in interpreting data obtained by more elaborate methods, and we shall show how this comes about by using the development of Jones.[2]

We try a solution of Bloch form

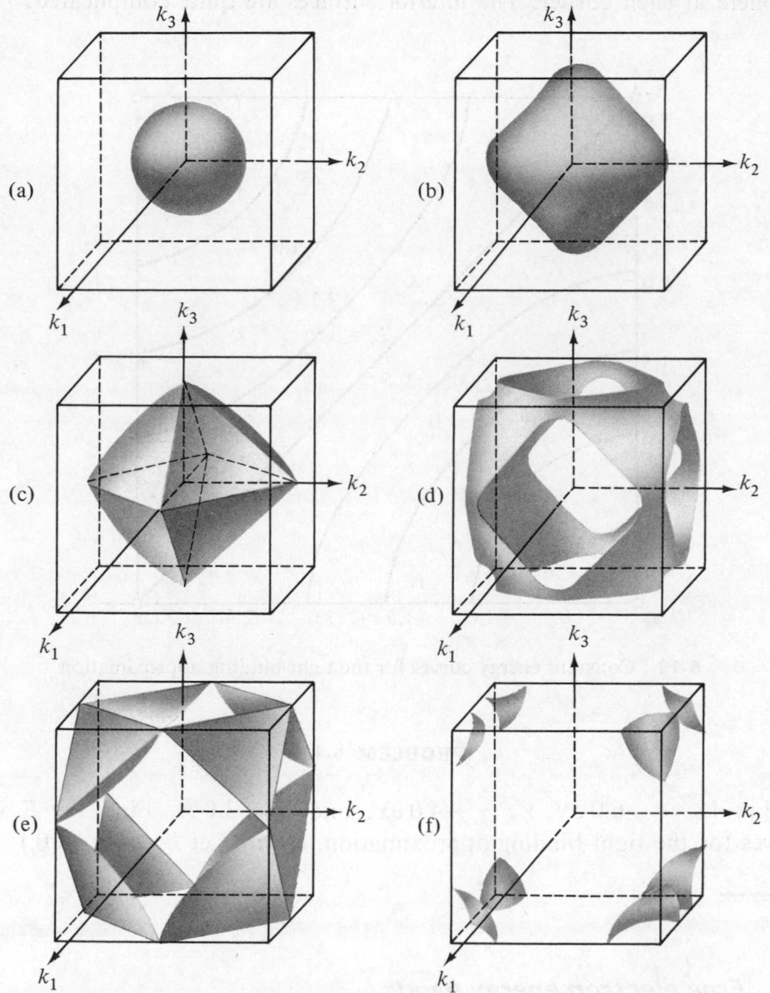

5-20 Constant energy surfaces for the tight-binding approximation

[2]H. Jones, *The Theory of Brillouin Zones and Electronic States in Crystals*, North-Holland Publishing Co., Amsterdam, 1960.

$$\psi(\mathbf{r}, \mathbf{k}) = u(\mathbf{r})e^{i\mathbf{k}\cdot\mathbf{r}} \tag{5-63}$$

where ψ is a function of both $\mathbf{r}$ and $\mathbf{k}$. Substituting (5-53) into (4-2), we obtain

$$\nabla^2 u + 2i\mathbf{k}\cdot\nabla u + \frac{2m}{\hbar^2}\left[E - \frac{\hbar^2 k^2}{2m} - V(\mathbf{r})\right]u = 0 \tag{5-69}$$

and for $V(\mathbf{r}) = 0$, this equation reduces to one of second order with constant coefficients, so that it has exponential solutions.

PROBLEM 5-13

Verify (5-69).

A convenient form in which to express the solutions to (5-69) when $V(\mathbf{r}) = 0$ comes by considering the geometry of the crystal lattice. As in (5-16), let $\mathbf{a}_1$, $\mathbf{a}_2$, $\mathbf{a}_3$ be three noncoplanar vectors specifying the direct lattice. Then the direct lattice is mapped out by the points $(n_1\mathbf{a}_1 + n_2\mathbf{a}_2 + n_3\mathbf{a}_3)$, where n_1, n_2, n_3 are integers which may be positive, negative, or zero. Each of the vectors $\mathbf{a}_i$ can be expressed in terms of its Cartesian coordinates as

$$\mathbf{a}_i = a_{ix}\mathbf{i} + a_{iy}\mathbf{j} + a_{iz}\mathbf{k}, \quad (i = 1, 2, 3)$$

so that the lattice points are given by the vectors

$$\begin{aligned} n_1\mathbf{a}_1 + n_2\mathbf{a}_2 + n_3\mathbf{a}_3 &= (n_1a_{1x} + n_2a_{2x} + n_3a_{3x})\mathbf{i} \\ &\quad + (n_1a_{1y} + n_2a_{2y} + n_3a_{3y})\mathbf{j} + (n_1a_{1z} + n_2a_{2z} + n_3a_{3z})\mathbf{k} \end{aligned}$$

or, in matrix form,

$$\begin{pmatrix} a_{1x} & a_{2x} & a_{3x} \\ a_{1y} & a_{2y} & a_{3y} \\ a_{1z} & a_{2z} & a_{3z} \end{pmatrix}\begin{pmatrix} n_1 \\ n_2 \\ n_3 \end{pmatrix} = \mathbf{An} \tag{5-70}$$

The reciprocal lattice vectors $\mathbf{b}_1$, $\mathbf{b}_2$, $\mathbf{b}_3$ are determined by the definition

$$\mathbf{a}_i \cdot \mathbf{b}_j = \delta_{ij} \tag{5-16}$$

from which it follows that a matrix $\mathbf{B}$ exists which satisfies the relation

$$(m_1 \quad m_2 \quad m_3)\begin{pmatrix} b_{1x} & b_{1y} & b_{1z} \\ b_{2x} & b_{2y} & b_{2z} \\ b_{3x} & b_{3y} & b_{3z} \end{pmatrix} = \mathbf{mB} \tag{5-71}$$

where **m** is the row matrix

$$\mathbf{m} = (m_1 \quad m_2 \quad m_3) \tag{5-72}$$

and m_1, m_2, m_3 are integers. The direct and reciprocal lattices are related by

$$\mathbf{BA} = \mathbf{I} \tag{5-73}$$

where **I** is the unit matrix. (See Problem 5-14.) The solutions to (5-69) can now be expressed in terms of these matrices. Since the Bloch theorem requires $u(\mathbf{r})$ to have the periodicity of the lattice, we see that we may impose the condition

$$u(\mathbf{r} + \mathbf{An}) = u(\mathbf{r})$$

The exponential function satisfying this condition is

$$u(\mathbf{r}) = e^{\pm 2\pi i \mathbf{mB}\cdot\mathbf{r}}$$

for
$$u(\mathbf{r} + \mathbf{An}) = e^{\pm 2\pi i \mathbf{mB}\cdot(\mathbf{r}+\mathbf{An})}$$

Since the multiplication of matrices is associative, then

$$\mathbf{mB}\cdot\mathbf{An} = \mathbf{m}\cdot(\mathbf{BA})\mathbf{n} = \mathbf{m}\cdot\mathbf{n}$$

by (5-73) and the quantity $\mathbf{m}\cdot\mathbf{n}$ reduces to a single integer. Hence

$$e^{\pm 2\pi i \mathbf{m}\cdot\mathbf{n}} = 1$$

and we are left with $\quad u(\mathbf{r} + \mathbf{An}) = e^{\pm 2\pi i \mathbf{mB}\cdot\mathbf{r}} = u(\mathbf{r})$

as we wished to show. It is convenient to use the exponential with the negative sign, so that

$$u(\mathbf{r}) = e^{-2\pi i \mathbf{mB}\cdot\mathbf{r}} \tag{5-74}$$

PROBLEM 5-14

(a) Use (5-16) to establish (5-73).
(b) Substitute (5-74) in (5-69) to show that

$$E = \frac{\hbar^2}{2m}(\mathbf{k} - 2\pi\mathbf{mB})^2 \tag{5-75}$$

Equation (5-75) gives the energies for the free electron and the corresponding wave functions are

$$\psi(\mathbf{r}, \mathbf{k}) = u(\mathbf{r})e^{i\mathbf{k}\cdot\mathbf{r}} = e^{i(\mathbf{k}-2\pi\mathbf{mB})\cdot\mathbf{r}} \tag{5-76}$$

Let us apply these results to a simple cubic crystal as an initial example. It is customary to place the primitive vectors of the cubic crystal parallel to the coordinate axes. Then, as Figs. 5-8 and 5-9 indicate, the corresponding vectors $\mathbf{k}_1$, $\mathbf{k}_2$, $\mathbf{k}_3$ in k-space are collinear with $\mathbf{a}_1$, $\mathbf{a}_2$, $\mathbf{a}_3$, respectively. This means that the vectors $\mathbf{k}_1$, $\mathbf{k}_2$, $\mathbf{k}_3$ which generate the Brillouin zone may also be designated as $\mathbf{k}_x$, $\mathbf{k}_y$, $\mathbf{k}_z$, and this could be done in Fig. 5-20. Equation (5-75) expresses the energy E as a function of k_x, k_y, k_z, so that we shall obtain constant energy surfaces which are paraboloid in shape, and hence simpler than those of Fig. 5-20. We can indicate this by plotting the E vs. k curves in the first Brillouin zone. Figure 5-10 indicates that the reciprocal lattice is also simple cubic and the first zone is bounded by the six planes $k_1 = \pm\pi/a$, $k_2 = \pm\pi/a$, $k_3 = \pm\pi/a$. It is customary to label points and axes of high symmetry in the Brillouin zone as shown in Fig. 5-21; this notation

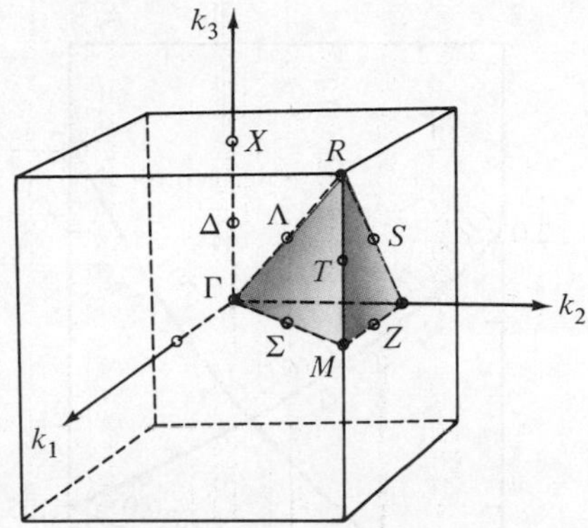

5-21 Brillouin zone for simple cubic lattice

comes from a paper usually referred to as BSW.[3] For example, the point Γ has coordinates $k_1 = 0$, $k_2 = 0$, $k_3 = 0$ and M has coordinates $k_1 = \pi/a$, $k_2 = \pi/a$, $k_3 = 0$; the Σ-axis joins these two points.

Let us next define

$$\xi = k_1 a, \qquad \eta = k_2 a, \qquad \zeta = k_3 a \tag{5-77}$$

Then the components of $\mathbf{k}$ may be written as ξ/a, η/a, ζ/a. Also

$$\mathbf{mB} = \frac{(m_1 \quad m_2 \quad m_3)}{a} \tag{5-78}$$

[3] L. P. Bouckaert, R. Smoluchowski, and E. Wigner, *Phys. Rev.*, **50**, 58 (1936).

again using row matrix notation. Equation (5-75) thus assumes the form

$$E = \frac{\hbar^2}{2ma^2}[(\xi - 2\pi m_1)^2 + (\eta - 2\pi m_2)^2 + (\zeta - 2\pi m_3)^2] \tag{5-79}$$

and (5-76) becomes

$$\psi = \exp\left[\frac{i\{(\xi - 2\pi m_1)x + (\eta - 2\pi m_2)y + (\zeta - 2\pi m_3)z\}}{a}\right] \tag{5-80}$$

To plot E as a function of $\mathbf{k}$ using (5-79) on a two-dimensional graph, we select various directions in the Brillouin zone. For example, along the Δ-axis of Fig. 5-21 we have $\xi = \eta = 0$, so that

$$E = \frac{\hbar^2}{2ma^2}[m_1^2 + m_2^2 + (\xi - m_3)^2] \tag{5-81}$$

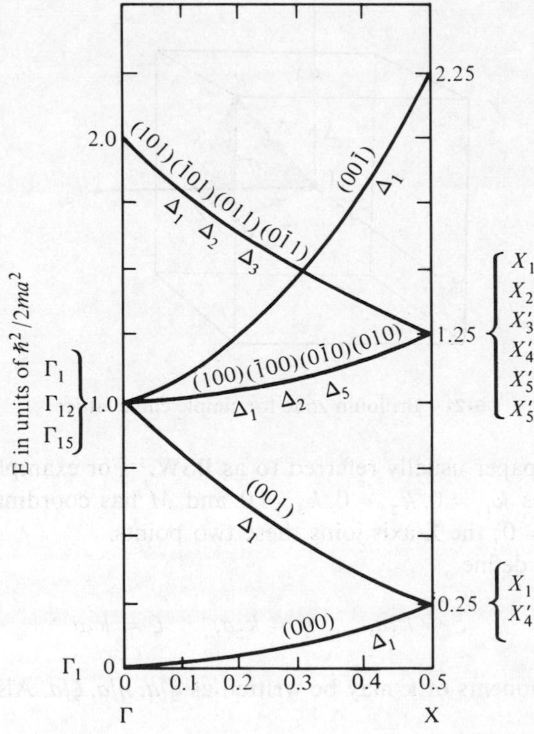

5-22 Free electron energy bands for the Δ-axis of a simple cubic crystal

and
$$\psi = \exp\left[2\pi i \frac{-m_1 x - m_2 y + (\xi - m_3)z}{a}\right] \tag{5-82}$$

where $\xi = \zeta/2\pi$, using a common convention. The curves of E vs. ξ are shown in Fig. 5-22, each curve being labelled with the values of (m_1, m_2, m_3) to which it corresponds. Note that two of these curves are four-fold degenerate, since the wave functions (5-82) are distinct for each triplet (m_1, m_2, m_3) shown on the parabolas of Fig. 5-22. We realize now that the dashed curve of Fig. 5-18(b)—the free electron energy—is simply a parabola through the origin.

PROBLEM 5-15

(a) Verify the curves of Fig. 5-22, the coordinates of the endpoints and the values of the triplets (m_1, m_2, m_3).

(b) Verify the dashed curve of Fig. 5-18(b) numerically by assuming that the wells coalesce; i.e., let $b = 0$.

The symmetry of the Brillouin zone and the direct lattice are both simple cubic; that is, the point groups are identical. In solving Problem 5-9, it is necessary to show that the Brillouin zone for a face-centered cubic crystal is body-centered, and the converse is also true. Thus, the translation groups need not be the same. However, because we can generally work with a reduced zone, only the point group symmetry in k-space is of importance. A cubic zone requires that we consider the 48-member holohedral group O_h. For our purposes, we shall regard this group as the combination of the octahedral group O with the inversion operator J. That is,

$$O_h = O \times J \tag{5-83}$$

where O has the symmetry operations of Table 1-4. We then generate Table 5-1. These operations are actually expressed in terms of a cube. As Fig. 1-11 shows, the two polyhedra have the same symmetry.

Since this group has five classes and 24 operations, we use the fact that we need five numbers whose squares add up to 24 in order to determine the dimensionality of the representations. Empirically, we see that

$$1^2 + 1^2 + 2^2 + 3^2 + 3^2 = 24$$

and we may again use Burnside's method to find the character system. Having done this, we can go from O to O_h by realizing that the group C_i composed simply of E and J will have the character system of Table 5-3, and that (5-83) is equivalent to

$$O_h = O \times C_i \tag{5-84}$$

TABLE 5-2 ***Symmetry Operations for the Octahedral Poiut Group O***

Operation	Multiplicity	Description
E	1	The identity.
C_4	6	$\pm 90°$ rotation about a coordinate axis.
C_4^2	3	$\pm 180°$ rotation about a coordinate axis.
C_2	6	$\pm 180°$ rotation about an axis of the type $x = y$, $z = 0$.
C_3	8	$\pm 120°$ rotation about a cube diagonal.
J	1	Inversion through the center.
JC_4	6	Combination of J and the rotations above.
JC_4^2	3	
JC_2	6	
JC_3	8	

TABLE 5-3 ***Character System for the Group C_i or $\bar{1}$***

C_i	E	J
A_g	1	1
A_u	1	-1

Hence, the character system of O_h is obtained from that of O by multiplying every element in the table for O three times by $+1$, once by -1, and arranging the results as shown in Table 5-4; that is, three of the four sections of this table are the original character system for O and the fourth one is its negative. This table shows the two types of notation used to designate the ten representations; the BSW[3] labels as used by physicists are on the right, and the Mulliken[4] symbols favored by physical chemists are on the left.

In our previous work using the coordinates x, y, z of an arbitrary point as a basis, we found that this basis was reducible. For example, when the z-axis was chosen as a three-fold axis C_3, then the reduction gave a nondegenerate representation associated with z and a doubly-degenerate representation for x and y, as the symmetry leads us to expect. For O_h, we see from Table 5-4 that we can have two triply-degenerate irreducible representations. Let us then consider the effect of the cubic symmetry operations on an arbitrary point with coordinates x, y, z. For C_4, for example, a 90° rotation about the OZ-axis will convert x, y, z into either $-y, x, z$ (which we denote by $\bar{y}xz$) or into $y\bar{x}z$, and the inversion J on xyz gives $\overline{xyz}$. The results for the group O are given in Table 5-5 and the additional twenty-four entries for O_h can be obtained by simply reversing all three signs. The notation of this table is due to Jones,[2] so that the combinations of coordinates shown are *Jones symbols.*

[4]Mulliken, *Phys. Rev.*, **43**, 279 (1933).

TABLE 5-4 ***Character System for the Group Q_h or $\frac{4}{m}\bar{3}\frac{2}{m}$***

O_h		E	$3C_4^2$	$6C_4$	$6C_2$	$8C_3$	J	$3JC_4^2$	$6JC_4$	$6JC_2$	$8JC_3$	Associated functions
A_{1g}	Γ_1	1	1	1	1	1	1	1	1	1	1	s
A_{2g}	Γ_2	1	1	−1	−1	1	1	1	−1	−1	1	
E_g	Γ_{12}	2	2	0	0	−1	2	2	0	0	−1	$(d_{z^2}, d_{x^2-y^2})$
F_{1g}	Γ'_{15}	3	−1	1	−1	0	3	−1	1	−1	0	(R_x, R_y, R_z)
F_{1u}	Γ'_{25}	3	−1	−1	1	0	3	−1	−1	1	0	(d_{xy}, d_{yz}, d_{zx})
A_{1u}	Γ'_1	1	1	1	1	1	−1	−1	−1	−1	−1	
A_{2u}	Γ'_2	1	1	−1	−1	1	−1	−1	1	1	−1	
E_u	Γ'_{12}	2	2	0	0	−1	−2	−2	0	0	1	
F_{1u}	Γ_{15}	3	−1	1	−1	0	−3	1	−1	1	0	(p_x, p_y, p_z)
F_{2u}	Γ_{25}	3	−1	−1	1	0	−3	1	1	−1	0	

TABLE 5-5 ***Jones Symbols for the Group O or 432***

E	xyz
$3C_4^2$	$\bar{x}\bar{y}z,\quad x\bar{y}\bar{z},\quad \bar{x}y\bar{z}$
$6C_4$	$yx\bar{z},\quad y\bar{x}z,\quad \bar{x}zy,\quad \bar{x}\bar{z}\bar{y},\quad \bar{z}\bar{y}\bar{x},\quad \bar{z}\bar{y}x$
$8C_3$	$zxy,\quad yzx,\quad z\bar{x}\bar{y},\quad \bar{y}\bar{z}x,\quad \bar{z}\bar{x}y,\quad \bar{y}z\bar{x},\quad \bar{z}x\bar{y},\quad y\bar{z}\bar{x}$

We can identify functions with irreducible representations since E has the form

$$\begin{pmatrix} 1 & 0 & 0 \\ 0 & 1 & 0 \\ 0 & 0 & 1 \end{pmatrix}\begin{pmatrix} x \\ y \\ z \end{pmatrix} = \begin{pmatrix} x \\ y \\ z \end{pmatrix}$$

with a character of 3 and that one matrix for C_4^2, for example, is

$$\begin{pmatrix} 1 & 0 & 0 \\ 0 & -1 & 0 \\ 0 & 0 & -1 \end{pmatrix}\begin{pmatrix} x \\ y \\ z \end{pmatrix} = \begin{pmatrix} x \\ \bar{y} \\ \bar{z} \end{pmatrix}$$

with a character of -1. The full set of characters is then 3, -1, 1, -1, 0, -3, 1, -1, 1, 0, which we see is identical to the representation Γ_{15} of Table 5-4. Since the orbitals p_x, p_y, p_z are proportional to x, y, z, respectively, the set of triply-degenerate p functions belongs to the representation Γ_{15}. Using the same arguments, the completely symmetric s orbital belongs to Γ_1, and by considering the behavior of the products, we can show that d_{xy}, d_{yz}, d_{zx} belong to Γ'_{25}, and $d_{z^2}, d_{x^2-y^2}$ belong to Γ_{12}. The Jones symbols also identify the basis functions quickly, since the three combinations associated with C_4^2, for example, represent one invariant coordinate and two which go negative, giving a net character of -1. We also see that the origin O in direct space (which is the same point as the origin Γ in k-space) is the invariant point of a 48-element group, explaining the use of the symbol Γ to denote the representations of this group. But when we move from Γ to X along the Δ-axis, which is equivalent to finding the energy and the wave function for some value of **k** not equal to zero, the symmetry is different. We recognize that an arbitrary point on the Δ-axis is left invariant by the five operations

$$E,\ C_4^2,\ C_4,\ JC_4^2,\ JC_2$$

where C_4 refers to a 90° rotation about the k_z-axis, whereas JC_4^2 is a 180° rotation about the k_x-axis or the k_y-axis, followed by the inversion. To specify this more precisely, we shall denote a rotation about the k_z-axis with a subscript $||$ and one about either of the other two axes with a subscript $\perp$, so that the above operations become

$$E,\ C_{4||}^2,\ C_{4||},\ JC_{4\perp}^2,\ JC_2$$

PROBLEM 5-16

Prove that the symmetry operations for a point on the Δ-axis form a group.

The character system, as taken from BSW, is shown in Table 5-6. The designation of the representations, which appears strange, will be considered shortly. The group of eight operations in this table is referred to as a *little group*. If **k** is an arbitrary vector joining Γ with some point (k_x, k_y, k_z) in the Brillouin zone, then the little group generates seven additional vectors **k** whose coordinates are indicated by the Jones symbols at the bottom of Table 5-6. These eight vectors are said to form the *star* of **k** for the little group associated with the Δ-axis.

At the point X, the symmetry does not return to that of Γ. That is, a rotation about the k_x-or k_y-axis (denoted by $\perp$) is not equivalent to a rotation

TABLE 5-6 ***Character System for the Δ-Axis of a Simple Cubic Crystal***

	E	$C_{4\parallel}^2$	$2C_{4\parallel}$	$2JC_{4\perp}^2$	$2JC_2$	Associated function
Δ_1	1	1	1	1	1	z^3
Δ_2	1	1	−1	1	−1	$x^2 - y^2$
Δ_2'	1	1	−1	−1	1	xy
Δ_1'	1	1	1	−1	−1	
Δ_5	2	−2	0	0	0	xz, yz
	xyz	$\bar{x}\bar{y}z$	$y\bar{x}z$ $\bar{y}xz$	$x\bar{y}z$ $\bar{x}yz$	$\bar{y}\bar{x}z$ yxz	

about the k_z-axis (denoted by $||$). To see why, consider the function ψ of Eq. (5-82) along the lowest curve of Fig. 5-22, for which $m_1 = 0$, $m_2 = 0$, $m_3 = 0$. Then

$$\psi = e^{2\pi i \xi z/a} = e^{ik_z z}$$

from (5-77). At X, this reduces to

$$\psi = e^{i\pi z/a}$$

since $k_z = \pi/a$ at this boundary of the Brillouin zone. Then a 90° rotation $C_{4\parallel}$ about the k_z-axis has no effect on ψ, but a 90° rotation $C_{4\perp}$ about either of the other two axis converts it into $e^{-i\pi z/a}$. Using arguments of this type, we may obtain the character system for the little group at X given in Table 5-7. Note that this table is also the Kronecker product of a smaller group with C_i.

As we have been developing the procedure for solving the Schroedinger equation, we have been concentrating our attention on the symmetry properties of the Brillouin zone, rather than on the direct lattice, and these properties have been considered in connection with the free electron solutions rather than the exact solutions. To show that what we have done so far does, in fact, take into account the behavior of the exact solutions in the direct lattice, let us go back to the original Bloch solution

$$\psi(\mathbf{r}, \mathbf{k}) = u(\mathbf{r})e^{i\mathbf{k}\cdot\mathbf{r}} \qquad \textbf{(5-63)}$$

Let us take $\mathbf{r}$ as a direct lattice vector, so that by (5-70), it may be written as

$$\mathbf{r} = n_1\mathbf{a}_1 + n_2\mathbf{a}_2 + n_3\mathbf{a}_3 \qquad \textbf{(5-85)}$$

TABLE 5-7 ***Character System for the Point X of a Simple Cubic Crystal***

	E	$2C_{4\perp}^2$	$C_{4\parallel}^2$	$2C_{4\parallel}$	$2C_2$	J	$2JC_{4\perp}^2$	$JC_{\parallel 4}^2$	$2JC_{4\parallel}$	$2JC_2$
X_1	1	1	1	1	1	1	1	1	1	1
X_2	1	1	1	−1	−1	1	1	1	−1	−1
X_3	1	−1	1	−1	1	1	−1	1	−1	1
X_4	1	−1	1	1	−1	1	−1	1	1	−1
X_5	2	0	−2	0	0	2	0	−2	0	0
X_1'	1	1	1	1	1	−1	−1	−1	−1	−1
X_2'	1	1	1	−1	−1	−1	−1	−1	1	1
X_3'	1	−1	1	−1	1	−1	1	−1	1	−1
X_4'	1	−1	1	1	−1	−1	1	−1	−1	1
X_5'	2	0	−2	0	0	−2	0	2	0	0
	xyz	$x\bar{y}\bar{z}$	$\bar{x}\bar{y}z$	$\bar{y}xz$	$yx\bar{z}$					
		$\bar{x}y\bar{z}$		$y\bar{x}z$	$\bar{x}\bar{y}\bar{z}$					

and let $\mathbf{k}$ be a reciprocal lattice vector, which by (5-71) is

$$\mathbf{r}_R = m_1\mathbf{b}_1 + m_2\mathbf{b}_2 + m_3\mathbf{b}_3 = \mathbf{mB}$$

or, in k-space,

$$\mathbf{r}_K = 2\pi(m_1\mathbf{b}_1 + m_2\mathbf{b}_2 + m_3\mathbf{b}_3) = m_1\mathbf{k}_1 + m_2\mathbf{k}_2 + m_3\mathbf{k}_3 \tag{5-86}$$

Consider the point $A(0, 0, \pi/c)$ at the top of the Brillouin zone. Let us perform a symmetry operation in direct space—such as J or $C_{4\perp}^2$—which converts the function ψ into its value at $A(0, 0, -\pi/c)$. Then $\mathbf{r}$ is converted into another lattice vector $\mathbf{r}'$ and $\mathbf{k}$ is changed to $\mathbf{k} + \mathbf{r}_K$, where the distance between the top and bottom of the k_z-axis is a reciprocal lattice vector $\mathbf{r}_K$. Hence, ψ becomes

$$\psi(\mathbf{r}', \mathbf{k} + \mathbf{r}_K) = u(\mathbf{r}')e^{i(\mathbf{k}+\mathbf{r}_K)\cdot\mathbf{r}'}$$

But

$$u(\mathbf{r}') = u(\mathbf{r})$$

since $u(\mathbf{r})$ is periodic. Also

$$e^{i\mathbf{r}_K\cdot\mathbf{r}'} = 1$$

for by (5-85) and (5-86), the terms in the exponent combine to give $2\pi i$ times an integer. Therefore, ψ is invariant under a direct space symmetry operation. Further, the two points $A(0, 0, \pi/c)$ and $A(0, 0, -\pi/c)$ are *equivalent*; although they are at opposite ends of the k_z-axis, their symmetry properties are identical. The same statement holds for any two points in k-space separated by a reciprocal lattice vector.

The character tables (Tables 5-4, 5-6, and 5-7) enable us to introduce the concept of *compatibility*, for which Jones[2] has the simplest explanation. As we go along the k_z-axis from Γ to X, we see that there is a summation relation among some of the characters. For example, the row Γ_{12} of Table 5-4 is the sum of the rows Δ_1 and Δ_2 of Table 5-6, and this is symbolized as

$$\Gamma_{12} \longrightarrow \Delta_1 + \Delta_2$$

where we compare characters for only those operations which are common to both tables; that is, for the column headings of Table 5-6. Similarly, we have

$$\Gamma'_{15} \longrightarrow \Delta'_1 + \Delta_5$$

and there are also direct relations such as

$$\Gamma'_1 \longrightarrow \Delta_1, \qquad \Gamma'_2 \longrightarrow \Delta'_2$$

The complete set of compatibility relations for the path ΓX is given in Table 5-8; the upper section shows the logic behind the choice of subscripts for the Γ representations.

The reasoning behind the compatibility relations is based on the requirement that the trace of a matrix be invariant under a similarity transformation; Eq. (2-60) verifies this for a matrix of Table 2-3. Consider as an example the triply-degenerate representation Γ_{15}, which is irreducible at the center of the Brillouin zone. As we move out along the axis, this representation may become reducible and a single similarity transformation can convert all the 3×3 matrices into purely diagonal ones or to block form. In either case, the new traces add up to the original trace so that the character systems are connected by compatibility relations of the type given. Another physical way of thinking about compatibility relations comes by considering the three p functions along some symmetry axis of a cubic crystal. At the point Γ, the functions p_x, p_y, p_z are completely equivalent in a cubic environment, and we have a three-fold degeneracy in our E vs. $\mathbf{k}$ curve. Suppose we move out along the k_z-axis. Then p_x and p_y are still equivalent, but p_z in general corresponds to a different energy. Hence, the three-fold degeneracy at the center of the zone has been split into a nondegenerate and a double degenerate energy curve; the E vs. $\mathbf{k}$ curves will touch at the center of the zone, but one will move away for $\mathbf{k} \neq 0$.

In order to compute the projection operators associated with the full cubic group O_h, we must determine the matrices comprising the two-dimen-

TABLE 5-8 ***Compatibility Relations for the Δ- and Σ-Axes of a Simple Cubic Lattice***

$\Gamma_1 \to \Delta_1$	$\Gamma'_1 \to \Delta'_1$
$\Gamma_2 \to \Delta_2$	$\Gamma'_2 \to \Delta'_2$
$\Gamma_{12} \to \Delta_1 + \Delta_2$	$\Gamma'_{12} \to \Delta'_1 + \Delta'_2$
$\Gamma'_{15} \to \Delta'_1 + \Delta_5$	$\Gamma_{15} \to \Delta_1 + \Delta_5$
$\Gamma'_{25} \to \Delta'_2 + \Delta_5$	$\Gamma_{25} \to \Delta_2 + \Delta_5$
$X_1 \to \Delta_1$	$X'_1 \to \Delta'_1$
$X_2 \to \Delta_2$	$X'_2 \to \Delta'_2$
$X_3 \to \Delta'_2$	$X'_3 \to \Delta_2$
$X_4 \to \Delta'_1$	$X'_4 \to \Delta_1$
$X_5 \to \Delta_5$	$X'_5 \to \Delta_5$
$\Gamma_1 \to \Sigma_1$	$\Gamma'_1 \to \Sigma_2$
$\Gamma_2 \to \Sigma_4$	$\Gamma'_2 \to \Sigma_3$
$\Gamma_{12} \to \Sigma_1$	$\Gamma'_{12} \to \Sigma_2 + \Sigma_3$
$\Gamma'_{15} \to \Sigma_2 + \Sigma_3 + \Sigma_4$	$\Gamma_{15} \to \Sigma_1 + \Sigma_3 + \Sigma_4$
$\Gamma'_{25} \to \Sigma_1 + \Sigma_2 + \Sigma_3$	$\Gamma_{25} \to \Sigma_1 + \Sigma_2 + \Sigma_4$
$M_1 \to \Sigma_1$	$M'_1 \to \Sigma_2$
$M_2 \to \Sigma_4$	$M'_2 \to \Sigma_3$
$M_3 \to \Sigma_1$	$M'_3 \to \Sigma_2$
$M_4 \to \Sigma_4$	$M'_4 \to \Sigma_3$
$M_5 \to \Sigma_2 + \Sigma_3$	$M'_5 \to \Sigma_1 + \Sigma_4$

sional and three-dimensional irreducible representations. The latter can be determined by inspection from the Jones symbols, but the two-dimensional matrices involve some calculation. Let us follow the procedure of Slater[5] in demonstrating how this may be done. We recall from Chapter 1 that the group T_d or $\bar{4}3m$ is obtained from the tetrahedral group T or 23 by incorporating mirror planes. Then O_h may be generated by combining T_d with J or O with J; the latter approach gave us Table 5-1, but Slater prefers to start with T_d. Let us define the 24 operations of this group in terms of the motion of the point labelled 1 in Fig. 5-23. The covering operations produce points 2, 3, . . . , 24. If we denote these operations by $R_1, R_2, \ldots, R_{24}$, the Cayley table may be compactly expressed in terms of the associated subscripts, as indicated

[5] J. C. Slater, *Quantum Theory of Molecules and Solids:* Vol. 2, *Symmetry and Energy Bands in Crystals*, McGraw-Hill Book Company, New York, 1965.

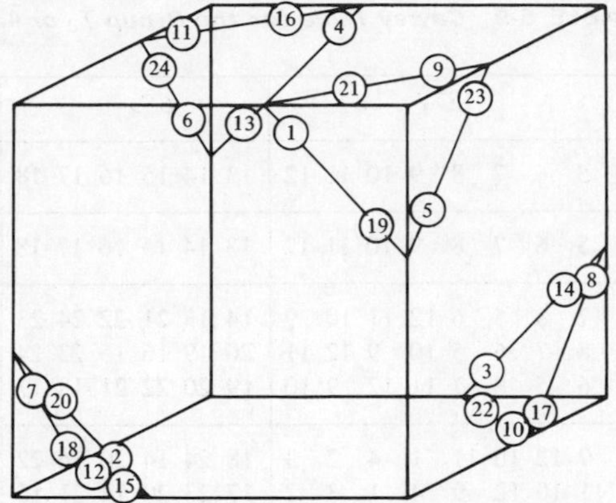

5-23 Symmetry operations for the group T_d or $\bar{4}3\,m$

in Table 5-9. The corresponding class designations are also shown to facilitate comparison of Slater's symbols with the Schoenflies notation.

Figure 5-23 indicates that the four operations R_5, R_6, R_7, R_8 represent rotations of $+120°$ about the body diagonals, and therefore correspond to the matrix

$$C_3 = \begin{pmatrix} -\frac{1}{2} & -\frac{\sqrt{3}}{2} \\ \frac{\sqrt{3}}{2} & -\frac{1}{2} \end{pmatrix}$$

as indicated in Table 5-9. In the same way, R_9, R_{10}, R_{11}, R_{12} are represented by

$$C_3^{-1} = \begin{pmatrix} -\frac{1}{2} & \frac{\sqrt{3}}{2} \\ -\frac{\sqrt{3}}{2} & -\frac{1}{2} \end{pmatrix}$$

and by Table 5-9, we see that

$$R_9 R_7 = R_2$$

so that

$$R_2 = \begin{pmatrix} 1 & 0 \\ 0 & 1 \end{pmatrix}$$

This result is given in columnar form in Table 5-10 in order to save space. The remaining matrices belong to Γ_{12} are computed in the same way.

Having found the complete representations for the group T_d, we use the procedure based on Eq. (5-84) to find the representations of the 48 member

TABLE 5-9 ***Cayley Table for the Group T_d or $\bar{4}3m$***

	E	$3C_4^2$			$8C_3$								$6S_4$						$6\sigma_d$					
	1	2	3	4	5	6	7	8	9	10	11	12	13	14	15	16	17	18	19	20	21	22	23	24
1	1	2	3	4	5	6	7	8	9	10	11	12	13	14	15	16	17	18	19	20	21	22	23	24
2	2	1	4	3	7	8	5	6	12	11	10	9	14	13	21	22	24	23	20	19	15	16	18	17
3	3	4	1	2	8	7	6	5	10	9	12	11	20	19	16	15	23	24	14	13	22	21	17	18
4	4	3	2	1	6	5	8	7	11	12	9	10	19	20	22	21	18	17	13	14	16	15	24	23
5	5	8	6	7	9	12	10	11	1	4	2	3	18	24	14	20	16	22	23	17	19	13	21	15
6	6	7	5	8	11	10	12	9	4	1	3	2	17	23	20	14	21	15	24	18	13	19	16	22
7	7	6	8	5	12	9	11	10	2	3	1	4	23	17	13	19	22	16	18	24	20	14	15	21
8	8	5	7	6	10	11	9	12	3	2	4	1	24	18	19	13	15	21	17	23	14	20	22	16
9	9	11	12	10	1	3	4	2	5	7	8	6	22	15	24	17	20	13	21	16	23	18	19	14
10	10	12	11	9	3	1	2	4	8	6	5	7	21	16	18	23	13	20	22	15	17	24	14	19
11	11	9	10	12	4	2	1	3	6	8	7	5	15	22	23	18	14	19	16	21	24	17	13	20
12	12	10	9	11	2	4	3	1	7	5	6	8	16	21	17	24	19	14	15	22	18	23	20	13
13	13	14	19	20	16	15	22	21	24	23	18	17	2	1	8	7	9	10	4	3	6	5	11	12
14	14	13	20	19	22	21	16	15	17	18	23	24	1	2	6	5	12	11	3	4	8	7	10	9
15	15	22	16	21	18	23	17	24	20	13	19	14	9	11	3	1	6	8	12	10	2	4	7	5
16	16	21	15	22	24	17	23	18	13	20	14	19	10	12	1	3	7	5	11	9	4	2	6	8
17	17	23	24	18	14	20	19	13	22	16	15	21	7	6	9	12	4	1	8	5	10	11	3	2
18	18	24	23	17	20	14	13	19	15	21	22	16	8	5	11	10	1	4	7	6	12	9	2	3
19	19	20	13	14	21	22	15	16	23	24	17	18	3	4	7	8	11	12	1	2	5	6	9	10
20	20	19	14	13	15	16	21	22	18	17	24	23	4	3	5	6	10	9	2	1	7	8	12	11
21	21	16	22	15	23	18	24	17	19	14	20	13	12	10	4	2	8	6	9	11	1	3	5	7
22	22	15	21	16	17	24	18	23	14	19	13	20	11	9	2	4	5	7	10	12	3	1	8	6
23	23	17	18	24	19	13	14	20	21	15	16	22	6	7	10	11	2	3	5	8	9	12	1	4
24	24	18	17	23	13	19	20	14	16	22	21	15	5	8	12	9	3	2	6	7	11	10	4	1

group O_h. That is, we take the Kronecker product of the representations in Table 5-10 with those in Table 5-3. We need not write the result out in full detail, since its structure will be similar to that of Table 5-4; there will be three subsections just like Table 5-10 and the lower right-hand corner will be the negative of the other three portions. Knowing the irreducible representations, we are in a position to use the projection operators to find the functions ϕ associated with the curves of Fig. 5-22. The lowest one corresponds to $\mathbf{m} = (000)$ and $\xi = 0$, so that (5-82) gives

$$\psi = 1$$

This symmetric function will obviously belong to Γ_1, as Fig. 5-22 indicates

TABLE 5-10 *Irreducible Representations for the Group O_h or $\frac{4}{m}3\frac{2}{m}$*

(see text for construction of lower half)

	R_1	R_2	R_3	R_4	R_5	R_6	R_7	R_8	R_9	R_{10}	R_{11}	R_{12}	R_{13}	R_{14}	R_{15}	R_{16}	R_{17}	R_{18}	R_{19}	R_{20}	R_{21}	R_{22}	R_{23}	R_{24}
Γ_1	1	1	1	1	1	1	1	1	1	1	1	1	1	1	1	1	1	1	1	1	1	1	1	1
Γ_2	1	1	1	1	1	1	1	1	1	1	1	1	−1	−1	−1	−1	−1	−1	−1	−1	−1	−1	−1	−1
$(\Gamma_{12})_{11}$	1	1	1	1	$-\frac{1}{2}$	$-\frac{1}{2}$	$-\frac{1}{2}$	$-\frac{1}{2}$	$-\frac{1}{2}$	$-\frac{1}{2}$	$-\frac{1}{2}$	$-\frac{1}{2}$	1	1	$-\frac{1}{2}$	$-\frac{1}{2}$	$-\frac{1}{2}$	$-\frac{1}{2}$	1	1	$-\frac{1}{2}$	$-\frac{1}{2}$	$-\frac{1}{2}$	$-\frac{1}{2}$
$(\Gamma_{12})_{21}$	0	0	0	0	$\frac{\sqrt{3}}{2}$	$\frac{\sqrt{3}}{2}$	$\frac{\sqrt{3}}{2}$	$\frac{\sqrt{3}}{2}$	$\frac{-\sqrt{3}}{2}$	$\frac{-\sqrt{3}}{2}$	$\frac{-\sqrt{3}}{2}$	$\frac{-\sqrt{3}}{2}$	0	0	$\frac{-\sqrt{3}}{2}$	$\frac{-\sqrt{3}}{2}$	$\frac{\sqrt{3}}{2}$	$\frac{\sqrt{3}}{2}$	0	0	$\frac{-\sqrt{3}}{2}$	$\frac{-\sqrt{3}}{2}$	$\frac{\sqrt{3}}{2}$	$\frac{\sqrt{3}}{2}$
$(\Gamma_{12})_{12}$	0	0	0	0	$\frac{-\sqrt{3}}{2}$	$\frac{-\sqrt{3}}{2}$	$\frac{-\sqrt{3}}{2}$	$\frac{-\sqrt{3}}{2}$	$\frac{\sqrt{3}}{2}$	$\frac{\sqrt{3}}{2}$	$\frac{\sqrt{3}}{2}$	$\frac{\sqrt{3}}{2}$	0	0	$\frac{-\sqrt{3}}{2}$	$\frac{-\sqrt{3}}{2}$	$\frac{\sqrt{3}}{2}$	$\frac{\sqrt{3}}{2}$	0	0	$\frac{-\sqrt{3}}{2}$	$\frac{-\sqrt{3}}{2}$	$\frac{\sqrt{3}}{2}$	$\frac{\sqrt{3}}{2}$
$(\Gamma_{12})_{22}$	1	1	1	1	$-\frac{1}{2}$	$-\frac{1}{2}$	$-\frac{1}{2}$	$-\frac{1}{2}$	$-\frac{1}{2}$	$-\frac{1}{2}$	$-\frac{1}{2}$	$-\frac{1}{2}$	−1	−1	$\frac{1}{2}$	$\frac{1}{2}$	$\frac{1}{2}$	$\frac{1}{2}$	−1	−1	$\frac{1}{2}$	$\frac{1}{2}$	$\frac{1}{2}$	$\frac{1}{2}$
$\chi(\Gamma_{12})$	2	2	2	2	−1	−1	−1	−1	−1	−1	−1	−1	0	0	0	0	0	0	0	0	0	0	0	0
$(\Gamma'_{25})_{11}$	1	1	−1	−1	0	0	0	0	0	0	0	0	−1	−1	0	0	0	0	1	1	0	0	0	0
$(\Gamma'_{25})_{21}$	0	0	0	0	1	−1	−1	1	0	0	0	0	0	0	0	0	1	−1	0	0	0	0	1	−1
$(\Gamma'_{25})_{31}$	0	0	0	0	0	0	0	0	1	−1	1	−1	0	0	−1	1	0	0	0	0	1	−1	0	0
$(\Gamma'_{25})_{12}$	0	0	0	0	0	0	0	0	1	−1	−1	1	0	0	0	0	−1	1	0	0	0	0	1	−1
$(\Gamma'_{25})_{22}$	1	−1	1	−1	0	0	0	0	0	0	0	0	0	0	−1	−1	0	0	0	0	1	1	0	0
$(\Gamma'_{25})_{32}$	0	0	0	0	1	1	−1	−1	0	0	0	0	1	−1	0	0	0	0	1	−1	0	0	0	0
$(\Gamma'_{25})_{13}$	0	0	0	0	1	−1	1	−1	0	0	0	0	0	0	1	−1	0	0	0	0	1	−1	0	0
$(\Gamma'_{25})_{22}$	0	0	0	0	0	0	0	0	1	1	−1	−1	−1	1	0	0	0	0	1	−1	0	0	0	0
$(\Gamma'_{25})_{33}$	1	−1	−1	1	0	0	0	0	0	0	0	0	0	0	0	0	−1	−1	0	0	0	0	1	1
$\chi(\Gamma'_{25})$	3	−1	−1	−1	0	0	0	0	0	0	0	0	−1	−1	−1	−1	−1	−1	1	1	1	1	1	1
$(\Gamma'_{15})_{11}$	1	1	−1	−1	0	0	0	0	0	0	0	0	1	1	0	0	0	0	−1	−1	0	0	0	0
$(\Gamma'_{15})_{21}$	0	0	0	0	1	−1	−1	1	0	0	0	0	0	0	0	0	−1	1	0	0	0	0	−1	1
$(\Gamma'_{15})_{31}$	0	0	0	0	0	0	0	0	1	−1	1	−1	0	0	1	−1	0	0	0	0	−1	1	0	0
$(\Gamma'_{15})_{12}$	0	0	0	0	0	0	0	0	1	−1	−1	1	0	0	0	0	1	−1	0	0	0	0	−1	1
$(\Gamma'_{15})_{22}$	1	−1	1	−1	0	0	0	0	0	0	0	0	0	0	1	1	0	0	0	0	−1	−1	0	0
$(\Gamma'_{15})_{32}$	0	0	0	0	1	1	−1	−1	0	0	0	0	−1	1	0	0	0	0	−1	1	0	0	0	0
$(\Gamma'_{15})_{13}$	0	0	0	0	1	−1	1	−1	0	0	0	0	0	0	−1	1	0	0	0	0	−1	1	0	0
$(\Gamma'_{15})_{23}$	0	0	0	0	0	0	0	0	1	1	−1	−1	1	−1	0	0	0	0	−1	1	0	0	0	0
$(\Gamma'_{15})_{23}$	1	−1	−1	1	0	0	0	0	0	0		0	0	0	0	0	1	1	0	0	0	0	−1	−1
$\chi(\Gamma'_{15})$	3	−1	−1	−1	0	0	0	0	0	0	0	0	1	1	1	1	1	1	−1	−1	−1	−1	−1	−1

on the left-hand edge. As we move from Γ to a point on the Δ-axis, the function becomes

$$\psi = e^{2\pi i \xi z/a}$$

which has the symmetry of z; that is, it should belong to Δ, but to no other representation. This also is indicated on the curve and by the compatibility relations of Table 5-8. As we reach $X, \xi = 0.5$ and

$$\psi = e^{i\pi z/a} \tag{5-87}$$

The Jones symbols at the bottom of Table 5-7 show that z goes negative under two of the first five operations: $C_{4\perp}^2$ and C_2. Only the representations X_1 and X'_4 have characters which are negative for these two operations and positive for the others. Thus, a function of z, alone, should belong to X_1 and to X'_4, and to no other irreducible representation. To verify this, we obtain the projection operators from (2-138), and the representations, except for X_5 and X'_5, from Table 5-7. To find the matrices for these two representations, we can use an argument based on our knowledge of the Jones symbols. The matrices which convert xyz into the other seven combinations shown are easily obtained. For example, $C_{4\perp}^2$ has the reducible representations

$$\begin{pmatrix} 1 & 0 & 0 \\ 0 & -1 & 0 \\ 0 & 0 & -1 \end{pmatrix} \begin{pmatrix} x \\ y \\ z \end{pmatrix} = \begin{pmatrix} x \\ -y \\ -z \end{pmatrix}$$

and

$$\begin{pmatrix} -1 & 0 & 0 \\ 0 & 1 & 0 \\ 0 & 0 & -1 \end{pmatrix} \begin{pmatrix} x \\ y \\ z \end{pmatrix} = \begin{pmatrix} -x \\ y \\ -z \end{pmatrix}$$

The lower right hand 1×1 block has already been identified with X'_4. Hence, the 2×2 section remaining must belong to X'_5, but not to X_5, since the signs for $\bar{x}\bar{y}\bar{z}$ belonging to J, and so on, are not correct. However, the matrices for X_5 follow immediately from those of X'_5; the first five are identical and the second five have the signs reversed. Let us employ the projection operator method to find those linear combinations ϕ, obtained from the original function ψ, which possess the proper symmetry to match the simple cubic Brillouin zone.

PROBLEM 5-17

Use (2-138) to show that the two nonlinear combinations generated from (5-82) are

$$X_1: \quad \phi = \frac{\cos \pi z}{a} \tag{5-88}$$

$$X'_4: \quad \phi = \frac{\sin \pi z}{a} \tag{5-89}$$

Again we note that the compatibility relations agree with the representations to which these functions belong. Although it was easy in this situation to deduce the complete representations, this procedure is generally not so simple for noncubic groups. Hence, it is worthwhile knowing that tables of irreducible representations for 20 of the most commonly encountered space groups are given by Slater.[5]

Returning to $\mathbf{\Gamma}$ along the curve connecting $E = 0.25$ (in reduced units) and $E = 1.0$, at the left-hand end there are six values for $\mathbf{m}$ corresponding to the same energy; namely, $\mathbf{m} = (100), (\bar{1}00), (0\bar{1}0), (010), (00\bar{1})$, and (001). Since this is the most complicated situation we have encountered, we shall work out the linear combinations for the curve connecting $E = 1.0$ with $E = 1.25$. For example,

$$\psi_{\mathrm{T}00} = \exp \frac{2\pi i(x + \xi z)}{a}$$

The projections result in

$$\begin{aligned}\Delta_1: \quad \phi = \Big[1 \exp \frac{2\pi i x}{a} &+ 1 \exp -\frac{2\pi i x}{a} + 1 \exp \frac{2\pi i y}{a} \\ &+ 1 \exp -\frac{2\pi i y}{a} + 1 \exp \frac{2\pi i x}{a} + 1 \exp -\frac{2\pi i x}{a} \\ &+ 1 \exp -\frac{2\pi i y}{a} + 1 \exp \frac{2\pi i y}{a} \Big] \exp \frac{2\pi \xi z}{a}\end{aligned}$$

where the exponential term in z is taken out, since z is invariant under the eight operations of the little group. Combining terms, the nonzero functions are

$$\begin{aligned}\Delta_1: \quad \phi &= \exp \frac{2\pi i \xi z}{a} \left[\cos \frac{2\pi x}{a} + \cos \frac{2\pi y}{a} \right] \\ \Delta_2: \quad \phi &= \exp \frac{2\pi i \xi z}{a} \left[\cos \frac{2\pi x}{a} - \cos \frac{2\pi y}{a} \right] \\ \Delta_5: \quad \phi &= \begin{cases} \exp \dfrac{2\pi i \xi z}{a} \sin \dfrac{2\pi x}{a} \\ \exp \dfrac{2\pi i \xi z}{a} \sin \dfrac{2\pi y}{a} \end{cases}\end{aligned} \tag{5-90}$$

We note that Δ_5 is two-fold degenerate, as indicated by the character table, and that the E vs. $\mathbf{k}$ curve is labelled to correspond with Eq. (5-90). A similar procedure at the value $E = 1.0$ gives the following associated functions:

$$\Gamma_1: \quad \phi = \cos\frac{2\pi x}{a} + \cos\frac{2\pi y}{a} + \cos\frac{2\pi z}{a}$$

$$\Gamma_{12}: \quad \phi = \begin{cases} \cos\dfrac{2\pi x}{a} - \cos\dfrac{2\pi y}{a} \\ \cos\dfrac{2\pi z}{a} - \dfrac{1}{2}\left[\cos\dfrac{2\pi x}{a} + \cos\dfrac{2\pi y}{a}\right] \end{cases} \tag{5-91}$$

$$\Gamma_{15}: \quad \phi = \begin{cases} \sin\dfrac{2\pi x}{a} \\ \sin\dfrac{2\pi y}{a} \\ \sin\dfrac{2\pi z}{a} \end{cases}$$

where the degeneracies again correspond to the representations.

Finally, for completeness—and for future use—we have given the corresponding energy band diagram (Fig. 5-24), character system (Table 5-11), and compatibility relations (lower half of Table 5-8) for the Σ-axis. Figures 5-22 and 5-24 have the same scale and the left-hand edges should, of course, match.

As stated at the beginning of this section, these complicated results appear to have no practical application, since using $V(\mathbf{r}) = 0$ is a very radical assumption. In order to appreciate the value of the free electron approximation, we must discuss the way in which the Schroedinger equation can be solved for more realistic approximations. This we shall do in the next section.

TABLE 5-11 ***Irreducible Representations for the Σ-Axis of a Simple Cubic Lattice***

	E	C_2	JC_4^2	JC_2	Associated d functions
Σ_1	1	1	1	1	xy, z^2
Σ_2	1	1	−1	−1	$yz - xz$
Σ_3	1	−1	−1	1	$yz + xz$
Σ_4	1	−1	1	−1	$x^2 - y^2$
	xyz	$yx\bar{z}$	$xy\bar{z}$	yxz	

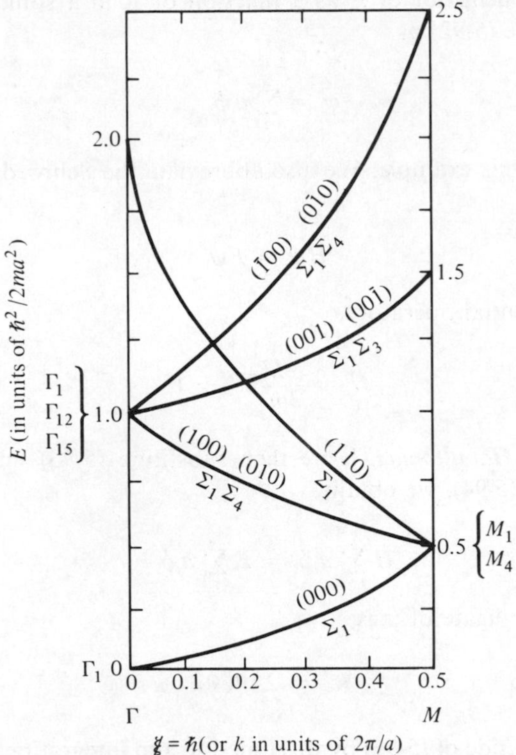

5-24 Free electron energy bands for the Σ-axis of a simple cubic crystal

5-5 *Approximate solutions of the Schroedinger equation*

As we have seen, the wave equation can be solved exactly for a relatively small number of physical systems. For most problems, it is necessary to make approximations, and we shall consider next how this may be done. To illustrate a method which works well in conjunction with group theory, let us return to the H_2^+ ion of Chapter 4. We discussed the exact solution there and also the approximation

$$\psi = a_1\phi_1 + a_2\phi_2 \tag{5-92}$$

where ϕ_1 and ϕ_2 were taken as $1s$ functions in Eq. (4-57). This type of solution is an example of the LCAO method. We can reach our previous conclusions about the behavior of E as a function of R in a somewhat different way. Let us write (5-92) as

$$\psi = \sum_{i=1}^{n} a_i \phi_i \tag{5-93}$$

where $n = 2$ in this example. We also abbreviate the Schroedinger equation (4-2) as

$$H\psi = E\psi \tag{5-94}$$

where the differential operator

$$H = \frac{-\hbar^2}{2m}\nabla^2 + V \tag{5-95}$$

is known as the *Hamiltonian*. If we then substitute (5-93) into the Schroedinger equation (5-94), we obtain

$$H \sum_{i=1}^{n} a_i \phi_i = E \sum_{i=1}^{n} a_i \phi_i \tag{5-96}$$

The complex conjugate of ψ is

$$\psi^* = \sum_{k=1}^{n} a_k^* \phi_k^* \tag{5-97}$$

Multiplying each side of (5-96) by this function and integrating over a volume τ (this symbol is used to avoid confusion with V for potential energy), we obtain

$$\int \sum_{i,k} a_k^* a_i \phi_k^* H \phi_i \, d\tau = E \int \sum_{i,k} a_k^* a_i \phi_k^* \phi_i \, d\tau \tag{5-98}$$

Letting

$$H_{ik} = \int \phi_k^* H \phi_i \, d\tau \qquad S_{ik} = \int \phi_k^* \phi_i \, d\tau \tag{5-99}$$

where the S_{ik} are overlap integrals like (4-52), we see that Eq. (5-98) becomes

$$\sum_{i,k} a_i^* a_k H_{ik} = E \sum_{i,k} a_i^* a_k S_{ik} \tag{5-100}$$

Differentiating these n equations with respect to a_k^* and using the fact that E can depend on the a_i^*, gives

$$\sum_k a_k H_{ik} = \frac{\partial E}{\partial a_k^*} \sum_k a_i^* a_k S_{ik} + E \sum_k a_k S_{ik} \tag{5-101}$$

PROBLEM 5-18

Verify (5-101).

Let us assume now that the quantum mechanical system is in its ground state, so that we are finding its lowest possible energy. This means that

$$\frac{\partial E}{\partial a_k^*} = 0 \tag{5-102}$$

for $k = 1, 2, \ldots, n$. Under these conditions, (5-101) reduces to n simultaneous equations which have a solution only if the determinant vanishes. That is,

$$\begin{vmatrix} H_{11} - S_{11}E & H_{12} - S_{12}E & \cdots & H_{1n} - S_{1n}E \\ H_{12} - S_{12}E & H_{22} - S_{22}E & \cdots & H_{2n} - S_{2n}E \\ \cdots & \cdots & \cdots & \cdots \\ \cdots & \cdots & \cdots & \cdots \\ H_{1n} - S_{1n}E & H_{2n} - S_{2n}E & \cdots & H_{nn} - S_{nn}E \end{vmatrix} = 0 \tag{5-103}$$

This will be recognized as a secular equation similar to those of Chapter 2. The use of the linear combination (5-92) and the consequent secular equation comprises what is known as the *method of linear variation functions.* When the functions ϕ_i are explicitly chosen as atomic orbitals, then we are using the LCAO method as a special case. On the other hand, the ϕ_i may be chosen as Bloch functions, as we shall see later.

To apply this to a simple example, consider the ground state of the H_2^+ ion. The secular equation is then

$$\begin{vmatrix} H_{11} - S_{11}E & H_{12} - S_{12}E \\ H_{21} - S_{21}E & H_{22} - S_{22}E \end{vmatrix} = 0 \tag{5-104}$$

where, by Problem 4-7, we use

$$\phi_1 = \frac{e^{-r_1/a_1}}{\pi^{1/2}a_1^{3/2}}, \qquad \phi_2 = \frac{e^{-r_2/a_1}}{\pi^{1/2}a_1^{3/2}} \tag{5-105}$$

These are $1s$ atomic orbitals with the radial attenuation explicitly included. Since they are also fully normalized, we see that

$$S_{11} = \int \phi_1^2 \, d\tau = 1 = S_{22}$$

We also have $$S_{12} = \frac{1}{\pi a_1^3}\int e^{(-r_1-r_2)/a_1}\,d\tau = S_{21}$$

which we shall denote by S.

PROBLEM 5-19

Why are ϕ_1 and ϕ_2 nonorthogonal?

In addition, $$H_{11} = H_{22}, \qquad H_{12} = H_{21}$$

The secular equation then becomes

$$\begin{vmatrix} H_{11} - E & H_{12} - SE \\ H_{12} - SE & H_{11} - E \end{vmatrix} = 0 \tag{5-106}$$

with roots $$E_1 = \frac{H_{11} + H_{12}}{1 + S}, \qquad E_2 = \frac{H_{11} - H_{12}}{1 - S} \tag{5-107}$$

Substituting E_1 into the two equations for the a_k from which the secular equation is derived, we find that

$$\pm[(H_{11}S - H_{12})a_1 + (H_{12} - SH_{11})a_2] = 0 \tag{5-108}$$

from which $$a_1 = a_2$$

By (5-92) $$\psi_1 = a_1(\phi_1 + \phi_2) \tag{5-109}$$

which is identical to (4-48). In a similar way, the root E_2 leads to

$$a_1 = -a_2 \tag{5-110}$$

and to $$\psi_2 = a_1(\phi_1 - \phi_2) \tag{5-111}$$

agreeing with (4-49). Hence, the use of a linear variation function gives results which confirm our previous intuitive approach. Using the normalization constants for the combinations as previously determined, we write

$$\psi_1 = \frac{\phi_1 + \phi_2}{\sqrt{2 + 2S}}, \qquad \psi_2 = \frac{\phi_1 - \phi_2}{\sqrt{2 - 2S}} \tag{5-112}$$

Eyring, Walter, and Kimball[6] have shown how to determine the integrals

[6]H. Eyring, J. Walter, and G. E. Kimball, *Quantum Chemistry*, John Wiley & Sons, Inc., New York, 1944.

H_{11}, H_{12}, and S in (5-111) and (5-112). To evaluate S, we need the volume element $d\tau$ in ellipsoidal coordinates, which they show to be

$$d\tau = \frac{R^3}{8}(\mu^2 - \nu^2)\, d\mu\, d\nu\, d\phi$$

Then
$$S = \frac{R^3}{8\pi a_1^3}\int_1^\infty \int_{-1}^1 \int_0^{2\pi} e^{-R\mu/a_1}(\mu^2 - \nu^2)\, d\mu\, d\nu\, d\phi \tag{5-113}$$

since
$$e^{-(r_1+r_2)/a_1} = e^{-Ru/a_1}$$

by (4-39). Integrating over ϕ and ν gives

$$S = \frac{R^3}{2a_1^3}\int_1^\infty \mu^2 e^{-R\mu/a_1}\, d\mu - \frac{R^3}{6a_1^3}\int_1^\infty e^{-R\mu}\, d\mu \tag{5-114}$$

The two integrals in (5-114) are particular examples of the standard integral

$$\int_1^\infty x^n e^{-ax}\, dx = \frac{n!\, e^{-a}}{a^{n+1}} \sum_{k=0}^{n} \frac{a^k}{k!} \tag{5-115}$$

from which
$$S = e^{-R}\left(1 + R + \frac{R^2}{3}\right) \tag{5-116}$$

In evaluating H_{11} and H_{12}, let us include the proton repulsion in the potential energy so that we are working with the total energy of the ion and not just the electron energy. In this case, the Hamiltonian becomes

$$H = -\left[\frac{\hbar^2}{2m}\nabla^2 + \frac{e^2}{4\pi\epsilon_0 r_1} + \frac{e^2}{4\pi\epsilon_0 r_2} - \frac{e^2}{4\pi\epsilon_0 R}\right] \tag{5-117}$$

We then see that
$$H\phi_1 = E_H\phi_1 + \frac{e^2}{4\pi\epsilon_0}\left[-\frac{1}{r_2} + \frac{1}{R}\right]\phi_1 \tag{5-118}$$

where E_H is the ground state energy of the hydrogen atom, since the first two terms on the right of (5-117) are the Hamiltonian for the atom. From (5-118), it follows that

$$H_{11} = \int \phi_1 H \phi_1\, d\tau = E_H + \frac{e^2}{4\pi\epsilon_0 R} - E_{11} \tag{5-119}$$

where
$$E_{11} = \frac{e^2}{4\pi\epsilon_0}\int \frac{\phi_1^2\, d\tau}{r_2} \tag{5-120}$$

and
$$H_{12} = \int \phi_1 H \phi_2\, d\tau \tag{5-121}$$

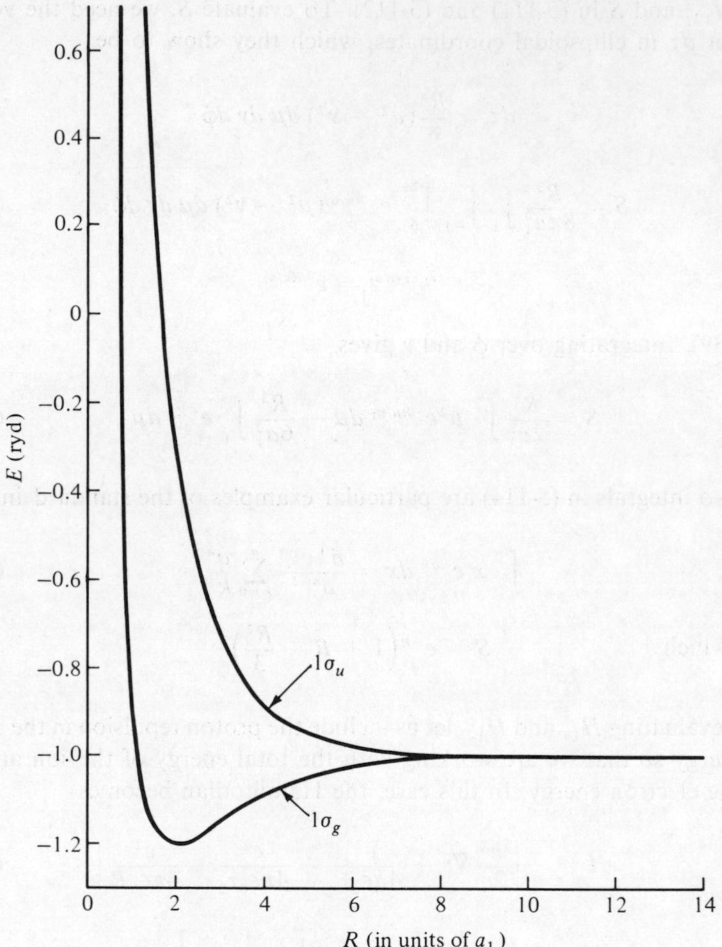

5-25 Energy as a function of proton spacing for the H_2^+ ion

or
$$H_{12} = E_H S + \frac{e^2}{4\pi\epsilon_0 R} S - E_{12} \tag{5-122}$$

where
$$E_{12} = \frac{e^2}{4\pi\epsilon_0} \int \frac{\phi_1 \phi_2}{r_1}\, d\tau \tag{5-123}$$

These two integrals may be evaluated just as S was, and we obtain

$$E_{11} = \frac{e^2}{4\pi\epsilon_0 R}\,[1 - e^{-2R}(1 + R)] \tag{5-124}$$

and
$$E_{12} = \frac{e^2}{4\pi\epsilon_0}\, e^{-R}(1 + R) \tag{5-125}$$

Substituting (5-121) and (5-125) into (5-107) gives

$$E_1 = E_H + \frac{e^2}{4\pi\epsilon_0 R} - \frac{E_{11} + E_{12}}{1 + S}$$
$$E_2 = E_H + \frac{e^2}{4\pi\epsilon R} - \frac{E_{11} - E_{12}}{1 - S} \tag{5-126}$$

Plotting E_1 and E_2 as a function of R, we obtain the curves of Fig. 5-25. We thus see that the symmetric bonding combination has a stable minimum at $R = 2.5a_1 = 1.25$ A, whereas the anti-bonding curve has no minimum.

PROBLEM 5-20

Figure 4-13 indicates that for $R = 2a_1$, the $1s\sigma$ curve gives an energy value of approximately

$$E = -2.2E_1$$

Show from this that the minimum in Fig. 5-25 has the right value.

Let us consider in more detail the integrals E_{11} and E_{12} which are known respectively as the *coulomb* and the *exchange* integral. Since $\phi_1^2\,d\tau$ is the probability of finding the electron in an orbit around proton 1, then $e^2\phi_1^2/4\pi\epsilon_0 r_2$ is simply a measure of the energy of attraction between proton 2 and the electron charge cloud around proton 1. The exchange integral, on the other hand, has no simple classical interpretation; it is purely quantum mechanical in character. If we were to make $\phi_2 = 0$, then we would have

$$E_1 = E_H + \frac{e^2}{4\pi\epsilon_0 R} - E_{11}$$

Hence, the presence of E_{12} lowers E_1 and, in fact, is responsible for the minimum; we ascribe this stable value to the *quantum mechanical exchange forces.*

The linear combinations

$$\psi = \frac{1}{\pi^{1/2}a_1^{3/2}}(e^{-r_1/a_1} \pm e^{-r_2/a_1}) \tag{5-127}$$

also permit us to look at the electron energy curves of Fig. 4-13 in a different way. A plot of these two functions along the internuclear axis would appear as shown in Fig. 5-26(a) (for a separation of $R = 8a_1$). As we let R go to zero, the LCAO method shows that the two functions approach the behavior

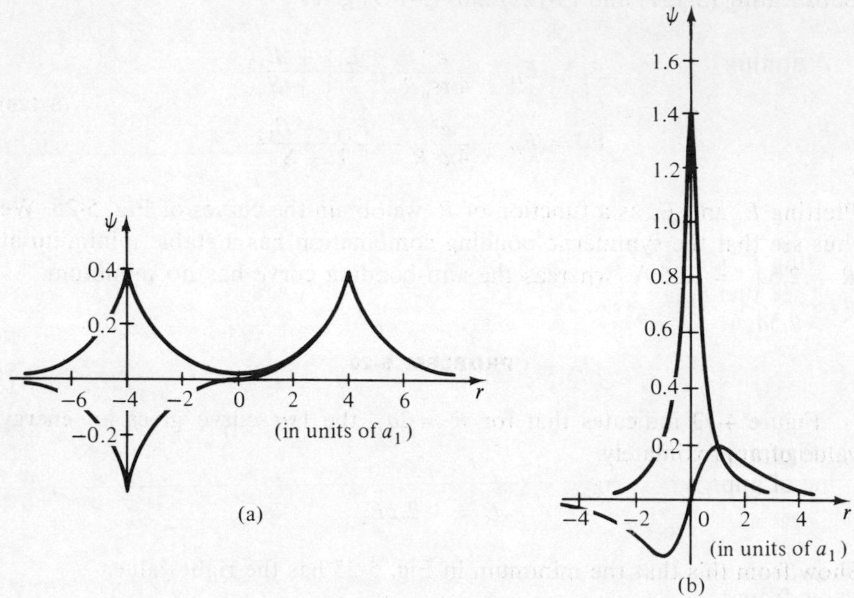

5-26 (a) The symmetric and antisymmetric LCAO functions for the H_2^+ ion; (b) the limiting functions as R goes to zero

indicated in Fig. 5-26(b). That is, the symmetric function becomes an s orbital which decays exponentially but is still symmetric, whereas the antisymmetric function assumes the lowest antisymmetric form available: the p function.

The LCAO method also provides some additional information on the correlation properties, as considered in connection with Fig. 4-13. Suppose we have a secular equation like (5-104) which was generated from a linear combination of orthonormal functions ϕ_1 and ϕ_2. Then

$$S_{11} = S_{22} = 1 \quad \text{and} \quad S_{12} = S_{21} = 0$$

so that (5-104) becomes

$$\begin{vmatrix} H_{11} - E & H_{12} \\ H_{12} & H_{22} - E \end{vmatrix} = 0 \tag{5-128}$$

Let us ask what the conditions are for the two roots of (5-128) to be equal. These are that both

$$H_{11} = H_{22} \tag{5-129}$$

and
$$H_{12} = 0 \tag{5-130}$$

Suppose that ϕ_1 is even and ϕ_2 is odd. Then $H\phi_2$ is also odd, so that $H_{12} =$

0. By adjusting R, it may also be possible to satisfy (5-129) and make the two roots equal, so that two corresponding curves in the correlation diagram can intersect. On the other hand, if ϕ_1 and ϕ_2 are of the *same* symmetry type, then H_{12} cannot vanish and the curves cannot intersect; this is the *non-crossing* rule. When used in conjunction with symmetry arguments concerning functions, it leads to a diagram of the type shown in Fig. 5-27, which is schematic rather than quantitative.

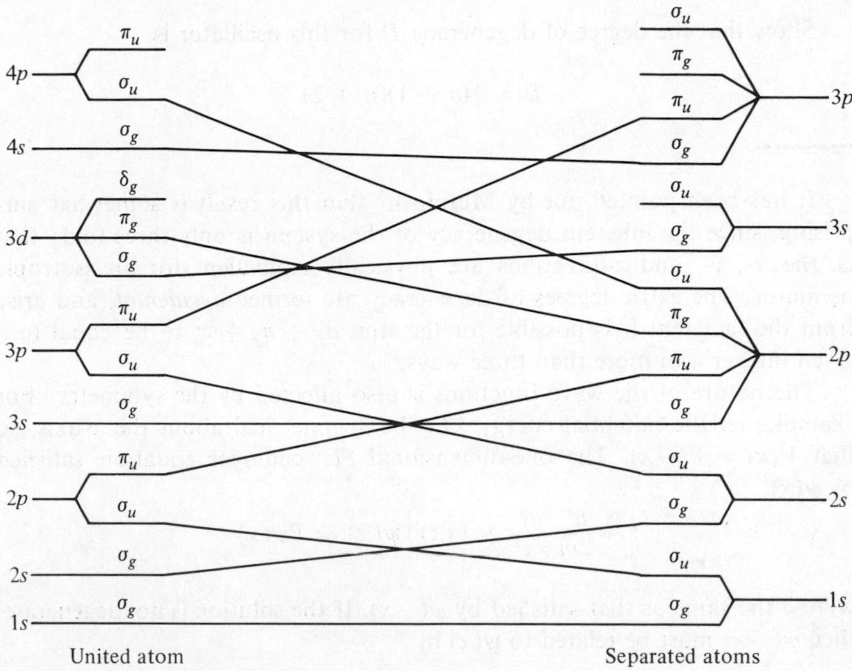

5-27 The correlation diagram for the homonuclear diatomic molecule

PROBLEM 5-21

Explain the notation and structure of the levels for $n = 3$.

5-6 The nearly-free electron approximation

It would be of value to consider in more detail the connection between symmetry and degeneracy. We have already stated—in connection with the two-dimensional isotropic oscillator of Sec. 3-4—that a degenerate energy E

corresponds to two or more solutions of the Schroedinger equation. Extending these considerations to a three-dimensional isotropic oscillator, the energy levels become

$$E = (n_x + n_y + n_z + \tfrac{3}{2})h\nu = (n + \tfrac{3}{2})h\nu$$

PROBLEM 5-22

Show that the degree of degeneracy D for this oscillator is

$$D = \tfrac{1}{2}(n + 1)(n + 2)$$

It has been pointed out by McIntosh[7] that this result is somewhat surprising, since the inherent degeneracy of the system is only three-fold; that is, the x-, y-, and z-directions are physically equivalent for an isotropic oscillator. The extra degrees of degeneracy are termed *accidental*, and arise from the fact that it is possible for the sum $n_x + n_y + n_z$ to be equal to a given integer n in more than three ways.

The nature of the wave functions is also affected by the symmetry. For example, let the potential energy $V(x)$ be symmetrical about the y-axis, so that $V(x) = V(-x)$. The one-dimensional Schroedinger equation satisfied by $\psi(x)$,

$$\left[-\frac{\hbar^2}{2m}\frac{d^2}{dx^2} + V(x)\right]\psi(x) = E\psi(x)$$

will be the same as that satisfied by $\psi(-x)$. If the solution is nondegenerate, then $\psi(-x)$ must be related to $\psi(x)$ by

$$\psi(-x) = c^2\psi(x)$$

where c is a constant. If the wave functions are normalized, then $c = \pm 1$, and $\psi(x)$ is either an even or an odd function. Hence, the simple symmetry for this example has imposed conditions on the solution, and we shall see in the more complicated cases to be discussed that the linear combinations of functions taken as solutions have symmetry properties consistent with the lattice to which they apply. This general principle is also true for single functions. For example, there are five d functions corresponding to the energy level E_3 for the hydrogen atom. If such an atom is placed inside a cubic crystal, so that the symmetry of the potential energy $V(\mathbf{r})$ is lowered from spherical to cubic, the character system of Table 5-4 shows that the five-fold

[7] H. V. McIntosh, *Am. J. Phys.*, **27**, 620 (1959).

degeneracy is split up into a three-fold energy belonging to Γ'_{25}, and a two-fold level for Γ_{12}. This phenomenon is called *crystal-field splitting*. On the other hand, the degeneracy of the E_2 level in hydrogen would be unaffected, since it goes in its original form with the Γ_{15} representation.

As another example of the connection between symmetry and degeneracy, consider the free electron energy bands of Figs. 5-22, and 5-24. This high degree of degeneracy should not be surprising, since $V(\mathbf{r}) = 0$ is a fully symmetric function. Hence, if we consider the effect of a potential energy with cubic symmetry, we would expect a reduction in the degeneracy. To see this in a very approximate way, let us explicitly treat the point at $k = 0.5$, $E = 0.25$ in Fig. 5-22. The free electron functions were found to be

$$X_1: \quad \phi = \cos\left(\frac{\pi x}{a}\right) \tag{5-88}$$

$$X'_4: \quad \phi = \sin\left(\frac{\pi z}{a}\right) \tag{5-89}$$

We shall use these to form a linear combination, obtaining a 2×2 secular determinant. Let

$$\begin{aligned} \phi_1 &= N_1 \cos\left(\frac{\pi z}{a}\right) \\ \phi_2 &= N_2 \sin\left(\frac{\pi z}{a}\right) \end{aligned} \tag{5-131}$$

Then
$$H_{11} = \int_0^a \phi_1 H \phi_1 \, dz$$

But
$$H\phi_1 = \left[-\frac{\hbar^2}{2m}\nabla^2 + V\right]\phi_1 = E_0\phi_1 + V\phi_1 \tag{5-132}$$

where E_0 is the energy when $V = 0$, Hence

$$H_{11} = E_0 \int_0^a \phi_1^2 \, dz + \int_0^a \phi_1^2 V \, dz = E_0 + V_{11}$$

where
$$V_{ij} = \int \phi_i V \phi_j \, dz \tag{5-133}$$

Similarly,
$$H_{22} = E_0 + V_{22}$$

But
$$H_{12} = \int_0^a \phi_1 H \phi_2 \, dz = E_0 \int \phi_1 \phi_2 \, dz + V_{12} = V_{12} = H_{21}$$

since ϕ_1 and ϕ_2 are orthogonal in the interval $(0, a)$. In the same way

$$S_{11} = S_{22} = 1, \qquad S_{12} = S_{21} = 0$$

so that the secular determinant becomes

$$\begin{vmatrix} E_0 + V_{11} - E & V_{12} \\ V_{12} & E_0 + V_{22} - E \end{vmatrix} = 0$$

or

$$\begin{aligned}(E_0 - E) &= \frac{(V_{11} + V_{22}) \pm \sqrt{(V_{11} + V_{22})^2 - 4(V_{11}V_{22} - V_{12}^2)}}{2} \\ &= \frac{(V_{11} + V_{22}) \pm \sqrt{(V_{11} - V_{22})^2 + 4V_{12}^2)}}{2}\end{aligned} \tag{5-134}$$

Since V_{11} involves $\cos^2 (\pi z/a)$, and V_{22} involves $\sin^2 (\pi z/a)$ but V_{12} involves the product $\sin (\pi z/a) \cos (\pi z/a)$, we would expect V_{12} to be smaller than either V_{11} or V_{22}.

PROBLEM 5-23

Establish this fact graphically.

This means that (5-134) becomes

$$(E_0 - E) = \begin{cases} V_{11} \\ V_{22} \end{cases}$$

or

$$E = \begin{cases} E_0 - V_{11} \\ E_0 - V_{22} \end{cases} \tag{5-135}$$

The lowest energy at the point X in the Brillouin zone now has two distinct values, separated by an amount $(V_{11} + V_{22})$, and the E vs. $\mathbf{k}$ curves appear as schematically indicated in Fig. 5-28. We say that the imposition of a cubic potential has removed the degeneracy, just as it does in the crystal field splitting example considered previously. In fact, this separation of the two levels is the energy gap we previously found as a consequence of a periodic potential. We note that this gap is small in comparison to the width of the allowed band; this is what we would expect if the potential energy is not significantly different from zero.

Our discussion in this section has been a mixture of qualitative and quantitative arguments. In order to see more precisely how the imposition of a triply periodic nonzero potential energy affects the degeneracies of the

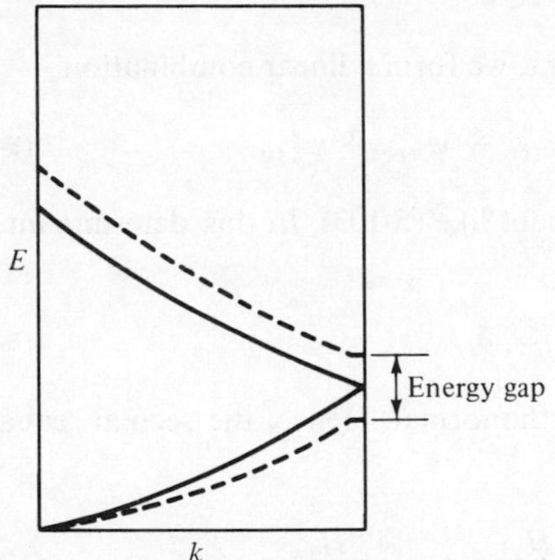

5-28 The effect of a non-zero potential on the lowest curves of Fig. 5-22

free electron energy bands, we should consider the secular equations in more detail. This is our next task.

5-7 *The method of orthogonalized plane waves*

The calculation of the previous section can be made more general by using a linear combination of functions of the form of (5-82), rather than the particular combinations given in (5-88) and (5-89). We recall that (5-82), or (5-76), follows from (5-63), which is a Bloch function. We may also call this a *Bloch plane wave*, since ψ has a periodic amplitude $u(\mathbf{r})$ and $e^{i\mathbf{k}\cdot\mathbf{r}}$ specifies a wave of propagation vector $\mathbf{k}$. If we use a linear combination of Bloch waves as the basis for producing a secular equation, and if we choose these functions to be orthogonal, then we are using the *orthogonalized plane wave* or *OPW* method to compute the energy band structure.

The free electron wave functions have the form

$$\psi(\mathbf{r}, \mathbf{k}) = e^{i(\mathbf{k} - 2\pi \mathbf{m}\mathbf{B}) \cdot \mathbf{r}} \tag{5-76}$$

Let us consider as an example the point M with an energy 0.5 in Fig. 5-24. There are four degenerate functions for this value of E, specified by

$$\mathbf{m} = (000), (100), (010), (110) \tag{5-136}$$

Since $\xi = \eta$ and $\zeta = 0$ along the Σ-axis, the corresponding functions are

$$\begin{aligned} \psi_{000} &= e^{i\pi(x+y)/a} \\ \psi_{100} &= e^{i\pi(-x+y)/a} \\ \psi_{010} &= e^{i\pi(x-y)/a} \\ \psi_{110} &= e^{i\pi(-x-y)/a} \end{aligned} \tag{5-137}$$

To find the actual energy at this point, we form a linear combination

$$\psi = \psi_{000} + \psi_{100} + \psi_{010} + \psi_{110} \tag{5-138}$$

and obtain a 4 × 4 secular determinant like (5-103). In this determinant, the overlap integrals become

$$S_{ij} = \delta_{ij} \tag{5-139}$$

since the functions in (5-138) are orthonormal. Hence, the secular equation simplifies to

$$\begin{vmatrix} H_{11} - E & H_{12} & H_{13} & H_{14} \\ H_{12} & H_{22} - E & H_{23} & H_{24} \\ H_{13} & H_{23} & H_{33} - E & H_{34} \\ H_{14} & H_{24} & H_{34} & H_{44} - E \end{vmatrix} = 0 \tag{5-140}$$

Any element H_{ij} has the form

$$H_{m_1m_2m_3;\, m'_1m'_2m'_3} = \int \psi^*_{m_1m_2m_3} \left[-\frac{\hbar^2}{2m}\nabla^2 + V \right] \psi_{m'_1m'_2m'_3}\, d\tau \tag{5-141}$$

where the indices m_1, m_2, m_3 are identical to m'_1, m'_2, m'_3 for $i = j$ and different if $i \neq j$. Using

$$\begin{aligned} \nabla^2 \psi_{m_1m_2m_3} &= \nabla^2 e^{i(\mathbf{k} - 2\pi m\mathbf{B}) \cdot \mathbf{r}} \\ &= -[(\mathbf{k} - 2\pi m\mathbf{B}) \cdot \mathbf{r}]^2\, \psi_{m_1m_2m_3} \end{aligned} \tag{5-142}$$

the integral in (5-141) becomes

$$H_{\mathbf{mm}'} = \int \psi^*_{\mathbf{m}} V \psi_{\mathbf{m}'}\, d\tau \tag{5-143}$$

when $\mathbf{m} \neq \mathbf{m}'$. This follows from the fact that $\psi^*_{\mathbf{m}}$ is orthogonal to $\psi_{\mathbf{m}'}$ and hence the part involving (5-142) vanishes. For $\mathbf{m} = \mathbf{m}'$, we get

$$H_{\mathbf{mm}'} = E_0 + \int \psi^*_{\mathbf{m}} V \psi_{\mathbf{m}'}\, d\tau \tag{5-144}$$

since

$$-\frac{\hbar^2}{2m}\nabla^2 \psi_{\mathbf{m}} = E_0 \psi_{\mathbf{m}}$$

where E_0 is the free electron energy.

To further simplify (5-140), we must consider the nature of the potential energy. It is well-known that a three-dimensional periodic function, such as the potential energy $V(x, y, z)$ in a crystal, can be expressed as a complex triple Fourier series of the form

$$V(x, y, z) = \sum_{-\infty}^{\infty} V_{n_1n_2n_3} e^{i\pi(n_1x+n_2y+n_3z)/a} \tag{5-145}$$

where n_1, n_2, and n_3 are integers and we are assuming a cubic region whose corners are at

$$(a, a, a) \quad (a, a, \bar{a}), \quad \ldots \quad (\bar{a}, \bar{a}, \bar{a})$$

The exponential functions in (5-145) are orthogonal over the interval $(-a, a)$. That is,

$$\int_{-a}^{a} e^{i\pi n_i x/a}\, e^{i\pi n_j x/a}\, dx = \delta_{ij} \tag{5-146}$$

Equation (5-145) expresses $V(x, y, z)$ in direct space. However, any cubic lattice has a reciprocal lattice which is also a cubic, so that $V(x, y, z)$ has a similar periodicity in reciprocal space. By (5-71), the reciprocal lattice points are specified as **mB**. Hence, we should write the Fourier series as

$$V(x, y, z) = \sum_{\mathbf{m}} V_{\mathbf{m}}\, e^{i\mathbf{mB}\cdot\mathbf{r}} \tag{5-147}$$

where $\mathbf{m} = (m_1, m_2, m_3)$ specifies triplets of integers and where

$$V_{\mathbf{m}} = V_{000},\ V_{100},\ V_{010}, \ldots \tag{5-148}$$

are the Fourier coefficients.

The elements in the secular determinant may now be determined. For the off-diagonal terms H_{ij}, we obtain

$$H_{ij} = \int_{-a}^{a}\int_{-a}^{a}\int_{-a}^{a} \psi_i^* V(x, y, z)\psi_j\, dx\, dy\, dz \tag{5-149}$$

where ψ_i and ψ_j are any of the four functions listed in (5-137). As a matter of notation, each of these functions should be specified by three indices, so that (5-149) becomes

$$H_{m_1m_2m_3;\, m'_1m'_2m'_3} = \int_{-a}^{a}\int_{-a}^{a}\int_{-a}^{a} \psi^*_{m_1m_2m_3}\{\sum_{\mathbf{m}} V_{\mathbf{m}}\, e^{i\mathbf{mB}\cdot\mathbf{r}}\}\psi_{m'_1m'_2m'_3}\, d\tau \tag{5-150}$$

Substituting any two functions of (5-137) into (5-150), the term $e^{-2\pi i(0.5z/a)}$ and its complex conjugate cancel, leaving only terms of the form $e^{-2\pi ix/a}$. The summation also consists of similar terms. For example, using ψ_{000} and ψ_{100}, we have

$$\begin{aligned} H_{000;\,100} = \int_{-a}^{a}\int_{-a}^{a}\int_{-a}^{a} &(1)[V_{000} + V_{100}\, e^{-2\pi ix/a} \\ &+ V_{010}\, e^{-2\pi iy/a} + \cdots]\,(e^{-2\pi ix})\, dx\, dy\, dz \end{aligned} \tag{5-151}$$

Since the only nonvanishing integral, by (5-146), is the one involving $(e^{-2\pi ix/a})^2$, it then follows that

$$H_{000;\,100} = V_{100}$$

In a similar way, we find relations such as

$$H_{000;\,110} = V_{110} \tag{5-152}$$

and

$$H_{100;\,110} = V_{010} \tag{5-153}$$

There are also relations among the Fourier coefficients which are a direct consequence of cubic symmetry. These are

$$V_{100} = V_{001} = V_{010} \tag{5-154}$$

By way of proof, consider a potential energy for which the coefficients in (5-154) are the only nonvanishing ones. Then

$$V(x, y, z) = V_{100}\, e^{-2\pi ix/a} + V_{010}\, e^{-2\pi iy/a} + V_{001}\, e^{-2\pi iz/a} \tag{5-155}$$

By symmetry

$$V(x, y, z) = V(y, x. z) \tag{5-156}$$

since V must conform to the cubic group. Combining this with (5-155) shows that

$$V_{100} = V_{010} \tag{5-157}$$

and so on.

Finally, we are ready to ask how group theory can be used to reduce the 4×4 secular determinant into smaller blocks. At the point M in the Brillouin zone, the little group is composed of E, C_2, JC_4^2, and JC_2. Using the four functions in (5-137) as a reducible basis, we obtain the character system

	E	C_2	JC_4^2	JC_2
χ_M	4	2	4	2

PROBLEM 5-24

Verify this character system.

Then by (2-63), we find

$$n_1 = 3, \qquad n_2 = 0, \qquad n_3 = 0, \qquad n_4 = 1 \tag{5-158}$$

Using the projection operators obtained from Table 5-10, we find the normalized linear combinations

$$\Sigma_1: \quad \psi = \begin{cases} \psi_{000} \\ \dfrac{1}{\sqrt{2}}(\psi_{100} + \psi_{010}) \\ \psi_{110} \end{cases} \tag{5-159}$$

and

$$\Sigma_4: \quad \psi = \frac{1}{\sqrt{2}}(\psi_{100} - \psi_{010}) \tag{5-160}$$

These functions are thus *symmetric combinations of plane waves* (SCPW) and when we use them, in place of the original basis, we find that any integral involving one function belonging to Σ_4 and any one of the three Σ_1 functions will vanish. For example,

$$\int (\psi^*_{100} - \psi^*_{010})V(\psi_{100} + \psi_{010})\, d\tau$$
$$= \int |\psi_{100}|^2 V\, d\tau - \int |\psi_{010}|^2 V\, d\tau = V_{100} - V_{010} = 0$$

by (5-154) and relations like (5-152). Working out the nonvanishing combinations, we find that the secular equation becomes

$$\begin{vmatrix} E_0+V_{000}-E & \sqrt{2}\,V_{100} & V_{110} & 0 \\ \sqrt{2}\,V_{100} & E_0+V_{000}-V_{110}-E & \sqrt{2}\,V_{100} & 0 \\ V_{110} & \sqrt{2}\,V_{100} & E_0+V_{000}-E & 0 \\ 0 & 0 & 0 & E_0+V_{000}+V_{110}-E \end{vmatrix} = 0 \tag{5-161}$$

where the 3×3 block belongs to Σ_1 and the 1×1 block to Σ_4.

The cubic corresponding to Σ_1 can be solved analytically. Let

$$V_{110} = g, \qquad V_{100} = d, \qquad E_0 + V_{000} = b$$

Then

$$(b - E)^2(b - E - g) + 4gd^2 - 4d^2(b - E) - g^2(b - E - g) = 0$$

or

$$(b - E - g)[(b - E)^2 - g^2 - 4d^2] = 0$$

from which

$$b - E - g = 0, \qquad b - E = \pm(g^2 + 4d^{21/2})$$

and

$$\left.\begin{aligned} E_1 &= b - g \\ E_2 &= b + (g^2 + 4d^2)^{1/2} \\ E_3 &= b - (g^2 + 4d^2)^{1/2} \end{aligned}\right\} \Sigma_1 \quad E = b + g\} \ \Sigma_4$$

Since

$$(V_{110}^2 + 4V_{100}^2)^{1/2} > |V_{110}| \tag{5-162}$$

we see that the root E_2 is the largest of the four and E_3 is the smallest. The intermediate Σ_1 root E_1 is then separated from the Σ_4 root E by an amount $2V_{110}$, and the potential $V(\mathbf{r})$ has removed the degeneracy. The resultant shape of the bands might appear as shown in Fig. 5-29.

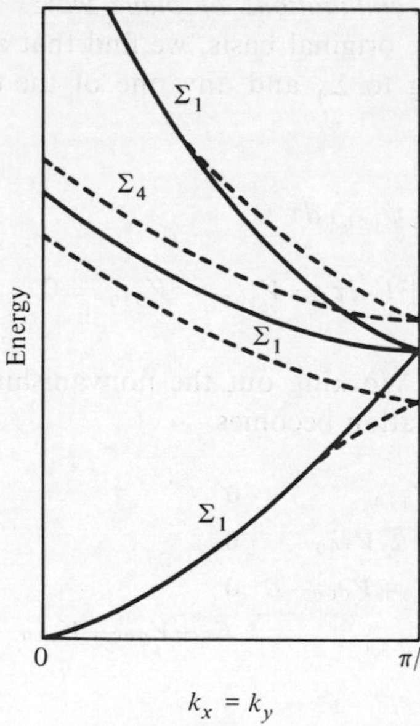

5-29 The effect of a non-zero potential on the lowest curves of Fig. 5-24

PROBLEM 5-25

Show that the degeneracies at the point M in Fig. 5-29 are incorrect, in general. However, it is possible to have a relation between V_{100} and V_{110} which does correspond to the situation of Fig. 5-29, as we shall see in a real material to be studied in Sec. 5-10. What is this relation?

5-8 *A two-dimensional example: the square lattice*

The transition from the free electron to the nearly-free approximation is beautifully demonstrated in an example given by Slater,[5] who considers a two-dimensional potential $V(x, y)$. The Fourier coefficients (in units of $\hbar^2/2ma^2$) are taken as

$$V_{00} = 0.0000$$

$$\left.\begin{aligned} V_{10} &= V_{01} = V_{\bar{1}0} = V_{0\bar{1}} = -0.0400 \\ V_{11} &= V_{\bar{1}1} = V_{1\bar{1}} = V_{\bar{1}\bar{1}} = -0.0250 \\ V_{20} &= V_{\bar{2}0} = V_{02} = V_{0\bar{2}} = -0.0100 \\ V_{21} &= V_{\bar{2}1} = V_{2\bar{1}} = V_{\bar{2}\bar{2}} = -0.0060 \\ V_{22} &= V_{\bar{2}2} = V_{2\bar{2}} = V_{\bar{2}\bar{2}} = -0.0015 \end{aligned}\right\} \tag{5-163}$$

and all other $V_{\mathbf{m}}$ are zero.

PROBLEM 5-26

Use the data of (5-163) to determine the shape of the potential well and show that it is physically reasonable.

The two-dimensional Brillouin zone is labelled as shown in Fig. 5-30, borrowing the symbols of Fig. 5-21. Slater's notation—which we shall use in a

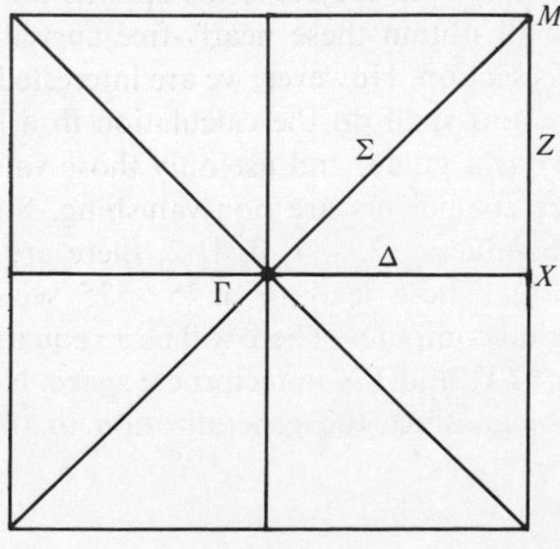

5-30 Brillouin zone for a square lattice

number of places in this chapter in order to assist in making the transition to his books—is expressed in terms of Jones symbols by Table 5-12. Using the methods discussed in Chapter 2, we readily obtain the character system shown in Table 5-13(a). The representation matrices for Γ_5 are also easy to obtain since the Jones symbols constitute a two-dimensional basis.

TABLE 5-12 ***Relation of Slater Symbols to Jones Symbols for a Square Lattice***

X_0	xy
X_1	$\bar{y}x$
X_{-1}	$y\bar{x}$
X_2	$\bar{x}\bar{y}$
Y_0	$x\bar{y}$
Y_1	yx
Y_{-1}	$\bar{y}\bar{x}$
Y_2	$\bar{x}y$

PROBLEM 5-27

Find the matrices for Γ_5 by using Table 5-12.

The same characters and representations apply to the point M; the little groups for Σ, Δ, and X are subgroups of the group for Γ. The compatibility relations for the points Γ, M, and X are shown in Table 5-14 while the free electron bands are given by Fig. 5-31(a). Figure 5-31(b) shows the energy bands obtained when the electron is not free but subject to the applied potential as specified by Eq. (5-163). We shall obtain these nearly-free curves by using the OPW method of the previous section. However, we are interested in the numerical results of this example and shall do the calculation in a less sophisticated way. Let us use (5-163) as a guide, and use only those values of m_1 and m_2 for which the Fourier coefficients are non-vanishing. Since either m_1 or m_2 can take on the five values -2, -1, 0, 1, 2, there are 25 different functions generated by (5-76). These lead to a 25×25 secular equation which can be solved on a digital computer. There will be an equation corresponding to each direction ΓM, MX, and ΓX in reciprocal space. Note that we refer to this as a cubic problem, since the generalization to three dimensions is fairly simple.

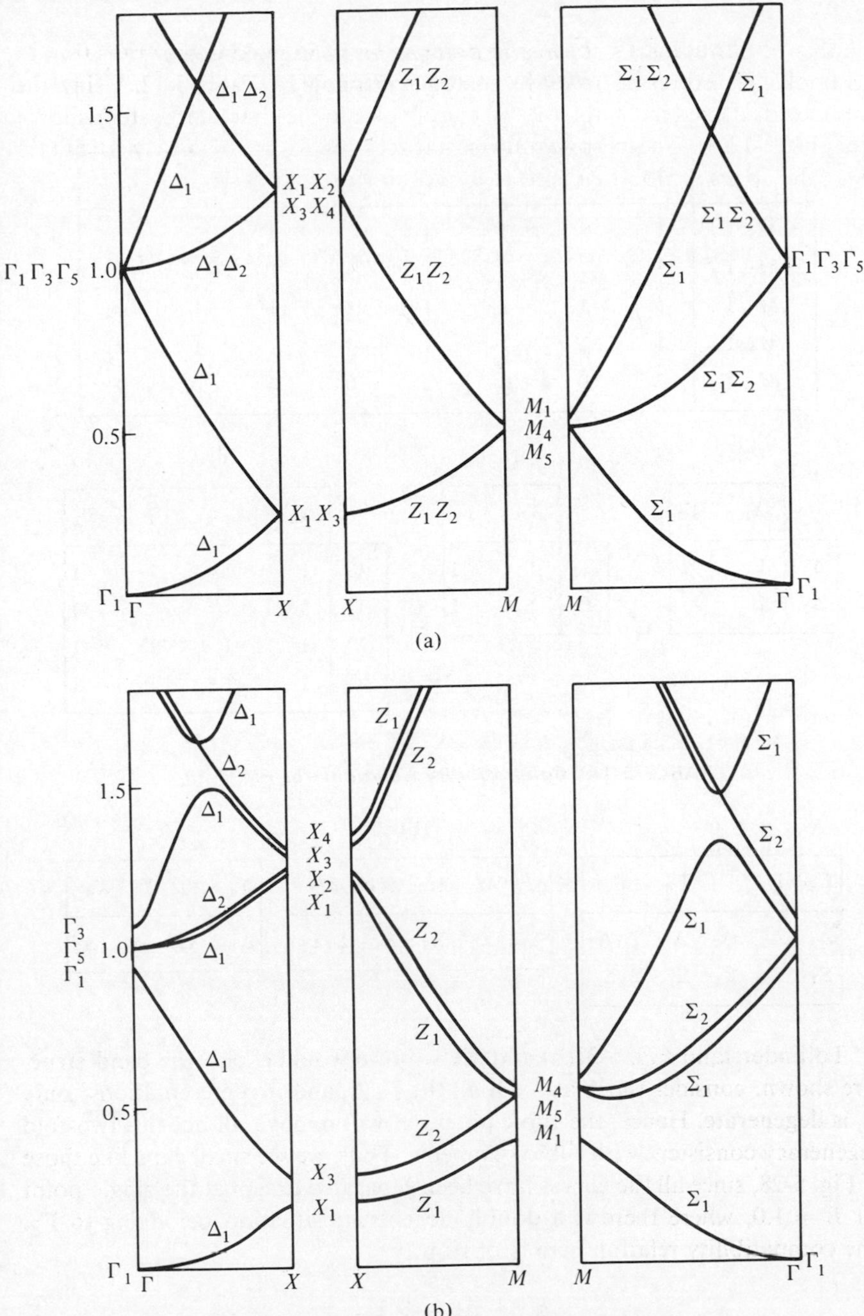

5-31 (a) Free electron energy curves for a square lattice; (b) corresponding curves for a non-zero potential

TABLE 5-13 *Character Systems for Points and Axes of High Symmetry in Fig. 5-30*

(a)

	X_0	X_1	X_{-1}	X_2	Y_0	Y_1	Y_{-1}	Y_2
M_1, Γ_1	1	1	1	1	1	1	1	1
M_2, Γ_2	1	1	1	1	−1	−1	−1	−1
M_3, Γ_3	1	−1	−1	1	1	−1	−1	1
M_4, Γ_4	1	−1	−1	1	−1	1	1	−1
M_5, Γ_5	2	0	0	−2	0	0	0	0

(b)

	X_0	Y_0
Δ_1	1	1
Δ_2	1	−1

(c)

	X_0	Y_1
Σ_1	1	1
Σ_2	1	−1

(d)

	X_0	X_2	Y_0	Y_2
X_1	1	1	1	1
X_2	1	1	−1	−1
X_3	1	−1	1	−1
X_4	1	−1	−1	1

TABLE 5-14 *Compatibility Relations for Fig. 5-30*

(a)

Γ_1	Γ_2	Γ_3	Γ_4	Γ_5
Δ_1	Δ_2	Δ_1	Δ_2	$\Delta_1\Delta_2$
Σ_1	Σ_2	Σ_2	Σ_1	$\Sigma_1\Sigma_2$

(b)

M_1	M_2	M_3	M_4	M_5
Σ_1	Σ_2	Σ_2	Σ_1	$\Sigma_1\Sigma_2$

(c)

X_1	X_2	X_3	X_4
Δ_1	Δ_2	Δ_1	Δ_2

To understand Fig. 5-31(b) and see why we would expect the band structure shown, consider the Δ-axis. Of all the Γ, Δ, and X representations, only Γ_5 is degenerate. Hence, the cubic potential will remove all but this two-fold degeneracy consistent with cubic symmetry. Thus, we see small gaps like those of Fig. 5-28, since all the curves have been separated except at the single point for $E = 1.0$, where there is a doubly-degenerate situation belonging to Γ_5. The compatibility relation here is obviously

$$\Delta_1 + \Delta_2 \longrightarrow \Gamma_5$$

as indicated in a slightly different way in Table 5-14. The lower energy curves behave much as we have already seen in simpler cases, but we notice something new in the upper portion of the figure. The intersection of the Δ_1 curve with the $\Delta_1\Delta_2$ curve at the top of Fig. 5-31(a) is converted into a maximum and a minimum in Fig. 5-31(b). This should not be surprising, since there really is no other way of removing the degeneracy in the Δ_1 levels which exists for $V = 0$. We could, in fact, have predicted the effect of the cubic potential by using some of the arguments just presented, and we shall do this later for a more complex problem.

PROBLEM 5-28

The two-dimensional form of (5-80) is

$$\psi = \exp\left[\frac{2\pi i}{a}\{\xi - m_1)x + (h - m_2)y\}\right]$$

For reasons of algebraic convenience, we shall take the corresponding form of (5-145) as

$$V(x, y) = \sum_{n_1, n_2} V_{n_1 n_2} e^{-2\pi i(n_1 x + n_2 y)/a}$$

To apply the OPW method, the solution to the Schroedinger equation should be taken as a linear combination of plane waves. Hence, we let this solution have the form

$$\psi = \sum_{m_1, m_2} f(\xi - m_1, h - m_2) \exp\left[\frac{2\pi i}{a}\{(\xi - m_1)x + (h - m_2)y\}\right]$$

where the quantities $f(\xi - m_1, h - m_2)$ are the Fourier coefficients of this expansion. Substitute the expressions for V and ψ into the Schroedinger equation, show that a 25×25 secular determinant is obtained, and determine the roots numerically.

5-9 The tight-binding or LCAO method

A brief introduction to the use of atomic orbitals to form symmetry-adapted functions was given in Sec. 5-5. In Sec. 5-3, we showed how the assumption of tightly bound electrons led to a specific family of E vs. $\mathbf{k}$ curves in the simple cubic lattice. We realize, of course, that the use of atomic orbitals is equivalent to the assumption of tight binding, for we are then saying that the crystal is formed from a set of atoms in their ground state, originally far apart, which are brought together with the proper spatial symmetry.

To treat the tight-binding or LCAO method in a more systematic way, following Dekker,[8] let us consider an atom with position-vector $\mathbf{R}_j$. An electron at $\mathbf{r}$ will have a coordinate $(\mathbf{r} - \mathbf{R}_j)$ with respect to this atom, and we will denote the atomic orbitals for this electron by $\phi(\mathbf{r} - \mathbf{R}_j)$. We then try a solution

$$\psi = \sum_j \exp(i\mathbf{k} \cdot \mathbf{R}_j)\phi(\mathbf{r} - \mathbf{R}_j) \tag{5-164}$$

where the sum is taken over all the atoms in the crystal. To show that this solution is in Bloch form, we let $\mathbf{r}_a$ denote the position-vector of a lattice point (which is not necessarily the same as an atomic position $\mathbf{R}_j$). Substituting $\mathbf{r} + \mathbf{r}_a$ for $\mathbf{r}$ in Eq. (5-164) and multiplying the right-hand side by exp $(i\mathbf{k} \cdot \mathbf{r}_a) \exp(-i\mathbf{k} \cdot \mathbf{r}_a)$ gives

$$\psi(\mathbf{r} + \mathbf{r}_a) = \exp(i\mathbf{k} \cdot \mathbf{r}_a) \sum_j \exp[i\mathbf{k} \cdot (\mathbf{R}_j - \mathbf{r}_a)]\phi[\mathbf{r} - (\mathbf{R}_j - \mathbf{r}_a)]$$

or

$$\psi(\mathbf{r} + \mathbf{r}_a) = \exp(i\mathbf{k} \cdot \mathbf{r}_a)\psi(\mathbf{r})$$

Now multiply the Schroedinger equation (4-2) for ψ by ψ^* and then integrate over all space, obtaining

$$\int \psi^* \left[-\frac{\hbar^2}{2m} \nabla^2 + V \right] \psi \, d\tau - E \int \psi^* \psi \, d\tau = 0 \tag{5-165}$$

It is convenient to decompose V into two terms, as follows

$$V(\mathbf{r}) = V_a(\mathbf{r} - \mathbf{R}_j) + V'(\mathbf{r} - \mathbf{R}_j) \tag{5-166}$$

where V_a is the potential energy which an electron would have in a single, isolated atom and V' is the additional potential energy it would acquire when the atom is incorporated into the crystal. Substituting this into Eq. (5-165)

$$\int \psi^* \left[-\frac{\hbar^2}{2m} \nabla^2 + V_a + V' \right] \psi \, d\tau - E \int \psi^* \psi \, d\tau = 0 \tag{5-167}$$

The second integral on the left we will evaluate by substituting from (5-164). This results in a sum of products of atomic orbitals, and if we use the assumption that the atoms are widely spaced, then integrals involving functions centered around two different atoms will vanish. That is, integrals of the form

[8] A. J. Dekker, *Solid State Physics*, Prentice-Hall, Inc., Englewood Cliffs, N.J., 1957.

$$\int \psi^*(\mathbf{r} - \mathbf{R}_i)\psi(\mathbf{r} - \mathbf{R}_j)\, d\tau$$

are negligible, but integrals like

$$\int \psi^*(\mathbf{r} - \mathbf{R}_j)\psi(\mathbf{r} - \mathbf{R}_j)\, d\tau$$

around a single atom reduce to unity for normalized functions. Since there are N such integrals, where N is the number of atoms, the second integral becomes EN.

The first term in (5-167) can be broken up into two parts, one of which is chosen as

$$\int \psi^* \left[-\frac{\hbar^2}{2m}\nabla^2 + V_a \right] \psi\, d\tau = NE_a \qquad \textbf{(5-168)}$$

where E_a is the energy of the electron in the free atom. This leaves only the term involving V' to be considered. We shall neglect all terms in the summations except those involving immediate neighbors. Define

$$\begin{aligned} \alpha &= -N \int \phi^*(\mathbf{r} - \mathbf{R}_j)V'\phi(\mathbf{r} - \mathbf{R}_j)\, d\tau \\ \beta &= -N \int \phi^*(\mathbf{r} - \mathbf{R}_m)V'\phi(\mathbf{r} - \mathbf{R}_j)\, d\tau \end{aligned} \qquad \textbf{(5-169)}$$

where j and m are neighbors and where the negative signs are used to make α and β positive. If we assume for the moment that ϕ is spherically symmetric, corresponding to the s function of Fig. (4-3), then the β integrals are the same for all nearest neighbors. Using (5-168) and (5-169), Eq. (5-167) becomes

$$E = E_a - \alpha - \beta \sum_m \exp\,[i\mathbf{k} \cdot (\mathbf{R}_j - \mathbf{R}_m)] \qquad \textbf{(5-170)}$$

where the sum is over the nearest neighbors of atom j.

The nearest neighbors of an atom in the simple cubic lattice are specified by

$$(\pm a, 0, 0), \qquad (0, \pm a, 0), \qquad (0, 0, \pm a)$$

so that the energy becomes

$$E = E_a - \alpha - 2\beta(\cos k_x a + \cos k_y a + \cos k_z a) \qquad \textbf{(5-171)}$$

For a two-dimensional square lattice, let $k_z = 0$ in this equation, obtaining

$$E - E_a + \alpha = -2\beta(\cos k_x a + \cos k_y a)$$

where E_a, α, and β are unknown constants. Since these constants do not affect the form of the relation between E and k, we can replace E by a reduced energy E_r, defined as

$$E_r = \frac{E - E_a + \alpha}{-2\beta} = (\cos k_x a + \cos k_y a) \tag{5-172}$$

This is identical to (5-68), which means that the results of Fig. 5-20 are valid for more than that specific numerical example.

PROBLEM 5-29

Find the expressions for the constant energy surfaces of a body-centered cubic lattice and of a face-centered cubic lattice.

Next, we shall consider the tight-binding approximation in the simple cubic lattice when the functions $\phi(\mathbf{r} - \mathbf{R}_j)$ are taken as one of the p functions of Fig. 4-3. The integrals of Eq. (5-169) are different for p functions from those for s functions, and further, the integrals are not the same for all nearest neighbors because of the directional properties of p_x, p_y, and p_z. For convenience, consider just the p_x function, and denote the integral corresponding to α by α_p. There will be a function p_x located at the origin, and six more located at each of six nearest neighbors in a simple cubic lattice. Then there are two kinds of integrals of the form of β in Eq. (5-169), denoted by β_{p1} and β_{p2}, respectively, and we have

$$E = E_{ap} - \alpha_p - 2\beta_{p1} \cos k_x a - 2\beta_{p2}(\cos k_y a + \cos k_z a)$$

where
$$\alpha_p = -N \int \phi^*(0) V' \phi(0)\, d\tau$$

$$\beta_{p1} = -N \int \phi^*(a, 0, 0) V' \phi(0)\, d\tau \tag{5-173}$$

$$\beta_{p2} = -N \int \phi^*(0, a, 0) V' \phi(0)\, d\tau$$

and where, $\phi(0)$ is a function located at the origin, $\phi(a, 0, 0)$ is located at $x = a$, and $\phi(0, a, 0)$ is located at $y = a$ (moving this function to $y = -a$ or to $z = \pm a$ gives the same integral).

Now let us extend these results to a crystal with several nonequivalent atoms per unit cell, and express ϕ_k as a linear combination of all three p functions. Then there will be an expression of the form (5-164) for each atom in the unit cell; that is,

$$\psi = \sum_j \exp(i\mathbf{k} \cdot \mathbf{R}_j)\phi_n(\mathbf{r} - \mathbf{R}_j) \tag{5-174}$$

where we have used the convention

$$\phi_1 = p_x, \qquad \phi_2 = p_y, \qquad \phi_3 = p_z$$

in order to utilize the summation notation. The general solution to the Schroedinger equation is then

$$\psi = \sum_{j,n,s} B_{ns} \exp(i\mathbf{k} \cdot \mathbf{R}_{js})\phi_n(\mathbf{r} - \mathbf{R}_{js}) \tag{5-175}$$

where the B_{ns} are constants to be determined, and the subscript s refers to the nonequivalent atoms. Again writing V as $V_a + V'$, introducing this $\phi_k(r)$ into the Schroedinger equation, and multiplying by each of the ϕ_n^* in turn, followed by an integration, we obtain a series of homogeneous equations in the unknowns B_{ns}. Requiring the determinant of these coefficients to vanish again yields a secular equation, and examples will be given shortly. For a crystal of arbitrary complexity, and making use of combinations of s, p, and d functions, we would expect the resulting calculations to involve a large number of integrals of the form of (5-175), which we shall denote by the symbol

$$(\phi_n^* \mid V' \mid \phi_m) = N \int \phi_n^* V' \phi_m \, d\tau \tag{5-176}$$

We shall find that the tight-binding method leads to high order secular equations involving these integrals, and one of the key applications of group theory to solid state physics concerns itself with the choice of linear combinations of atomic orbitals which automatically reduce the degree of the secular determinant.

Slater and Koster[9] have shown how to conveniently handle the geometrical considerations involved in the tight-binding method. To introduce their results, choose some atom in a crystal as the origin O of a spherical coordinate system and consider the vector $\mathbf{r}$ determined by any one of its neighbors located at the point P (Fig. 5-32). The direction cosines l, m, and n of $\mathbf{r}$ with respect to the rectangular coordinates are given by

$$l = \frac{x}{r} = \cos\phi \sin\theta, \qquad m = \frac{y}{r} = \sin\phi \sin\theta, \qquad n = \frac{z}{r} = \cos\theta \tag{5-177}$$

We wish to set up a second coordinate system $OX'Y'Z'$ with the same origin at $OXYZ$ and with the OZ'-axis lying along OP. The transformation matrix will be designated by

$$\begin{array}{c} \\ OX' \\ OY' \\ OZ' \end{array} \begin{array}{c} \begin{array}{ccc} OX & OY & OZ \end{array} \\ \begin{pmatrix} a_{11} & a_{12} & a_{13} \\ a_{21} & a_{22} & a_{23} \\ a_{31} & a_{32} & a_{33} \end{pmatrix} \end{array}$$

[9] J. C. Slater and G. F. Koster, *Phys. Rev.*, **94**, 1498 (1954).

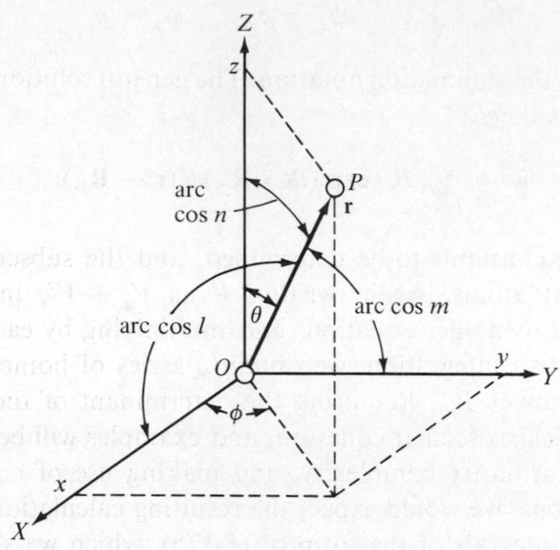

5-32 Coordinates used for two-center integrals

so that $a_{13} = l, \qquad a_{23} = m, \qquad a_{33} = n$ **(5-178)**

Consider now an integral of the form (5-176) with ϕ_n^* and ϕ_m both taken, for example, as p'_x, the function p_x in the new coordinate system. Writing p'_x simply as x', in accordance with the discussion of Sec. 4-4, we may, by Table 4-5, express the transformed functions as

$$\begin{pmatrix} x' \\ y' \\ z' \end{pmatrix} = \begin{pmatrix} a_{11} & a_{12} & a_{13} \\ a_{21} & a_{22} & a_{23} \\ a_{31} & a_{32} & a_{33} \end{pmatrix} \begin{pmatrix} \frac{\sqrt{3}}{2\sqrt{\pi}} \sin\theta\cos\phi \\ \frac{\sqrt{3}}{2\sqrt{\pi}} \sin\theta\sin\phi \\ \frac{\sqrt{3}}{2\sqrt{\pi}} \cos\theta \end{pmatrix} \tag{5-179}$$

However, it is more convenient for purposes of manipulation to use the functions of Table 4-4, although the linear combinations are easier to interpret physically. To reintroduce these functions, we take the three equations of the form:

$$x' = a_{11}x + a_{12}y + a_{13}z$$

and rewrite them as

$$\begin{aligned} x' &= \tfrac{1}{2}[(a_{11} + ia_{12})(x - iy)] + \tfrac{1}{2}[(a_{11} - ia_{12})(x + iy)] + a_{13}z \\ y' &= \tfrac{1}{2}[(a_{21} + ia_{22})(x - iy)] + \tfrac{1}{2}[(a_{21} - ia_{22})(x + iy)] + a_{23}z \end{aligned} \tag{5-180}$$

(with a similar expression for z'). We note that

$$\begin{aligned} x + iy &= \sqrt{2}\, p_{+1} \propto \sin\theta\, e^{i\phi} \\ x - iy &= \sqrt{2}\, p_{-1} \propto \sin\theta\, e^{-i\phi} \\ z &= p_0 \propto \cos\theta \end{aligned} \tag{5-181}$$

The integral we are interested in is then

$$\begin{aligned} (x'^* \,|\, V' \,|\, x') = (\{&\tfrac{1}{2}[(a_{11} - ia_{12})(x + iy)] \\ &+ \tfrac{1}{2}[(a_{11} + ia_{12})(x - iy)] + a_{13}z\} \,|\, V' \,| \\ &\{\tfrac{1}{2}[(a_{11} + ia_{12})(x - iy)] \\ &+ \tfrac{1}{2}[(a_{11} - ia_{12})(x + iy)] + a_{12}z\}) \end{aligned} \tag{5-182}$$

If we expand this expression and work out each of the nine integrals, we find that some of them will immediately vanish, since

$$\int_0^{2\pi} \exp(im_l\phi)\, d\phi = 0 \tag{5-183}$$

when $m_l = \pm 1, \pm 2, \ldots$. The nonvanishing integrals in this example are of two types, and the following special symbols are used for them:

$$\begin{aligned} (p_0^* \,|\, V' \,|\, p_0) &= \frac{3}{4\pi}(\cos\theta \,|\, V' \,|\, \cos\theta) = (pp\sigma) \\ (p_{\pm 1}^* \,|\, V' \,|\, p_{\pm 1}) &= \frac{3}{8\pi}(\sin\theta\, e^{\mp i\phi} \,|\, V' \,|\, \sin\theta\, e^{\pm i\phi}) \\ &= \frac{3}{8\pi}(\sin\theta \,|\, V' \,|\, \sin\theta) = (pp\pi) \end{aligned} \tag{5-184}$$

where the symbols σ and π in the notation $(pp\sigma)$ and $(pp\pi)$, respectively, refer to the quantum number λ of the H_2^+ ion, discussed in Sec. 4-6. The situation of Fig. 5-32 is physically equivalent to a single electron bound to two protons.

Using Eqs. (5-181), (5-183), and (5-184), Eq. (5-182) becomes

$$\begin{aligned} (x'^* \,|\, V' \,|\, x) &= \tfrac{1}{2}[(a_{11}^2 + a_{12}^2)](\sqrt{2}\, p_{+1} \,|\, V' \,|\, \sqrt{2}\, p_{-1}) \\ &\quad + a_{13}^2(p_0 \,|\, V' \,|\, p_0) = (1 - l^2)(pp\pi) + l^2(pp\sigma) \end{aligned} \tag{5-185}$$

where we have used the well-known property of an orthogonal matrix:

$$a_{11}^2 + a_{12}^2 + a_{13}^2 = 1 \tag{5-186}$$

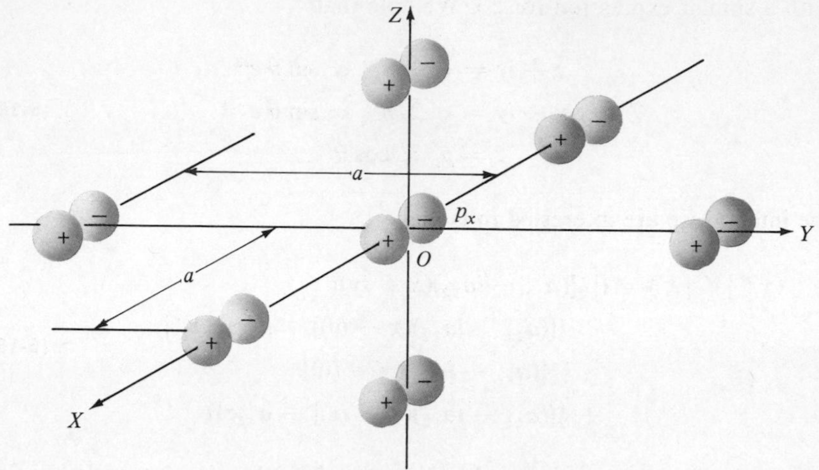

5-33 Arrangement of nearest-neighbor p_x functions

A somewhat more involved example is the integral $(x'^* \mid V' \mid x'y')$, since the normalization constants must be considered. We shall need the additional symbols:

$$(p_0^* \mid V' \mid d_0) = \frac{\sqrt{15}}{8\pi} (\cos\theta \mid V' \mid 3\cos^2\theta - 1) = (pd\sigma)$$
$$(p_{+1}^* \mid V' \mid d_{+1}) = \frac{\sqrt{45}}{8\pi} (\sin\theta \mid V' \mid \sin\theta\cos\theta) = (pd\pi) \tag{5-187}$$

From (5-180), we have that

$$\begin{aligned} x'y' = {} & \tfrac{1}{2}[(a_{11}a_{21} + a_{12}a_{22})] + \tfrac{1}{2}[(a_{11} + ia_{12})(x - iy)a_{23}z] \\ & + \tfrac{1}{2}[(a_{11} - ia_{12})(x + iy)a_{23}z] \\ & + \tfrac{1}{2}[(a_{21} + ia_{22})(x - iy)a_{13}z] \\ & + \tfrac{1}{2}[(a_{21} - ia_{22})(x + iy)a_{13}z] + a_{13}a_{23}z^2 \end{aligned} \tag{5-188}$$

The first and last terms of (5-188) can be combined to give:

$$\begin{aligned} -\frac{lm}{2}(x^2 + y^2) + lmz^2 &= lm[z^2 - \tfrac{1}{2}(x^2 + y^2)] \\ &= \frac{lm}{2}[3z^2 - r^2] \end{aligned} \tag{5-189}$$

since
$$x^2 + y^2 + z^2 = r^2$$

This term is the sole contributor to the σ integral, since all other terms in

(5-188) involve $e^{\pm i\phi}$. The π contribution comes from the remaining terms of (5-180); multiplying by x'^* and integrating gives

$$\begin{aligned}\tfrac{1}{2}[(a_{11}^2 + a_{12}^2)a_{23} + (a_{11}a_{21} + a_{12}a_{22})a_{13}(x + iy \,|\, V' \,|\, \{x - iy\}z)] \\ = \tfrac{1}{2}[(1 - 2l^2)m](x + iy \,|\, V' \,|\, \{x - iy\}z)\end{aligned} \tag{5-190}$$

The complete integral is then

$$\begin{aligned}(x'^* \,|\, V' \,|\, x'y') &= \frac{l^2 m}{2}(z \,|\, V' \,|\, 3z^2 - r^2) \\ &\quad + \tfrac{1}{2}[(1 - 2l^2)m](x + iy \,|\, V' \,|\, \{x - iy\}z) \\ &= \frac{l^2 m}{2}(\cos\theta \,|\, V' \,|\, 3\cos^2\theta - 1) \\ &\quad + \tfrac{1}{2}[(1 - 2l^2)m](\sin\theta \,|\, V' \,|\, \sin\theta\cos\theta)\end{aligned} \tag{5-191}$$

This can be written in terms of symbols as

$$(x'^* \,|\, V' \,|\, x'y') = \sqrt{3}\, l^2 m(pd\sigma) + (1 - 2l^2)(pd\pi) \tag{5-192}$$

Similar expressions have been worked out in the Slater and Koster[9] paper for all combinations of s, p, and d functions, and the results are given in Table 5-15. In this table, the simpler notation

$$(x'^* \,|\, V' \,|\, x'y') = (x, zy) \tag{5-193}$$

is used, and any entries not given can be found by simple permutation of the coordinates and direction cosines.

The formulas of Table 5-15 can now be applied to actual lattices and one example which we shall use shortly is the simple cubic. The nearest neighbors have already been stated to have coordinates $(\pm a, 0, 0)$, $(0, \pm a, 0)$, and $(0, 0, \pm a)$, and we are interested in evaluating terms corresponding to the last term on the right of (5-170). As we have already shown, for nearest neighbors, this gives

$$(ss\sigma)_1 \sum \exp[ik \cdot (R_j - R_m)] = 2(ss\sigma)_1(\cos\xi + \cos\eta + \cos\zeta) \tag{5-194}$$

where the subscript 1 denotes first-nearest neighbors, $(ss\sigma)_1$ corresponds to what we have previously called β, and we use the abbreviations

$$\xi = k_x a, \qquad \eta = k_y a, \qquad \zeta = k_z a \tag{5-77}$$

There are twelve second-nearest neighbors in a simple cubic lattice, located at points like $(a, a, 0)$ and the summation term is

$$(ss\sigma)_2\,[\exp[i(k_x+k_y)a]+\exp[i(k_x-k_y)a] + (\text{similar combinations})]$$
$$= 2(ss\sigma)_2[\cos(\xi+\eta)+\cos(\xi-\eta)+\cos(\eta+\zeta) + \cos(\eta-\zeta)+\cos(\zeta+\xi)+\cos(\zeta-\xi)] \tag{5-195}$$
$$= 4(ss\sigma)_2[\cos\xi\cos\eta+\cos\eta\cos\zeta+\cos\zeta\cos\xi]$$

Similarly, for eight third-nearest neighbors, located at the points ($\pm a$, $\pm a$, $\pm a$), the corresponding term is

$$2(ss\sigma)_3[\cos(\xi+\eta+\zeta)+\cos(\xi-\eta-\zeta) + \cos(-\xi+\eta-\zeta)+\cos(-\xi-\eta+\zeta)] = 8(ss\sigma)_3[\cos\xi\cos\eta\cos\zeta] \tag{5-196}$$

Denoting the integral for $j=m$ by s_0, and the sum of all integrals out to third-nearest neighbors by $(s|s)$, we obtain the first entry of Table 5-16, which shows how to obtain the nearest-neighbor integrals from the two-center integrals of Table 5-15.

This process for the p states is somewhat more involved since we have integrals of both the σ type and π type to consider, corresponding respectively to β_{p1} and β_{p2} of (5-173). To indicate how the entry $(x|x)$ of Table (5-16) is calculated, we realize from Fig. 5-33 that a function p_x located at O will give rise to two π integrals with its first-nearest neighbors on the y- and z-axes. The summation term for first-nearest neighbors is then

$$(pp\sigma)_1[\exp(ik_xa)+\exp(-ik_xa)]+(pp\pi)_1[\exp(ik_ya) + \exp(-ik_ya)+\exp(ik_za)+\exp(-k_za)] = 2(pp\sigma)_1\cos\xi+2(pp\pi)_1(\cos\eta+\cos\zeta) \tag{5-197}$$

The geometrical arrangement for second-nearest neighbors is indicated by Fig. 5-34, where we have sketched the p_x orbitals for the four neighbors lying in the xy-plane. These orbits do not form simple σ integrals and π integrals, as for nearest neighbors, but instead, we must go back to Table 5-15, using the entry

$$(x,x)=l^2(pp\sigma)+(1-l^2)(pp\pi) \tag{5-198}$$

which becomes
$$(x,x)_2=\tfrac{1}{2}[(pp\sigma)_2+(pp\pi)_2] \tag{5-199}$$

where $l=\sqrt{2}/2$ for a 45° angle. The summation (5-170) is

$$\exp[i(\xi+\eta)]+\exp[i(-\xi+\eta)]+\exp[i(-\xi-\eta)] + \exp[i(\xi-\eta)] = 4\cos\xi\cos\eta \tag{5-200}$$

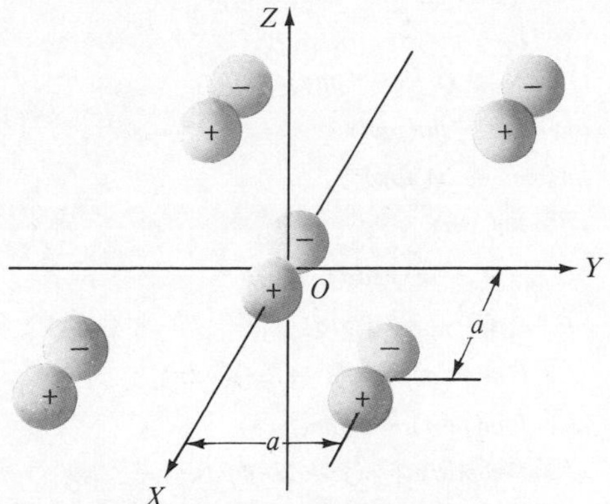

5-34 Arrangement of second-nearest p_x functions

so that the xy-plane gives a contribution $2 \cos \xi \cos \eta[(pp\sigma)_2 + (pp\pi)_2]$ to the complete integral. The xz-plane has a similar contribution but the yz-plane produces pure π integrals, giving a contribution $4 \cos \eta \cos \xi (pp\pi)_2$. The total result gives the second-nearest neighbor term in the table.

The third-nearest neighbors are shown in Fig. 5-35 and all eight integrals involve σ and π terms. By Table 5-15

$$(x, x)_3 = \tfrac{1}{3}(pp\sigma)_3 + \tfrac{2}{3}(pp\pi)_3 \tag{5-201}$$

and combining this with the summation as previously evaluated in (5-196) gives the last term of the $(x \mid x)$ entry in Table 5-16.

In this discussion, we have stressed primarily the manipulative features of the Slater-Koster method. Our subsequent considerations will involve some of the applications.

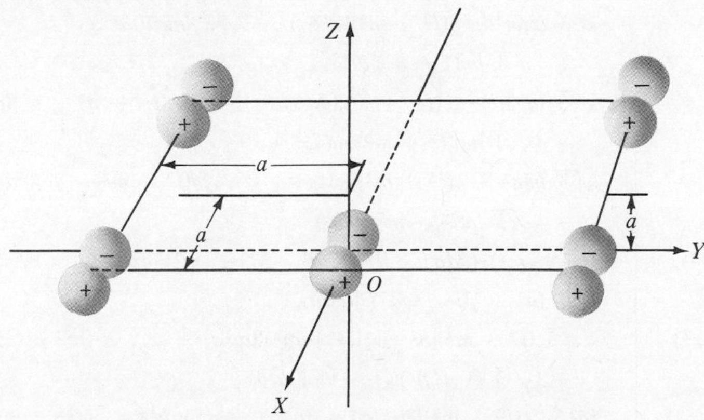

5-35 Arrangement of third-nearest p_x functions

TABLE 5-15 *Two-Center Integrals in Terms of Atomic-Orbital Integrals*

(s, s)	$(ss\sigma)$
(s, x)	$l(sp\sigma)$
(x, x)	$l^2(pp\sigma) + (1 - l^2)(pp\pi)$
(x, y)	$lm(pp\sigma) - lm(pp\pi)$
(x, z)	$ln(pp\sigma) - ln(pp\pi)$
(s, xy)	$\sqrt{3}\, lm(sd\sigma)$
$(s, x^2 - y^2)$	$\frac{1}{2}\sqrt{3}\,(l^2 - m^2)(sd\sigma)$
$(s, 3z^2 - r^2)$	$[n^2 - \frac{1}{2}(l^2 + m^2)](sd\sigma)$
(x, xy)	$\sqrt{3}\, l^2m(pd\sigma) + m(1 - 2l^2)(pd\pi)$
(x, yx)	$\sqrt{3}\, lmn(pd\sigma) - 2lmn(pd\pi)$
(x, zx)	$\sqrt{3}\, l^2n(pd\sigma) + n(1 - 2l^2)(pd\pi)$
$(x, x^2 - y^2)$	$\frac{1}{2}\sqrt{3}\, l(l^2 - m^2)(pd\sigma) + l(1 - l^2 + m^2)(pd\pi)$
$(y, x^2 - y^2)$	$\frac{1}{2}\sqrt{3}\, m(l^2 - m^2)(pd\sigma) - m(1 + l^2 - m^2)(pd\pi)$
$(z, x^2 - y^2)$	$\frac{1}{2}\sqrt{3}\, n(l^2 - m^2)(pd\sigma) - n(l^2 - m^2)(pd\pi)$
$(x, 3z^2 - r^2)$	$l[n^2 - \frac{1}{2}(l^2 + m^2)](pd\sigma) - \sqrt{3}\, ln^2(pd\pi)$
$(y, 3z^2 - r^2)$	$m[n^2 - \frac{1}{2}(l^2 + m^2)](pd\sigma) - \sqrt{3}\, mn^2(pd\pi)$
$(z, 3z^2 - r^2)$	$n[n^2 - \frac{1}{2}(l^2 + m^2)](pd\sigma) + \sqrt{3}\, n(l^2 + m^2)(pd\pi)$
(xy, xy)	$3l^2m^2(dd\sigma) + (l^2 + m^2 - 4l^2m^2)(dd\pi) + (n^2 + l^2m^2)(dd\delta)$
(xy, yz)	$3lm^2n(dd\sigma) + ln(1 - 4m^2)(dd\pi) + ln(m^2 - 1)(dd\delta)$
(xy, zx)	$3l^2mn(dd\sigma) + mn(1 - 4l^2)(dd\pi) + mn(l^2 - 1)(dd\delta)$
$(xy, x^2 - y^2)$	$\frac{3}{2}lm(l^2 - m^2)(dd\sigma) + 2lm(m^2 - l^2)(dd\pi) + \frac{1}{2}lm(l^2 - m^2)(dd\delta)$
$(yz, x^2 - y^2)$	$\frac{3}{2}mn(l^2 - m^2)(dd\sigma) - mn[1 + 2(l^2 - m^2)](dd\pi)$ $+ mn[1 + \frac{1}{2}(l^2 - m^2)](dd\delta)$
$(zx, x^2 - y^2)$	$\frac{3}{2}nl(l^2 - m^2)(dd\sigma) + nl[1 - 2(l^2 - m^2)](dd\pi)$ $- nl[1 - \frac{1}{2}(l^2 - m^2)](dd\delta)$
$(xy, 3z^2 - r^2)$	$\sqrt{3}\, lm[n^2 - \frac{1}{2}(l^2 + m^2)](dd\sigma) - 2\sqrt{3}\, lmn^2(dd\pi)$ $+ \frac{1}{2}\sqrt{3}\, lm(1 + n^2)(dd\delta)$
$(yz, 3z^2 - r^2)$	$\sqrt{3}\, mn[n^2 - \frac{1}{2}(l^2 + m^2)](dd\sigma) + \sqrt{3}\, mn(l^2 + m^2 - n^2)(dd\pi)$ $- \frac{1}{2}\sqrt{3}\, mn(l^2 + m^2)(dd\delta)$
$(xz, 3z^2 - r^2)$	$\sqrt{3}\, ln[n^2 - \frac{1}{2}(l^2 + m^2)](dd\sigma) + \sqrt{3}\, ln(l^2 + m^2 - n^2)(dd\pi)$ $- \frac{1}{2}\sqrt{3}\, ln(l^2 + m^2)(dd\delta)$
$(x^2 - y^2, x^2 - y^2)$	$\frac{3}{4}(l^2 - m^2)^2(dd\sigma) + [l^2 + m^2 - (l^2 - m^2)^2](dd\pi)$ $+ [n^2 + \frac{1}{4}(l^2 - m^2)^2](dd\delta)$
$(x^2 - y^2, 3z^2 - r^2)$	$\frac{1}{2}\sqrt{3}\,(l^2 - m^2)[n^2 - \frac{1}{2}(l^2 + m^2)](dd\sigma) + \sqrt{3}\, n^2(m^2 - l^2)(dd\pi)$ $+ \frac{1}{4}\sqrt{3}\,(1 + n^2)(l^2 - m^2)(dd\delta)$
$(3z^2 - r^2, 3z^2 - r^2)$	$[n^2 - \frac{1}{2}(l^2 + m^2)]^2(dd\sigma) + 3n^2(l^2 + m^2)(dd\pi) + \frac{3}{4}(l^2 + m^2)^2(dd\delta)$

TABLE 5-16 ***Nearest-Neighbor Integrals in Terms of Two-Center Integrals***

$(s\,\vert\, s)$	$s_0 + 2(ss\sigma)_1(\cos\xi + \cos\eta + \cos\zeta)$ $+ 4(ss\sigma)_2(\cos\xi\cos\eta + \cos\xi\cos\zeta + \cos\eta\cos\zeta)$ $+ 8(ss\sigma)_3\cos\xi\cos\eta\cos\zeta$
$(s\,\vert\, x)$	$2i(sp\sigma)_1\sin\xi + 2\sqrt{2}\,i(sp\sigma)_2(\sin\xi\cos\eta + \sin\xi\cos\zeta)$ $+ (8/\sqrt{3})i(sp\sigma)_3\sin\xi\cos\eta\cos\zeta$
$(s\,\vert\, xy)$	$-2\sqrt{3}\,(sd\sigma)_2\sin\xi\sin\eta - (8/\sqrt{3})(sd\sigma)_3\sin\xi\sin\eta\cos\zeta$
$(s\,\vert\, x^2 - y^2)$	$\sqrt{3}\,(sd\sigma)_1(\cos\xi - \cos\eta) + \sqrt{3}\,(sd\sigma)_2(\cos\xi\cos\zeta - \cos\eta\cos\zeta)$
$(s\,\vert\, 3z^2 - r^2)$	$(sd\sigma)_1(-\cos\xi - \cos\eta + 2\cos\zeta)$ $+ (sd\sigma)_2(-2\cos\xi\cos\eta + \cos\xi\cos\zeta + \cos\eta\cos\zeta)$
$(x\,\vert\, x)$	$p_0 + 2(pp\sigma)_1\cos\xi + 2(pp\pi)_1(\cos\eta + \cos\zeta)$ $+ 2(pp\sigma)_2(\cos\xi\cos\eta + \cos\xi\cos\zeta)$ $+ 2(pp\pi)_2(\cos\xi\cos\eta + \cos\xi\cos\zeta + 2\cos\eta\cos\zeta)$ $+ [(\frac{8}{3})(pp\sigma)_3 + (\frac{16}{3})(pp\pi)_3]\cos\xi\cos\eta\cos\zeta$
$(x\,\vert\, y)$	$-2[(pp\sigma)_2 - (pp\pi)_2]\sin\xi\sin\eta$ $- (\frac{8}{3})[(pp\sigma)_3 - (pp\pi)_3]\sin\xi\sin\eta\cos\zeta$
$(x\,\vert\, xy)$	$2i(pd\pi)_1\sin\eta + (\sqrt{6})i(pd\sigma)_2\cos\xi\sin\eta + 2\sqrt{2}\,i(pd\pi)_2\sin\eta\cos\zeta$ $+ [(\frac{8}{3})(pd\sigma)_3 + (8/3\sqrt{3})(pd\pi)_3]i\cos\xi\sin\eta\cos\zeta$
$(x\,\vert\, yz)$	$[-(\frac{8}{3})(pd\sigma)_3 + (16/3\sqrt{3})(pd\pi)_3]i\sin\xi\sin\eta\sin\zeta$
$(x\,\vert\, x^2 - y^2)$	$\sqrt{3}\,(pd\sigma)_1 i\sin\xi + (\frac{3}{2})^{1/2}(pd\sigma)_2 i\sin\xi\cos\zeta$ $+ 2\sqrt{2}\,(pd\pi)_2 i[\sin\xi\cos\eta + \frac{1}{2}\sin\xi\cos\zeta] + (8/\sqrt{3})(pd\pi)_3 i$ $\times \sin\xi\cos\eta\cos\zeta$
$(x\,\vert\, 3z^2 - r^2)$	$-(pd\sigma)_1 i\sin\xi - \sqrt{2}\,(pd\sigma)_2 i[\sin\xi\cos\eta - \frac{1}{2}\sin\xi\cos\zeta]$ $- (\sqrt{6})(pd\pi)_2 i\sin\xi\cos\zeta - (\frac{8}{3})(pd\pi)_3 i\sin\xi\cos\eta\cos\zeta$
$(z\,\vert\, 3z^2 - r^2)$	$2i(pd\sigma)_1\sin\zeta + [(1/\sqrt{2})(pd\sigma)_2 + (\sqrt{6})(pd\pi)_2]i$ $\times [\cos\xi\sin\zeta + \cos\eta\sin\zeta] + (\frac{16}{3})(pd\pi)_3 i\cos\xi\cos\eta\sin\zeta$
$(xy\,\vert\, xy)$	$d_0 + 2(dd\pi)_1(\cos\xi + \cos\eta) + 2(dd\delta)_1\cos\zeta + 3(dd\sigma)_2\cos\xi\cos\eta$ $+ 2(dd\pi)_2(\cos\xi\cos\zeta + \cos\eta\cos\zeta) + (dd\delta)_2$ $\times (\cos\xi\cos\eta + 2\cos\xi\cos\zeta + 2\cos\eta\cos\zeta)$ $+ [(\frac{8}{3})(dd\sigma)_3 + (\frac{16}{9})(dd\pi)_3 + (\frac{32}{9})(dd\delta)_3]\cos\xi\cos\eta\cos\zeta$
$(xy\,\vert\, xz)$	$2[-(dd\pi)_2 + (dd\delta)_2]\sin\eta\sin\zeta$ $+ [-(\frac{8}{3})(dd\sigma)_3 + (\frac{8}{9})(dd\pi)_3 + (\frac{16}{9})(dd\delta)_3]\cos\xi\sin\eta\sin\zeta$
$(xy\,\vert\, x^2 - y^2)$	zero
$(xy\,\vert\, 3z^2 - r^2)$	$\sqrt{3}\,[(dd\sigma)_2 - (dd\delta)_2]\sin\xi\sin\eta + (16/3\sqrt{3})[(dd\pi)_3 - (dd\delta)_3]$ $\sin\xi\sin\eta\cos\zeta$
$(xz\,\vert\, x^2 - y^2)$	$-\frac{3}{2}[(dd\sigma)_2 - (dd\delta)_2]\sin\xi\sin\zeta - (\frac{8}{3})[(dd\pi)_3 - (dd\delta)_3]$ $\times \sin\xi\cos\eta\sin\zeta$
$(xz\,\vert\, 3z^2 - r^2)$	$\frac{1}{2}\sqrt{3}\,(-(dd\sigma)_2 + (dd\delta)_2]\sin\xi\sin\zeta - (8/3\sqrt{3})[(dd\pi)_3 - (dd\delta)_3]$ $\times \sin\xi\cos\eta\sin\zeta$
$(x^2 - y^2\,\vert\, x^2 - y^2)$	$d_0 + \frac{3}{2}(dd\sigma)_1(\cos\xi + \cos\eta) + (dd\delta)_1(\frac{1}{2}\cos\xi + \frac{1}{2}\cos\eta + 2\cos\zeta)$ $+ 4(dd\pi)_2\cos\xi\cos\eta + [\frac{3}{4}(dd\sigma)_2 + (dd\pi)_2 + (\frac{9}{4})(dd\delta)_2]$ $\times (\cos\xi\cos\zeta + \cos\eta\cos\zeta) + [(\frac{16}{3})(dd\pi)_3 + (\frac{8}{3})(dd\delta)_3]$ $\cos\xi\cos\eta\cos\zeta$
$(3z^2 - r^2\,\vert\, 3z^2 - r^2)$	$d_0 + (dd\sigma)_1(\frac{1}{2}\cos\xi + \frac{1}{2}\cos\eta + 2\cos\zeta) + \frac{3}{2}(dd\delta)_1$ $\times (\cos\xi + \cos\eta) + (dd\sigma)$ $\times (\cos\xi\cos\eta + \frac{1}{4}\cos\xi\cos\zeta + \frac{1}{4}\cos\eta\cos\zeta) + 3(dd\pi)_2$ $\times (\cos\xi\cos\zeta + \cos\eta\cos\zeta) + 3(dd\delta)_2$ $\times (\cos\xi\cos\eta + \frac{1}{4}\cos\xi\cos\zeta + \frac{1}{4}\cos\eta\cos\zeta)$ $+ [(\frac{16}{3})(dd\pi)_3 + (\frac{8}{3})(dd\delta)_3]\cos\xi\cos\eta\cos\zeta$
$(x^2 - y^2\,\vert\, 3z^2 - r^2)$	$\frac{1}{2}\sqrt{3}\,[-(dd\sigma)_1 + (dd\delta)_1](\cos\xi - \cos\eta)$ $+ [\frac{1}{4}\sqrt{3}\,(dd\sigma)_2 - \sqrt{3}\,(dd\pi)_2$ $+ \frac{3}{4}\sqrt{3}\,(dd\delta)_2](\cos\xi\cos\zeta - \cos\eta\cos\zeta)$

5-10 A simple cubic symmorphic space group: beta-brass

An example which illustrates the usefulness of group theory as applied to the LCAO method is the study of beta-brass made by Amsterdam.[10] This alloy in its ideal form consists of equal numbers of copper and zinc atoms, with a unit cell as shown in Fig. 5-36. This unit cell contains one zine atom and the equivalent of one copper atom, and the lattice is simple cubic, with an atom of each kind assigned to each lattice point.

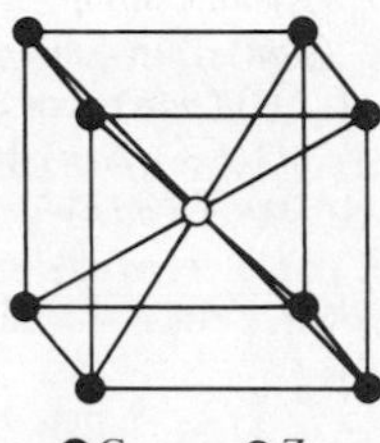

5-36 The unit cell of beta-brass

The electron configuration of copper is $1s^2\, 2s^2\, 2p^6\, 3s^2\, 3p^6\, 3d^{10}\, 4s^1$ and zinc has the outer subshell $3d^{10}\, 4s^2$. Amsterdam's calculation is then based on consideration of these twenty-three valence electrons. According to Pauling,[11] Cu has an ionic radius of 0.96 A and Zn has a radius of 0.74 A. It seems reasonable, therefore, to base a tight-binding approximation on interactions between Zn nearest neighbors and Cu next-nearest neighbors of Cu atoms. If we take the edge of the cubic unit cell all as $2a$, then the Cu-Zn interaction corresponds to third-nearest neighbor interaction in a simple cubic lattice with edge a, and the Cu-Cu interaction corresponds to nearest neighbor interaction in a simple cubic lattice of edge $2a$. Hence, we can use the results of Slater and Koster presented in Table 5-16. It is assumed that the bands are due to the ten $3d$ electrons which lie just beyond the argon-type core of Cu and Zn.

Let us establish the form of the secular equation which results from the use of the solution given in (5-175). The ϕ_n are the d functions, denoted as ϕ_1 through ϕ_5. The sum on the right of (5-175) involves ten terms—five each for Cu and for Zn—since the unit cell contains one copper and one zinc atom. Substituting in the Schroedinger equation, multiplying by each of the ten complex conjugates ψ^*, and integrating gives a 10×10 secular equation. A typical diagonal term in this determinant has the form $[(\phi_{1C} | V' | \phi_{1C}) - E]$,

[10]M. F. Amsterdam, "The Band Structure of Beta-Brass," Ph.D. dissertation, Temple University, Philadelphia, Penna. 1958.

[11]L. Pauling, *Nature of the Chemical Bond*, Cornell University Press, Ithaca, N.Y., 1945.

where ϕ_{1C} is one of the d functions for the Cu atom and V' is defined by (5-166). A typical off-diagonal term would be $(\phi_{1C}|\, V' \,|\phi_{2Z})$, but many of these terms vanish and the secular equation for the Σ-axis of Fig. 5-21 can be written as

$$
\begin{array}{ll}
 & \overbrace{\begin{array}{ccccc} xy & yz & zx & x^2-y^2 & z^2 \end{array}}^{\text{Cu}} \quad \overbrace{\begin{array}{ccccc} xy & yz & zx & x^2-y^2 & z^2 \end{array}}^{\text{Zn}} \\
\begin{array}{r} \text{Cu}\left\{\begin{array}{l} xy \\ yz \\ zx \\ x^2-y^2 \\ z^2 \end{array}\right. \\ \text{Zn}\left\{\begin{array}{l} xy \\ yz \\ zx \\ x^2-y^2 \\ z^2 \end{array}\right. \end{array} &
\left|\begin{array}{ccccc|ccccc}
G-E & 0 & 0 & 0 & 0 & M & 0 & 0 & 0 & Q \\
0 & H-E & 0 & 0 & 0 & 0 & M & P & 0 & 0 \\
0 & 0 & H-E & 0 & 0 & 0 & P & M & 0 & 0 \\
0 & 0 & 0 & K-E & 0 & 0 & 0 & 0 & N & 0 \\
0 & 0 & 0 & 0 & L-E & Q & 0 & 0 & 0 & N \\
\hline
M & 0 & 0 & 0 & Q & D-E & 0 & 0 & 0 & 0 \\
0 & M & P & 0 & 0 & 0 & D-E & 0 & 0 & 0 \\
0 & P & M & 0 & 0 & 0 & 0 & D-E & 0 & 0 \\
0 & 0 & 0 & N & 0 & 0 & 0 & 0 & D-E & 0 \\
Q & 0 & 0 & 0 & N & 0 & 0 & 0 & 0 & D-E
\end{array}\right| = 0
\end{array}
\tag{5-202}
$$

where the order of the rows and columns is arbitrary, and where the non-vanishing elements of the determinant are defined by

$$
\begin{aligned}
(xy_C|\, V' \,|xy_C) &= G \\
(yz_C|\, V' \,|yz_C) = (xz_C|\, V' \,|xz_C) &= H \\
(x^2 - y^2_C|\, V' \,|x^2 - y^2_C) &= K \\
(z^2_C|\, V' \,|z^2_C) &= L \\
(xy_Z|\, V' \,|xy_Z) = (yz_Z|\, V' \,|yz_Z) &= (zx_Z|\, V' \,|zx_Z) \\
&= (x^2 - y^2_Z)|\, V' \,|x^2 - y^2_Z) = (z^2_Z|\, V' \,|z^2_Z) = D \\
(xy_C|\, V' \,|xy_Z) = (xy_Z|\, V' \,|xy_C) &= (yz_C|\, V' \,|yz_Z) \\
&= (zx_C|\, V' \,|zx_Z) = M \\
(x^2 - y^2_C|\, V' \,|x^2 - y^2_Z) &= (z^2_C|\, V' \,|z^2_Z) = N \\
(yz_C|\, V' \,|zx_Z) &= P \\
(xy_C|\, V' \,|z^2_Z) &= Q
\end{aligned}
\tag{5-203}
$$

To show why the determinant of Eq. (5-202) has the form indicated, we turn to Table 5-16 for an evaluation of the various elements. For example, the lower right-hand quadrant has only diagonal integrals D, and these are all the same. This follows from the fact that all the entries in Table 5-16 containing the same two d functions are the sum of d_0 and terms involving $(dd\sigma)_1$, $(dd\sigma)_2$, $(dd\pi)_2$, etc. But we are neglecting Zn-Zn interactions, so that all these two-center integrals reduce to $d_0 = D$.

For the Cu-Cu terms, we use $\xi = \eta$ and $\zeta = 0$ to obtain as an example

$$G = (xy\,|\,xy) = d_0 + 2(dd\pi)_1 2\cos 2\xi + 2(dd\delta)_1 \cos 0$$

or

$$G = d_0 + 4(dd\pi)_1 \cos 2\xi + 2(dd\delta)_1$$

where we have replaced ξ by 2ξ to incorporate the fact that the copper atoms are separated by a distance $2a$, and where only nearest-neighbor terms have been retained. Similarly,

$$H = (yz\,|\,yz) = d_0 + 2(dd\pi)_1(\cos 2\xi + 1) + 2(dd\delta)_1 \cos 2\xi = (zx\,|\,zx)$$

TABLE 5-17 ***Terms in the Secular Equation for the Σ-Axis of Beta-Brass***

Zn-Zn
$D = d_0$
Cu-Cu
$(xy\,|\,xy) = G = d_0 + 4(dd\pi)_1 \cos 2\xi + 2(dd\delta)_1$
$(yz\,|\,yz) = (zx\,|\,zx) = H = d_0 + 2(dd\pi)_1(1 + \cos 2\xi) + 2(dd\delta)_1 \cos 2\xi$
$(x^2 - y^2\,|\,x^2 - y^2) = K = d_0 + 3(dd\sigma)_1 \cos 2\xi + (dd\delta)_1(\cos 2\xi + 2)$
$(3z^2 - r^2\,|\,3z^2 - r^2) = L = d_0 + (dd\sigma)_1(\cos 2\xi + 2) + 3(dd\delta)_1 \cos 2\xi$
Cu-Zn
$(xy\,|\,xy) = (yz\,|\,yz) = (zx\,|\,zx) = M$
$= [(\frac{8}{3})(dd\sigma)_3 + (\frac{16}{9})(dd\pi)_3 + (\frac{32}{9})(dd\delta)_2] \cos^2 \xi$
$(x^2 - y^2\,|\,x^2 - y^2) = (3z^2 - r^2\,|\,3z^2 - r^2) = N$
$= [(\frac{16}{3})(dd\pi)_3 + (\frac{8}{3})(dd\delta)_2] \cos^2 \xi$
$(yz\,|\,zx) = P = [-(\frac{8}{3})(dd\sigma)_3 + (\frac{8}{9})(dd\pi)_3 + (\frac{16}{9})(dd\delta)_3] \sin^2 \xi$
$(xy\,|\,3z^2 - r^2) = Q = (16/3\sqrt{3})[(dd\pi)_3 - (dd\delta)_3] \sin^2 \xi$

Carrying on in this fashion we obtain Table 5-17, which lists all the nonvanishing elements. Note that for the Cu-Zn entries, the lattice constant is a (rather than $2a$) and all terms except those for third-nearest neighbors have been dropped.

The secular determinant can now be rearranged to get it in the block form, so that

$$\begin{array}{cccccccccc} xy_C & xy_Z & z^2_C & z^2_Z & x^2-y^2_C & x^2-y^2_Z & yz_C & yz_Z & xz_C & xz_Z \end{array}$$

$$\left|\begin{array}{cccccccccc}
G-E & M & 0 & Q & & & & & & \\
M & D-E & Q & 0 & & & & & & \\
0 & Q & L-E & N & & & & & & \\
Q & 0 & N & D-E & & & & & & \\
 & & & & K-E & N & & & & \\
 & & & & N & D-E & & & & \\
 & & & & & & H-E & M & 0 & P \\
 & & & & & & M & D-E & P & 0 \\
 & & & & & & 0 & P & H-E & M \\
 & & & & & & P & 0 & M & D-E
\end{array}\right| = 0 \tag{5-204}$$

The labelling of the rows is the same as that of the columns and all elements not written are zero.

We can use the methods of group theory to reduce the 4×4 subdeterminant in the lower right-hand corner. Using the information at the bottom of Table 5-11, we calculate the way in which the five d functions will be affected by the symmetry operations associated with the Σ-axis, and this information is summarized in Table 5-18. This table lists the effect of each symmetry operation on each function. Using the character system of Table 5-10 gives for the representation Σ_1 the following linear combinations:

$$\begin{aligned}
xy&: \quad 1(xy) + 1(yx) + 1(xy) + 1(yx) = 4xy \\
yz&: \quad 1(yz) + 1(-xz) + 1(-yz) + 1(xz) = 0 \\
xz&: \quad 1(xz) + 1(-yz) + 1(-xz) + 1(yz) = 0 \\
x^2 - y^2&: \quad 1(x^2 - y^2) + 1(y^2 - x^2) + 1(x^2 - y^2) + 1(y^2 - x^2) = 0 \\
z^2&: \quad 1(z^2) + 1(z^2) + 1(z^2) + 1(z^2) = 4z^2.
\end{aligned}$$

Hence, xy and z^2 belong to Σ_1, but this representation is *not* degenerate (this point will be clarified below). Similarly, we find the other results in the last column of Table 5-10 and we note that group theory has converted our five individual d functions into five linear combinations. Further, the functions belonging to different irreducible representations do not produce any elements in the secular determinant; that is, they do not *mix*.

PROBLEM 5-30

Prove that

$$(yz - xz_C \,|\, V' \,|\, yz + xz_C) = 0 \tag{5-205}$$

$$(yz + xz_C \,|\, V' \,|\, yz + xz_C) = 2H \tag{5-206}$$

$$(yz + xz_C \,|\, V' \,|\, yz + xz_Z) = 2M + 2P \tag{5-207}$$

TABLE 5-18 ***The Effect of Σ-Axis Symmetry Operations on the d-Functions***

	E	C_2	JC_4^2	JC_2
xy	xy	yx	xy	yx
yz	yz	$-xz$	$-yz$	xz
xz	xz	$-yz$	$-xz$	yz
$x^2 - y^2$	$x^2 - y^2$	$-(x^2 - y^2)$	$x^2 - y^2$	$-(x^2 - y^2)$
z^2	z^2	z^2	z^2	z^2

The result expressed by (5-205) is a general property of integrals involving symmetrized linear combinations; we have seen an analogous behavior in connection with infrared selection rules, as discussed in Sec. 3-5.

Dropping the factors of 2, the secular equation now becomes

$$\begin{array}{cccccccccc} xy_C & xy_Z & z^2_C & z^2_Z & x^2-y^2_C & x^2-y^2_Z & xz+yz_C & xz+yz_Z & xz-yz_C & xz-yz_Z \end{array}$$

$$\left|\begin{array}{cccc|cc|cc|cc} G-E & M & 0 & Q & & & & & & \\ M & D-E & Q & 0 & & & & & & \\ 0 & Q & L-E & N & & & & & & \\ Q & 0 & N & D-E & & & & & & \\ \hline & & & & K-E & N & & & & \\ & & & & N & D-E & & & & \\ \hline & & & & & & H-E & M+P & & \\ & & & & & & M+P & D-E & & \\ \hline & & & & & & & & H-E & M-P \\ & & & & & & & & M-P & D-E \end{array}\right| = 0$$

$$\underbrace{\qquad\qquad}_{\Sigma_1}\quad\underbrace{\qquad}_{\Sigma_4}\quad\underbrace{\qquad}_{\Sigma_3}\quad\underbrace{\qquad}_{\Sigma_2} \tag{5-208}$$

The application of group theory has thus reduced one of the 4×4 subdeterminants into a pair of 2×2 determinants, and this is as far as the theory takes us in this direction.

We note that this example is in accord with (2-63). The five d functions form a reducible representation, with characters as follows:

$$\begin{array}{|cccc} E & C_2 & JC_4^2 & JC_2 \\ \hline 5 & 1 & 1 & 1 \end{array}$$

so that

$$n_1 = 2, \qquad n_2 = 1, \qquad n_3 = 1, \qquad n_4 = 1$$

Then the irreducible representation Σ_1 is contained twice in the reducible representation and the other three are contained once. Since the character for E in any irreducible representation automatically gives the degeneracy, then in the case of Σ_1, we must have two nondegenerate states, rather than a single, doubly-degenerate one. This example also demonstrates an important point that has not yet been emphasized. The original secular determinant we used to show the value of group theory, Eq. (2-107), was reduced from 6×6 size to the very limit possible: two 1×1 subdeterminants and two 2×2 subdeterminants. In the present case, since one zinc and one copper function serve as an irreducible basis for Σ_2, Σ_3, and Σ_4, we would expect 2×2 blocks on the main diagonal. But the representation Σ_1 has two sets belonging to it and, furthermore, $n_1 = 2$. Hence, the best we expect from group theory is a 4×4 block belonging to Σ_1. If this representation were doubly-degenerate, like Γ_3 in (2-144), then we would have a right to expect two 2×2 blocks.

For the Δ-axis of Fig. 5-21, using $\xi = \eta = 0$, we obtain the secular equation

$$\begin{array}{ccccc|ccccc} \multicolumn{5}{c}{\overbrace{\qquad\qquad\qquad\qquad}^{\text{Cu}}} & \multicolumn{5}{c}{\overbrace{\qquad\qquad\qquad\qquad}^{\text{Zn}}} \\ xy & yz & zx & x^2-y^2 & z^2 & xy & yz & zx & x^2-y^2 & z^2 \end{array}$$

$$\left\| \begin{array}{ccccc|ccccc} K-E & & & & & Z & & & & \\ & J-E & & & & & Z & & & \\ & & J-E & & & & & Z & & \\ & & & M-E & & & & & W & \\ & & & & R-E & & & & & W \\ \hline Z & & & & & D-E & & & & \\ & Z & & & & & D-E & & & \\ & & Z & & & & & D-E & & \\ & & & W & & & & & D-E & \\ & & & & W & & & & & D-E \end{array} \right\| = 0 \tag{5-209}$$

which, upon rearrangement, becomes

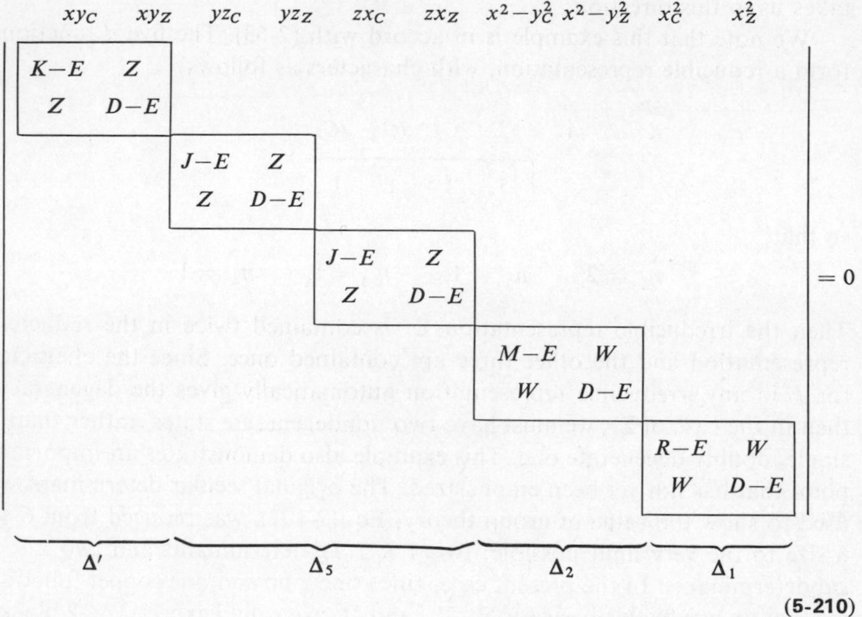

(5-210)

The elements of this determinant are listed in Table 5-19 and the *d* functions associated with the irreducible representations along the Δ-axis are given in the right-hand column of Table 5-6. This information was obtained by using the bottom row to set up a table analogous to Table 5-18 and then forming the projection operators belonging to each irreducible representation.

The determinant corresponding to the representation Δ_2' in (5-210) can be expanded to give

$$2E = K + D \pm [(D - K)^2 + 4Z^2]^{1/2} \tag{5-211}$$

and substituting the expressions of Table 5-19 yields the relation

$$\begin{aligned} 2E = {} & d_{0C} + 4(dd\pi)_1 + 2(dd\delta)_1 \cos 2\zeta + d_{0Z} \\ & \pm [\{d_{0Z} - d_{0C} - 4(dd\pi)_1 - 2(dd\delta)_1 \cos 2\zeta\}^2 \\ & + (\tfrac{32}{3})\{(dd\delta)_3 + (\tfrac{2}{3})(dd\pi)_3 + (\tfrac{4}{3})(dd\delta)_3\}^2]^{1/2} \end{aligned} \tag{5-212}$$

To assign values to the nearest-neighbor integrals in (5-212), Amsterdam[10] refers to an argument of Slater and Koster,[9] which in turn, is based on calculations by Howarth[12] for copper and Fletcher[13] for nickel. In the first

[12]D. J. Howarth, *Proc, Roy. Soc.*, **A220**, 513 (1953).

[13]G. C. Fletcher, *Proc. Phys. Soc.* (London), **A65**, 192 (1952).

TABLE 5-19 ***Terms in the Secular Equation for the Δ-Axis of Beta-Brass***

<u>Zn-Zn</u>
$D = d_0$
<u>Cu-Cu</u>
$(xy\,\vert\,xy) = K = d_0 + 4(dd\pi)_1 + 2(dd\delta)_1 \cos 2\zeta$
$(yz\,\vert\,yz) = J = (zx\,\vert\,zx) = d_0 + 2(dd\pi)_1(1 + 2\cos 2\zeta) + 2(dd\delta)_1$
$(x^2 - y^2\,\vert\,x^2 - y^2) = M = d_0 + d(dd\sigma)_1 + (dd\delta)_1(1 + 2\cos 2\zeta)$
$(3z^2 - r^2\,\vert\,3z^2 - r^2) = R = d_0 + (dd\sigma)_1(1 + 2\cos 2\zeta) + 3(dd\delta)_1$
<u>Cu-Zn</u>
$(xy\,\vert\,xy) = (yz\,\vert\,yz) = (xz\,\vert\,xz) = Z$
$= [(\frac{8}{3})(dd\sigma)_3 + (\frac{16}{9})(dd\pi)_3 + (\frac{32}{9})(dd\delta)_3]\cos\zeta$
$(x^2 - y^2\,\vert\,x^2 - y^2) = W = (3z^2 - r^2\,\vert\,3z^2 - r^2)$
$= [(\frac{16}{3})(dd\pi)_3 + (\frac{8}{3})(dd\delta)_3]\cos\zeta$

place, it is assumed that $(dd\sigma)$, $(dd\pi)$, and $(dd\delta)$ bear an identical relation to one another in copper, nickel, and beta-brass, using as a justification the fact that we are dealing only with valence electrons and the specific nature of the core is not too important. Now Fletcher was able to evaluate integrals of this type by using an analytical expression for $V'(r)$ obtained from a Hartree self-consistent calculation on Cu, (see Slater[5] for a discussion of the Hartree method), and Amsterdam quotes the ratios as

$$(dd\pi) = -0.54(dd\sigma), \qquad (dd\delta) = 0.82(dd\sigma) \tag{5-213}$$

Substituting (5-213) into (5-212), we obtain

$$\begin{aligned} E_r = {} & (dd\sigma)_1(0.16\cos 2\zeta - 2.16) \\ & \pm [\{(d_{0Z} - d_{0C}) + (dd\sigma)_1(2.16 - 0.16\cos 2\zeta)\}^2 \\ & + \{4(dd\sigma)_3\cos\zeta\}^2]^{1/2} \end{aligned} \tag{5-214}$$

where $E_r = 2E - d_{0Z} - d_{0C}$ is the reduced energy for this calculation. To estimate the term $(d_{0Z} - d_{0C})$ inside the radical, we use the $3d$ atomic levels of -10.6 and -3.3 eV and work functions of 4.3 and 5.6 eV for Zn and Cu, respectively, to obtain an energy difference of

$$[-14.9 - (-8.9)]\text{ eV} = -6.0\text{ eV} = -0.44\text{ ryd} \tag{5-215}$$

Assuming that the energy difference in the brass lattice is only slightly different from that for the free atoms permits us to use this value for $(d_{0Z} - d_{0C})$.

Finally, to estimate $(dd\sigma)_1$ and $(dd\sigma)_3$, we use the value

$$(dd\sigma) = -0.025 \text{ ryd}$$

as determined for Ni, and then use another Slater-Koster argument; namely that these integrals for other metals should vary inversely as the distance between the ion-core "surfaces." These distances, in turn, are given as follows:

Ni 2.5 A, Cu 3.0 A, CuZn 2.6 A

Hence $$(dd\sigma)_1 = -0.030 \text{ A}, \qquad (dd\sigma)_3 = -0.026 \text{ A} \qquad \textbf{(5-216)}$$

The values of (5-215) and (5-216) when substituted into (5-214) show that (1) $E_r = -0.41$ at $\zeta = 0$ and that (2) E_r is essentially constant as ζ goes from 0 to $\pi/2$, since the terms in $\cos \zeta$ are negligible. The curve corresponding to the positive sign is plotted in Fig. 5-37. Going to the next deter-

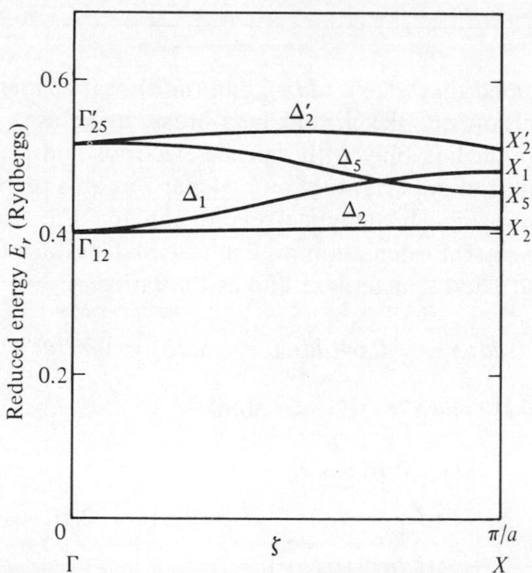

5-37 Tight-binding energy curves for the Δ-axis of beta-brass

minant, the one for Δ_5, we find in the same way that E_r starts at 0.55 but falls to about 0.41 at $\zeta = \pi/2$. The figure also shows the Δ_2 and Δ_1 curves, and we note the agreement with the compatibility relations. Figure 5-38 shows the corresponding results for the Σ-axis, and in this case, the fourth-order determinant was worked out on a computer.

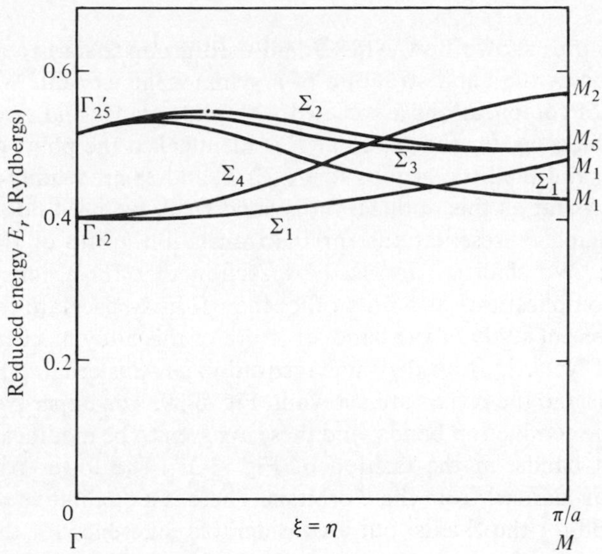

5-38 Tight-binding energy curves for the Σ-axis of beta-brass

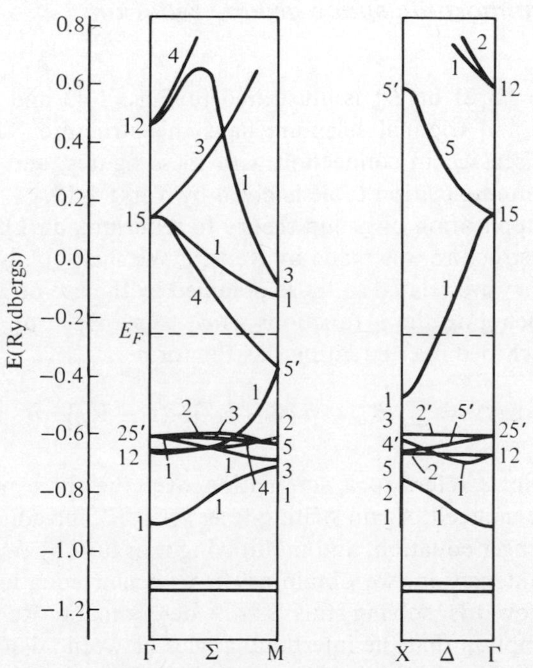

5-39 Beta-brass energy bands after Johnson and Amar[14] (the numerals indicate the subscripts of the corresponding representation)

We have thus shown how Amsterdam used group theory to assist in the determination of the band structure of a symmorphic crystal. We note that the absence of complications associated with glide planes and screw axes has been a real blessing, for the group at $\mathbf{\Gamma}$ is identical to the point group O_h in direct space, and the little groups for $\mathbf{\Delta}$, $\mathbf{\Sigma}$, X, and M are readily-determined subgroups. As far as this application is concerned, we need not even worry about irreducible representations for the translation group or the complete space group. We shall see in the next section that for a nonsymmorphic group, the complications at points other than $\mathbf{\Gamma}$ in k-space are tremendous.

A more recent study of the band structure of this alloy has been made by Johnson and Amar,[14] from the same institution as Amsterdam. Their results for the $\mathbf{\Sigma}$-axis and the $\mathbf{\Delta}$-axis are shown in Fig. 5-39. The upper parts of these figures are the conduction bands, and these are seen to be modifications of the free electron bands, in the fashion of Fig. 5-31. The lower parts are the valence bands, derived from the d orbitals. There is a qualitative resemblance with Fig. 5-38 for the $\mathbf{\Sigma}$-axis, but a considerable difference for the $\mathbf{\Delta}$ bands.

5-11 *A nonsymmorphic space group: tellurium*

The space group $P3_121$ or D_3^4 is illustrated in Figs. 1-43 and 1-44; single-crystal tellurium and trigonal selenium have this structure. The symmetry operations are discussed in connection with these figures, and a portion of the space group multiplication table is given by Table 1-19.

Prior to the application of group theory to tellurium, an LCAO calculation of the band structure was made by Reitz.[15] We shall briefly consider his results because they are related to those obtained by the use of group-theoretical methods. Denoting the p functions as $\phi_1 = p_x$, $\phi_2 = p_y$, $\phi_3 = p_z$, the solution to the Schroedinger equation has the form

$$\psi(\mathbf{r}) = \sum_{j,n,s} B_{n8} \exp\{i\mathbf{k} \cdot \mathbf{R}_{js}\}\phi_n(\mathbf{r} - \mathbf{R}_{js}) \tag{5-175}$$

where the subscript s refers to a summation over the three nonequivalent lattice sites in the unit cell. Again writing V as $V_a + V'$, introducing ψ above into the Schroedinger equation, and multiplying it in turn by ϕ_1^*, ϕ_2^*, and ϕ_3^*, followed by an integration, we obtain a 9×9 secular equation. As a first approximation towards solving this 9×9 determinant, Reitz made the simplifying assumption that the interbond angles between adjacent atoms in a chain were 90° instead of the true value 102.6°. Referring to Fig. 5-40, we

[14] K. H. Johnson and H. Amar, *Phys. Rev.*, **139**, A760 (1965).

[15] J. R. Reitz, *Phys. Rev.*, **105**, 1233 (1957).

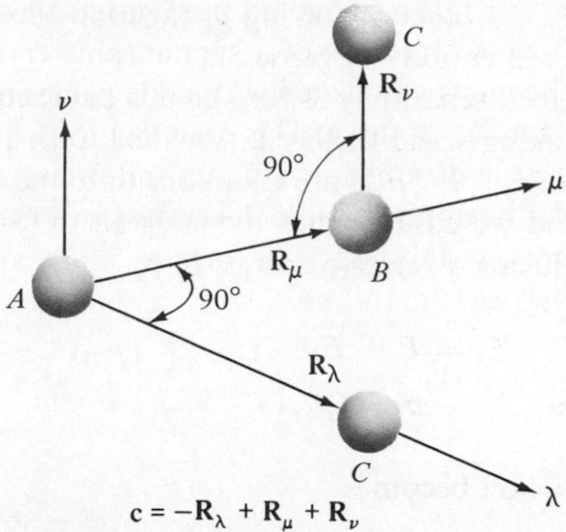

5-40 An approximation to the spiral chain structure of tellurium

label the three atoms in a unit cell as A, B, and C and let $\mathbf{R}_\lambda$, $\mathbf{R}_\mu$, and $\mathbf{R}_\nu$ denote the three nearest-neighbor directions, so that the lattice constant along the trigonal axis can be expressed

$$\mathbf{c} = -\mathbf{R}_\lambda + \mathbf{R}_\mu + \mathbf{R}_\nu$$

Considering only nearest-neighbor integrals and denoting the location of a function by a subscript A, B, or C, the integrals involving p_λ functions are

$$(p_{\lambda A} \,|\, V' \,|\, p_{\lambda A}) = (p_{\lambda C} \,|\, V' \,|\, p_{\lambda C}) \tag{5-217a}$$

$$(p_{\lambda B} \,|\, V' \,|\, p_{\lambda B}) \tag{5-217b}$$

$$(p_{\lambda A} \,|\, V' \,|\, p_{\lambda B}) = (p_{\lambda B} \,|\, V' \,|\, p_{\lambda A}) \tag{5-217c}$$

$$(p_{\lambda B} \,|\, V' \,|\, p_{\lambda C}) = (p_{\lambda C} \,|\, V' \,|\, p_{\lambda B}) \tag{5-217d}$$

$$(p_{\lambda C} \,|\, V' \,|\, p_{\lambda A}) = (p_{\lambda A} \,|\, V' \,|\, p_{\lambda C}) \tag{5-217e}$$

The relations among these integrals come from the fact that their values depend upon the relative orientation of the two p functions in each integrand. Since two different p functions are orthogonal, all terms of the form $(p_{\lambda A} \,|\, V' \,|\, p_{\mu A})$ vanish in this approximation, and the 9×9 determinant reduces to the form

$$\begin{vmatrix} \boxed{\lambda} & 0 & 0 \\ 0 & \boxed{\mu} & 0 \\ 0 & 0 & \boxed{\nu} \end{vmatrix} = 0$$

where λ denotes a 3×3 block involving p_λ functions only and 0 denotes a 3×3 block composed entirely of zeroes.

Equating the λ block separately to zero gives a cubic equation in E, which can be solved by making some further approximations. It is estimated that integrals of the form $(p_{\lambda A} | V' | p_{\lambda B})$ are about one third the magnitude of those like $(p_{\lambda A} | V' | p_{\lambda C})$ and we assume that the integrals in (5-217a) and (5-217b) are the same. Introducing a relative energy E_r by

$$E_r = E - E_a - (p_{\lambda A} | V' | p_{\lambda A})$$

and the abbreviation $$\sigma = (p_{\lambda A} | V' | p_{\lambda C})$$

then the secular equation becomes

$$-E_r^3 + (\tfrac{11}{9})E_r\sigma^2 + (\tfrac{2}{9})\sigma^3 \cos k_z c = 0 \quad \textbf{(5-218)}$$

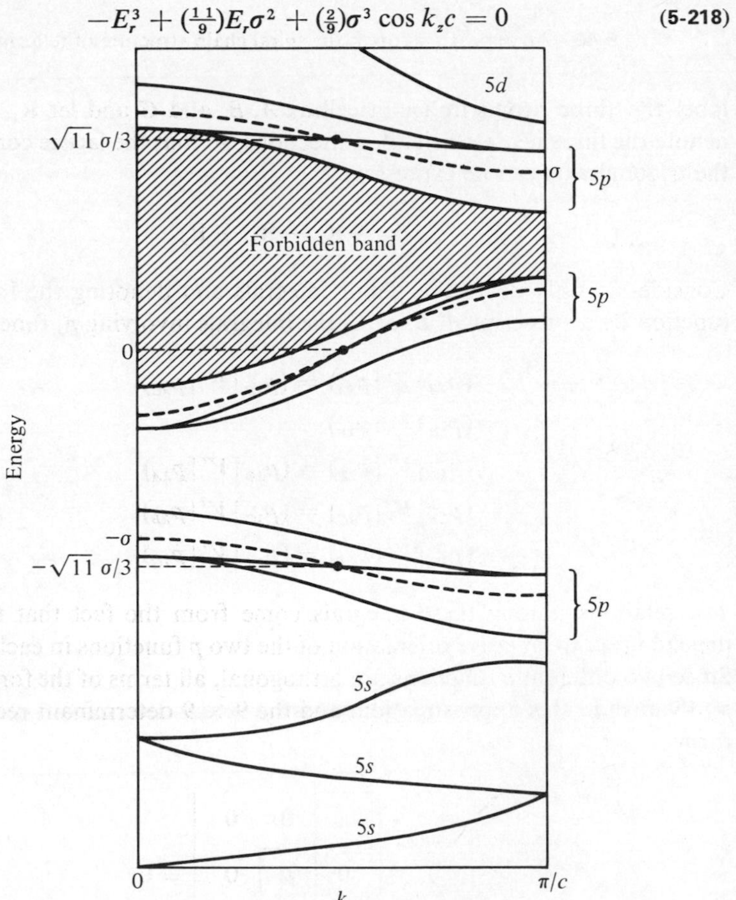

5-41 Tight-binding energy bands for tellurium

The dotted curves of Fig. 5-41 show the three roots of this equation as a function of k_z. The same set of curves will also come from the μ and ν equations and the bands are triply-degenerate. This degeneracy can be removed by stretching the bond angles to their true value of 102.6°, which results in the introduction of additional integrals of the form $(p_{\lambda A}\,|\,V'\,|\,p_{\mu B})$ into the secular equation. Reitz has solved this 9×9 equation by making further approximations and obtained the solid curves of Fig. 5-41. In this diagram, we have added the s bands based on the free electron approximation, which we shall consider next.

PROBLEM 5-31

Show that Eq. (5-218) comes from the secular determinant

$$\begin{vmatrix} (p_{\lambda A}\,|\,V'\,|\,p_{\lambda A})+E_a-E & (p_{\lambda A}\,|\,V'\,|\,p_{\lambda B})e^{i\mathbf{k}\cdot\mathbf{R}_\mu} & (p_{\lambda A}\,|\,V'\,|\,p_{\lambda C})e^{i\mathbf{k}\cdot\mathbf{R}_\lambda} \\ (p_{\lambda B}\,|\,V'\,|\,p_{\lambda A})e^{-i\mathbf{k}\cdot\mathbf{R}_\mu} & (p_{\lambda B}\,|\,V'\,|\,p_{\lambda B})+E_a-E & (p_{\lambda B}\,|\,V'\,|\,p_{\lambda C})e^{-i\mathbf{k}\cdot\mathbf{R}_\mu} \\ (p_{\lambda C}\,|\,V'\,|\,p_{\lambda A})e^{-i\mathbf{k}\cdot\mathbf{R}_\lambda} & (p_{\lambda C}\,|\,V'\,|\,p_{\lambda B})e^{-i\mathbf{k}\cdot\mathbf{R}_\mu} & (p_{\lambda C}\,|\,V'\,|\,p_{\lambda C})+E_a-E \end{vmatrix}$$

and that this determinant is the λ block of the complete 9×9 determinant.

PROBLEM 5-32

(a) Show that the Brillouin zone for tellurium has the shape indicated in Fig. 5-42 and that the matrices specifying the direct and reciprocal lattices are

$$\mathbf{A} = \begin{bmatrix} a & -\dfrac{a}{2} & 0 \\ 0 & \dfrac{\sqrt{3}\,a}{2} & 0 \\ 0 & 0 & c \end{bmatrix}, \qquad \mathbf{B} = \begin{bmatrix} \dfrac{1}{a} & \dfrac{1}{\sqrt{3}\,a} & 0 \\ 0 & \dfrac{2}{\sqrt{3}\,a} & 0 \\ 0 & 0 & \dfrac{1}{c} \end{bmatrix} \tag{5-219}$$

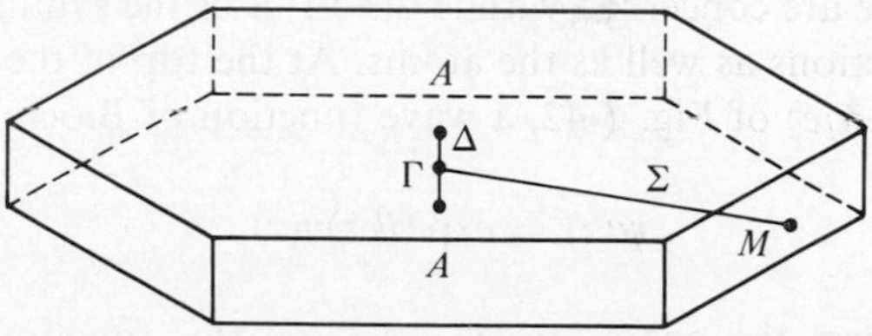

5-42 Brillouin zone of tellurium

(b) Find the coordinates of A and M.

(c) Show that the free electron wave functions and energies are

$$\psi = \exp\left\{\frac{2\pi i}{a}\left[(\xi - m_1)x + \left\{\frac{\eta - (m_1 + 2m_2)}{\sqrt{3}}\right\}y + \frac{a}{c}(\zeta - m_3)z\right]\right\} \tag{5-220}$$

$$E = \frac{h^2}{2ma^2}\left[(\xi - m_1)^2 + \left\{\frac{\eta - (m_1 + 2m_2)}{\sqrt{3}}\right\}^2 + \left(\frac{a}{c}\right)^2(\zeta - m_3)^2\right] \tag{5-221}$$

where
$$\mathbf{k} = \left(\frac{2\pi\xi}{a}, \frac{2\pi\eta}{a}, \frac{2\pi\zeta}{c}\right)$$

Equation (5-221) permits us to plot the E vs. $\mathbf{k}$ curves for the free electron approximation. For example, along the Δ-axis of Fig. 5-43, $\xi = \eta = 0$ and (5-221) reduces to

$$E = \frac{h^2}{2ma^2}\left[m_1^2 + \frac{(m_1 + 2m_2)^2}{3} + \left(\frac{a}{c}\right)^2(\zeta - m_3)^2\right] \tag{5-222}$$

For tellurium, the value of (a/c) is very close to 0.75, and we can plot E vs. ξ for various combinations of (m_1, m_2, m_3), as shown in Fig. (5-43). Note that the curve for which $E = \frac{4}{3}$ at $\zeta = 0$ has six possible sets of m values, the notation $(0\bar{1}0)$ being used for $m_1 = 0, m_2 = -1, m_3 = 0$, and this energy curve is therefore six-fold degenerate. The valence electrons of tellurium have the configuration $5s^2 5p^4$, and since there are three nonequivalent atoms in a single turn of the spiral chain, this corresponds to eighteen valence electrons. Considering spin allows two electrons per energy level so that the ninth level from the bottom represents the valence band, thus determining the position of the forbidden gap which has been shaded in. A similar calculation gives the E vs. $\mathbf{k}$ curves along the ΓM-direction, shown in Fig. 5-44.

In our introductory discussion on symmetry, we considered the effect of various operations on the atoms which lie in direct space. In reciprocal space, however, what we are concerned with is the effect of the symmetry operations on the wave functions as well as the atoms. At the top of the Brillouin zone, the point $A(0, 0, \pi/c)$ of Fig. 5-42, a wave function of Bloch form will be

$$\psi(z) = \exp(ikz)u(z) \tag{5-223}$$

If we now perform the operation $(C_3 \mid c/3)$ on this function three times, z becomes $z + c$, and

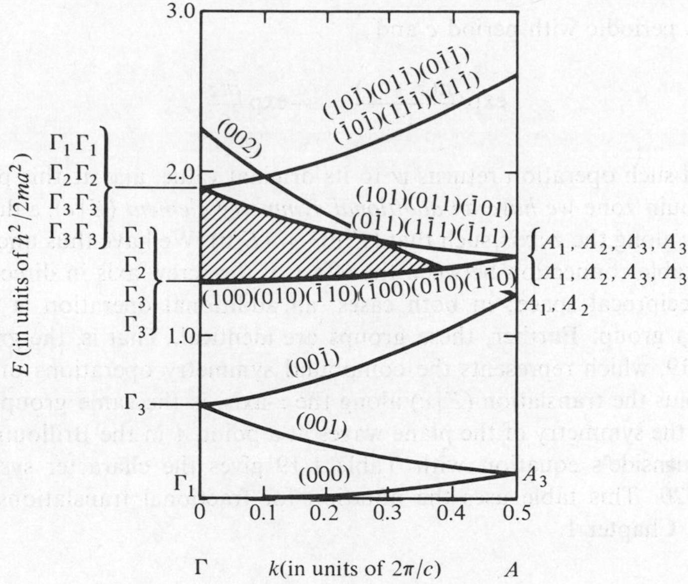

5-43 Free electron energy bands along the Δ-axis of tellurium

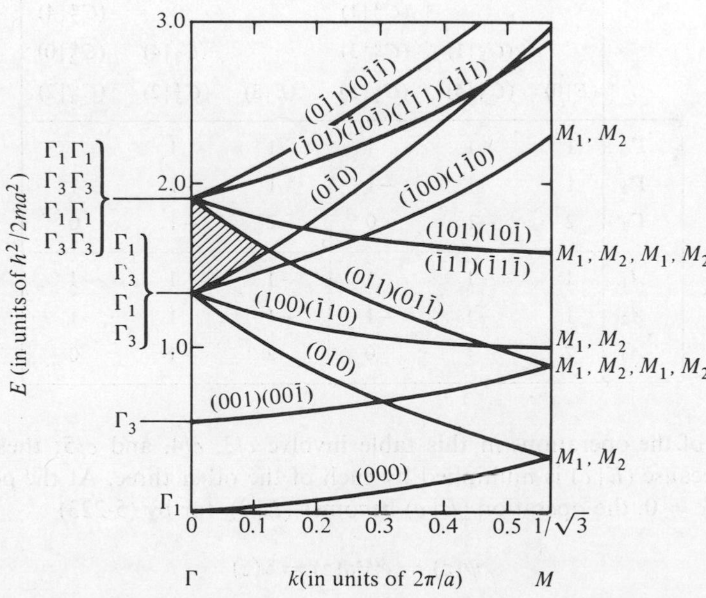

5-44 Free electron energy bands along the Σ-axis of tellurium

$$\psi = \exp[ik(z + c)]u(z) = -\psi$$

since u is periodic with period c and

$$\exp\frac{i\pi(z+c)}{c} = -\exp\frac{i\pi z}{c}$$

A second such operation returns ψ to its original value, and at this point in the Brillouin zone *we have an additional symmetry element* $(E\,|\,c)$, a displacement of c along the z-axis, such that $(E\,|\,c)^2 = (E\,|\,0)$. We have thus uncovered a remarkable connection between the effect of the screw axis in direct space and in reciprocal space; in both cases, an additional operation is needed to form a group. Further, these groups are identical. That is, the group of Table 1-19, which represents the compound symmetry operations of D_3^4 or $P3_1\,21$ plus the translation $(E\,|\,c)$ along the z-axis, is the same group which specifies the symmetry of the plane waves at a point A in the Brillouin zone. Using Burnside's equation with Table 1-19 gives the character system of Table 5-20. This table uses the notation for fractional translations introduced in Chapter 1.

TABLE 5-20 ***Character System for the Points Γ and A of Tellurium***

	$(E\|0)$	$(C_3\|1)$ $(C_3^2\|5)$	$(C_2^a\|1)$ $(C_2^b\|3)$ $(C_2^c\|5)$	$(E\|3)$	$(C_3\|4)$ $(C_3^2\|2)$	$(C_2^a\|4)$ $(C_2^b\|0)$ $(C_2^c\|2)$
Γ_1	1	1	1	1	1	1
Γ_2	1	1	−1	1	1	−1
Γ_3	2	−1	0	2	−1	0
A_1	1	−1	1	−1	1	−1
A_2	1	−1	−1	−1	1	1
A_3	2	1	0	−2	−1	0

Six of the operations in this table involve $c/3$, $c/4$, and $c/5$; these exist only because $(E\,|\,c)$ is multiplied by each of the other three. At the point Γ, where $k = 0$, the operation $(E\,|\,c)$ becomes $(E\,|\,0)$, for by (5-223)

$$\psi(z) = e^{ikz}u(z) = u(z)$$

Hence, $(C_3^2\,|\,5)$ is equivalent to $(C_3^2\,|\,2)$ for such a function, and the same is true for $(C_2^c|5)$, $(C_3|4)$, etc. The table contains only six distinct operations arranged in three classes rather than twelve operations comprising six classes. The character table for the point Γ is then the upper left-hand block of Table 5-20 (this is the same as the system for the point group D_3, which we call the *underlying point group* of D_3^4) and the character system at A is the full 6×6 table. Thus we see why the first three representations are labelled Γ_1, Γ_1, and Γ_3; in shortened form they belong to Γ and in their long form, they belong to A. We also have demonstrated the rather odd fact that the "little" group at the point A is twice the size of the group at Γ; this is a feature of nonsymmorphic groups.

For the character system along the Δ-axis, let us go back to the group C_3, composed of the elements

E, the identity
C_3, a 120° rotation
C_3^2, a 240° rotation

Since this group is Abelian, it has three classes and hence three one-dimensional representations. The character table then appears as follows:

	E	C_3	C_3^2
Δ_1	1	1	1
Δ_2	1	w	x
Δ_3	1	y	z

where w, x, y, and z are to be determined. Since

$$(C_3)^2 = C_3^2$$

then

$$w^2 = x$$

Using

$$C_3 C_3^2 = E$$

then

$$wx = w^3 = 1$$

or

$$w = 1^{1/3} = 1, e^{2\pi i/3}, e^{4\pi i/3}$$

Thus, the representation Δ_2 may be written as

	E	C_3	C_3^2
Δ_2	1	ω	ω^2

where

$$\omega = e^{2\pi i/3}$$

Finally, since any two representations must act like orthogonal vectors (remembering the modification for complex quantities introduced in Problem 5-11), we find that

$$y = \omega^2, z = \omega$$

and the complete character system is given by Table 5-21.

TABLE 5-21

	E	C_3	C_3^2
Δ_1	1	1	1
Δ_2	1	ω	ω^2
Δ_3	1	ω^2	ω

$\omega = e^{2\pi i/3}$

This table refers only to the rotational part of the compound operations associated with a screw axis. To include the translational properties, we go back to Table 5-1, where we saw that the translation group has representations which may be written as

$$\begin{aligned} e^{2\pi il/N} &= e^{2\pi ia(l/Na)} \\ &= e^{ika} \end{aligned} \tag{5-224}$$

where l is an integer, a is the lattice constant along the translation direction, and

$$k = \frac{2\pi l}{Na} \tag{5-225}$$

To show that the values of k are correctly specified by this equation, we use the cyclic boundary conditions to obtain

$$u(x + Na)e^{ik(x+Na)} = u(x)e^{ikx}$$

from which $$\exp(ikNa) = 1$$

and (5-225) immediately follows.

From this result, we would expect that the representations corresponding to screw operations $(E\,|\,0)$, $(C_3\,|\,c/3)$, and $(C_3^2\,|\,2c/3)$ along the z-axis have the form

$$e^{ik(0)} \quad e^{ik(c/3)} \quad e^{ik(2c/3)}$$

or $$1 \quad \delta \quad \delta^2$$

where $$\delta = e^{ikc/3}$$

Incorporating this result into the character system for the group C_3 or 3 gives Table 5-22.

TABLE 5-22

	$(E\|0)$	$(C_3\|c/3)$	$(C_3^2\|2c/3)$
Δ_1	1	δ	δ^2
Δ_2	1	$\omega\delta$	$\omega^2\delta^2$
Δ_3	1	$\omega^2\delta$	$\omega\delta^2$

$\omega = e^{2\pi i/3}$, $\delta = e^{ikc/3}$,

To find the compatibility relations, we realize that $\delta = 1$ for $k = 0$. Adding the characters for Δ_2 and Δ_3 and using

$$\omega + \omega^2 = -1 \tag{5-226}$$

shows that $$\Gamma_3 \longrightarrow \Delta_2 + \Delta_3$$

At the point A, $k = \pi/c$, so that

$$\delta = e^{i\pi/3} \tag{5-227}$$

Adding the rows for Δ_1 and Δ_3 in Table 5-22, we obtain

	$(E\|0)$	$(C_3\|c/3)$	$(C_3^2\|2c/3)$
$\Delta_1 + \Delta_3$	2	$\delta(1 + \omega^2)$	$\delta^2(1 + \omega)$

From (5-226) and (5-227), we find that

$$\delta(1 + \omega^2) = \delta(-\omega) = -e^{i\pi}$$
$$= 1$$

and similarly $$\delta^2(1 + \omega) = -1$$

so that $$\Delta_1 + \Delta_3 \longrightarrow A_3$$

from Table 5-20. Similar considerations produce the balance of Table 5-23, which also shows the compatibility relations for the Σ-axis.

The above discussion should also answer any questions about why the factor δ, which shows up in the representations for the Δ-axis, is absent at

TABLE 5-23 ***Compatibility Relations for Tellurium***

$\Gamma_1 \to \Delta_1$	$A_1 \to \Delta_2$
$\Gamma_2 \to \Delta_1$	$A_2 \to \Delta_2$
$\Gamma_3 \to \Delta_2 + \Delta_3$	$A_3 \to \Delta_1 + \Delta_3$
$\Gamma_1 \to \Sigma_1 \to M_1$	
$\Gamma_2 \to \Sigma_2 \to M_2$	
$\Gamma_3 \to \Sigma_1 + \Sigma_2$	

Γ and at A. At the center of the zone, δ reduces to unity; at the top of the zone, it combines with the rotational contribution to the representation, producing a real number. We have admittedly introduced δ in a rather empirical way; a more logical treatment will be considered later.

In order to compute projection operators, we need the complete representations rather than just the character tables. At the point Γ, the space group has the same representations as the point group, and these have been given in Chapter 2. Along the Δ-axis, there is no problem, since the three representations are one-dimensional. At the top of the zone, we again need matrices for Γ_3 and A_3, which are two-dimensional. These can be expressed in the same form as those for Γ or, as we have just seen, it is convenient to have them in the equivalent complex form. The matrices for six of the elements are given in Table 5-24; the others are the negative of these six.

TABLE 5-24 ***Two-Dimensional Irreducible Representations at the Point A for Tellurium***

	$(E\|0)$	$(C_3\|c/3)$	$(C_3^2\|2c/3)$	$(C_2^a\|c/3)$	$(C_2^b\|0)$	$(C_2^c\|2c/3)$
A_3	$\begin{pmatrix}1 & 0\\0 & 1\end{pmatrix}$	$\begin{pmatrix}-\omega^2 & 0\\0 & -\omega\end{pmatrix}$	$\begin{pmatrix}\omega & 0\\0 & \omega^2\end{pmatrix}$	$\begin{pmatrix}0 & -\omega^2\\-\omega & 0\end{pmatrix}$	$\begin{pmatrix}0 & 1\\1 & 0\end{pmatrix}$	$\begin{pmatrix}0 & \omega\\\omega^2 & 0\end{pmatrix}$

$$\omega = \exp(2\pi i/3)$$

We are now in a position to generate the symmetrized free electron wave functions. As an example, consider the functions at Γ for $\mathbf{m} = (001)$ or $(00\bar{1})$, which are degenerate with an energy of $\frac{9}{16}$ in units of $h^2/2ma^2$. We find, as we would expect, that Γ_1 and Γ_2 project to zero and Γ_3 gives two linear combinations, as indicated in Fig. 5-43.

We can obtain a qualitative notion of the band structure of tellurium by starting with the free electron energy bands of Figs. 5-43 and 5-44 and using the procedure discussed in connection with Figs. 5-32 (a) and (b). This was done by Nussbaum and Hager,[16] who made a qualitative determination of the tellurium band structure starting with the free electron scheme shown in Figs. 5-43 and 5-44. The presence of the crystalline field in tellurium will remove all the degeneracies at the center and edges of the Brillouin zone except those inherent in the irreducible representations and obeying the compatibility relations. By following these principles, and using Figs. 5-43 and 5-44 as a rough guide, we can sketch the bands as shown in Fig. 5-45. We see, for example, that the crystal field removes the degeneracy at the intersection of the (001) and (002) curve for the point A. Also, Table 5-23 shows that $\boldsymbol{\Gamma}_3 \longrightarrow \boldsymbol{\Delta}_2 + \boldsymbol{\Delta}_3$ and $A_3 \longrightarrow \boldsymbol{\Delta}_1 + \boldsymbol{\Delta}_3$, so that the curve joining $\boldsymbol{\Gamma}_3$ and A_3 can only belong to $\boldsymbol{\Delta}_3$.

The energy bands for tellurium given in this figure are seen to differ considerably from those of Reitz. For example, we show the extrema in both the valence and conduction bands as lying about halfway along the $\boldsymbol{\Delta}$-axis, rather than at the top. A material like this, for which the maximum point in the valence band does not have the same k-value as the minimum energy in the conduction band, is said to be an *indirect gap* semiconductor. The physical significance of a direct vs. an indirect gap semiconductor in connection with laser action is explained in another book.[17] The indirect gap can be used to account for some of the unusual electrical properties of tellurium.[18]

The energy band structure of tellurium has been considered by a number of people. The 9×9 secular equation of Reitz,[15] which he solved approximately to obtain the solid curves of Fig. 5-41, was symmetrized using the projection operator method by Asendorf.[19] As we would expect, he obtained essentially the same results. A more sophisticated method was used by Beissner,[20] whose E vs. $\mathbf{k}$ curves along the $\boldsymbol{\Delta}$-axis are somewhat like Fig. 5-45 with regard to the valence band, but differ significantly for the conduction band. He finds, in fact, that there is both a direct gap at $k_z = \pi/c$ and an indirect gap not too far away; the magnitudes are identical, agreeing quite well with the experimental value of 0.33 eV (0.024 ryd). However, the nature of the band structure is still controversial, since a complete OPW calculation (as we shall describe for silicon and germanium) has not been done. A summary of all the studies on the band structure of tellurium and selenium has recently

[16] A. Nussbaum and R. J. Hager, *Phys. Rev.*, **123**, 1958 (1961).

[17] A. Nussbaum, *Electromagnetic and Quantum Properties of Materials*, Prentice-Hall, Inc., Englewood Cliffs, N. J., 1966.

[18] J. S. Blakemore *et al.*, *Tellurium*, "Progress in Semiconductors," Vol. 6, John Wiley & Sons, Inc., New York, 1962.

[19] R. H. Asendorf, "A Group Theoretical Approach to the Band Structure of Tellurium," Ph.D. dissertation, University of Pennsylvania, Philadelphia, Penna. 1956; also *J. Chem. Phys.* **27**, 11 (1957).

[20] R. E. Beissner, *Phys. Rev.*, **145**, 479, (1966).

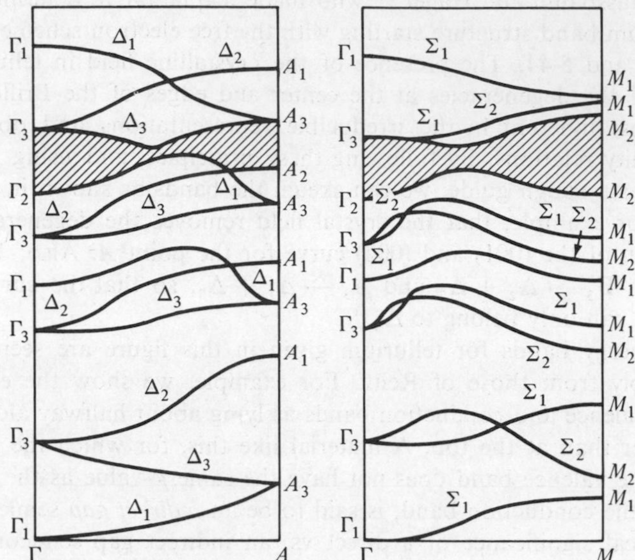

5-45 The nearly-free approximation energy bands for tellurium

been published; this article[21] also discusses the correlation of these calculations with experimental results.

PROBLEM 5-33

(a) Consider the effect of the group C_3 on the basis of Fig. 2-10. The corresponding projection operators will produce a symmetry-adapted basis of three vectors which we can label $\mathbf{e}_{ij}^{(k)}$. Show by example that the effect of some operation R of the group on any of these projected vectors is to simply multiply it by the associated matrix element. That is,

$$R\mathbf{e}_{ij}^{(k)} = \Gamma_{ij}^{(k)}\,\mathbf{e}_{ij}^{(k)}$$

(b) Using the approach of Sec. 5-13, express the Cayley table for D_{3v} in Slater's notation and compare it with the Schoenflies notation.

(c) Also use the method of Sec. 5-13 to determine the spacegroup multiplication table for D_3^4 or $P3_121$.

(d) Use the results of (a) and (c) to verify Slater's character tables for the tellurium space group.

---

[21] W. Becker, V. A. Johnson, and A. Nussbaum, "Electrical and Optical Properties of Tellurium" in *Tellurium-Selenium Conference*, Reinhold Publishing Corp., New York, 1971.

5-12 *Energy band structure as a function of lattice spacing*

The simplest group theoretical treatment of tellurium is due to Gáspár,[22] who was concerned with a problem in energy band determination different from those we have so far considered. Returning to the transcendental equation (5-59), we can extract from it another very important characteristic of periodic structures.

PROBLEM 5-34

Show from Eq. (5-59) that when $b/a = 10$, the energy bands of Fig. 5-17 reduce approximately to four distinct levels and that when $b/a = \frac{1}{240}$, the bands spread out, forming a continuum of levels.

The results of this problem, as well as the ranges of the bands for intermediate values of b/a, are shown in Fig. 5-46. This figure is essentially a plot of band structure as a function of lattice spacing. It shows how the Schroedinger equation predicts the broadening of the discrete energy levels associated with an electron in an isolated atom—those at the right-hand end of the

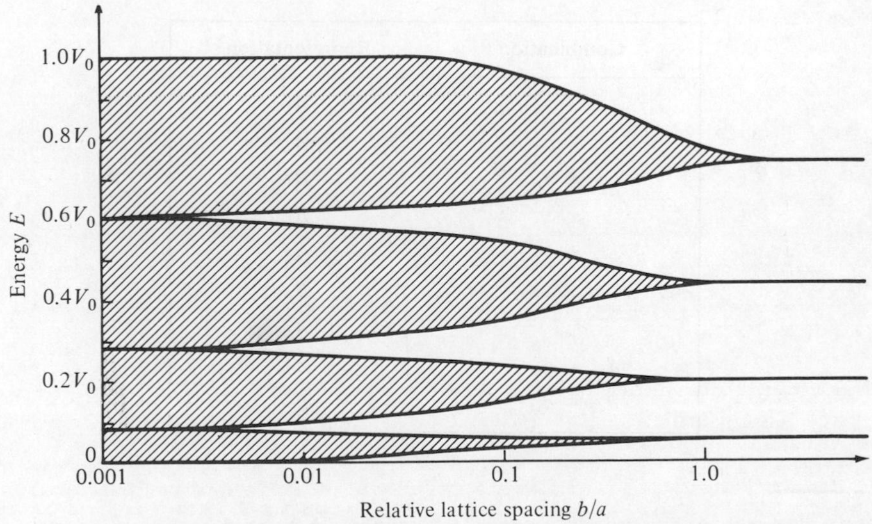

5-46 Effect of lattice spacing on band structure for the one-dimensional hydrogen crystal

[22] R. Gáspár, *Acta Phys. Acad. Sci. Hung.*, 7, 289 and 313 (1957).

diagram—into allowed and forbidden bands and eventually into a single continuous band as the interatomic spacing decreases.

The question arises as to the usefulness of such information. For many semiconductors, experiments have been done whose purpose is to study the change in physical or electrical properties with lattice spacing, usually accomplished by the application of pressure in an oil bath. Tellurium is particularly interesting in this respect because it *expands* along the *c*-axis under pressure. Referring to Fig. 1-44, if we regard the chains as helical springs, we can picture a simultaneous contraction in diameter and lengthening along the chain under pressure. To construct an energy band diagram which would correlate the observed electrical properties with this anomalous physical behavior, Gáspár proposed that the three atoms in the tellurium unit cell be regarded as a planar V-shaped molecule with the C_{2v} symmetry of Fig. 2-20. The bond angle is 102.6° and the coordinate system is oriented as shown in Fig. 5-47, with σ_v taken as a reflection in the xy plane. Using Table 2-13, we may identify s, p, and d combinations by the use of projection operators.

PROBLEM 5-35

Show that the linear combinations of atomic orbitals which are associated with the atom at the origin of Fig. 5-47 are

Combination	Representation
s	A_1
$p_x + p_y$	A_1
$p_x - p_y$	B_1
p_z	B_1
d_{z^2}	A_2
$d_{x^2-y^2}$	B_1
d_{xy}	A_1
$d_{xz} - d_{yz}$	A_2
$d_{xz} + d_{yz}$	B_2

The discussion given in connection with Fig. 4-9 brought out the point that the geometry of atomic orbitals provides only for bonds at right angles; when we deal with angles such as the one of 102.6° in Fig. 5-47, then we must consider the possibility of hybrid orbitals. Gáspár has proposed that the s

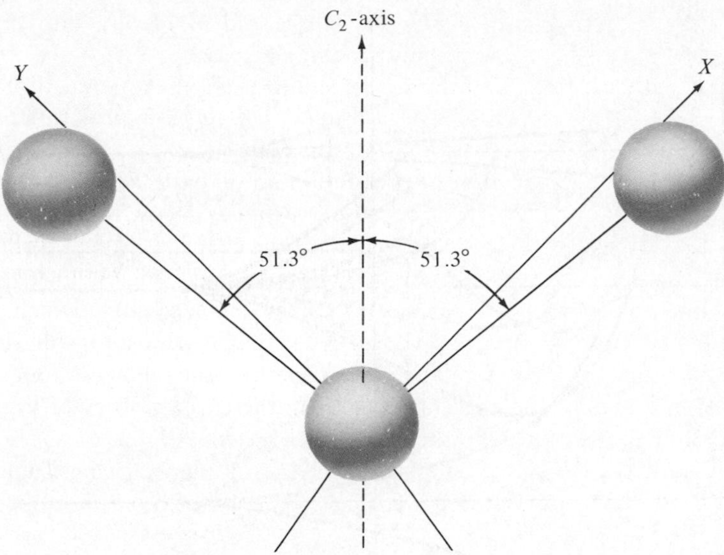

5-47 Tellurium regarded as a molecule with C_{2v} symmetry

orbital and the $(p_x + p_y)$ combination, both belonging to A_1, would combine to form two hybrids. One has the form

$$s_h = s + c(p_x + p_y)$$

where $c < 1$. That is, the original s orbital is modified by the addition of a small contribution from $(p_x + p_y)$ to form a band which may be regarded as originating from an atomic s level. This band is the lowest of the three shown in Fig. 5-48. Since the s functions are spherically symmetrical and have a very small radius, the band arising from them, even when hybridized with a small p function contribution, should be so low in energy that it lies below the valence band. Hence, the s band plays no role in the electrical properties of the material.

The other hybrid formed from A_1 orbitals is taken to have the form

$$p_h = c's + (p_x + p_y)$$

where, again, $c' < 1$. For this combination, the p-like characteristics dominate, and the long lobes associated with sp hybrids have a heavy overlap. This leads to the rather wide band labelled p_h which, as we shall see, forms a considerable portion of the valence band. As in the case of the water molecule of Fig. 4-9, the lobes of the hybrids lie along the lines joining the central atoms with the two end ones (that is, they lie along the chains of the real crystal), and explains why the p_h overlap is so large.

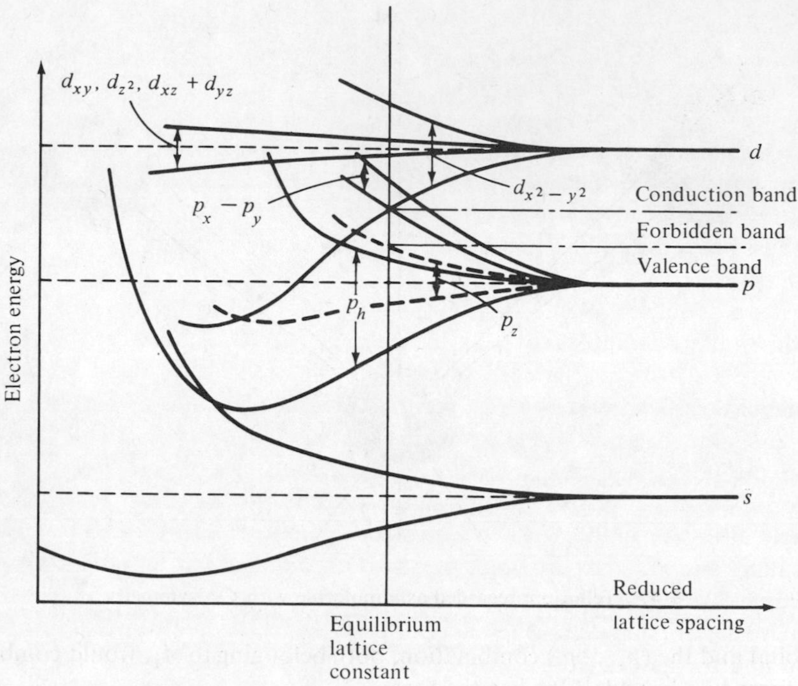

5-48 Energy band structure as a function of lattice spacing for tellurium, assuming C_{2v} symmetry

On the other hand, the combination $(p_x - p_y)$ will lie in a direction normal to the C_2-axis of Fig. 5-47. This is a direction which is roughly parallel to the helical axis. Hence, the positive lobe of a function centered on a given atom overlaps the negative lobe of the function for a neighbor on the chain, and we obtain an antibonding orbital, as shown in Fig. 4-6(b). Further, we realize from Fig. 4-13 that the curves representing energy as a function of atomic spacing turn upwards as R decreases (as we saw for the $2p\sigma$ orbital). This explains why the $(p_x - p_y)$ band of Fig. 5-48 is narrow, lies well above the p_h band, and shows no minimum.

Finally, we consider the p_z-band. These functions, being normal to the plane of Fig. 5-47 are somewhat like the π bonds of Fig. 4-15(c), although not precisely the same, because the orbitals are not parallel to one another as we move along the chain. Hence, the band which is formed (shown dotted in Fig. 5-48 for clarity) is not very wide but does contribute to the bonding. It is shown as overlapping with the top of the p_h-band to indicate that it adds somewhat to the total bonding.

Going on to the d functions, the greatest overlap exists for the $d_{x^2-y^2}$

orbitals, which have their lobes along the x- and y-axes; these are almost collinear with the internuclear directions. This band is taken to be very wide and the other d functions combine to form a narrower band which lies too high to be an active portion of the conduction band. It is assumed, however, that the conduction band is formed by the overlapping of the $(p_x - p_y)$ contribution from the p levels and the $d_{x^2-y^2}$ contribution of the d-levels. That is, there is no way of knowing just where on a diagram like that of Fig. 5-48 to place the equilibrium lattice spacing; it is located as shown in order to fit observed electrical and optical properties. The particular choice shown creates a double conduction band which consists of the $d_{x^2-y^2}$ band combined with an almost complete overlap by the $(p_x - p_y)$ band coming up from the p levels. Gáspár felt the need for this double conduction band to explain some electrical properties which we shall not consider here.

A slightly more elaborate approach was taken by Callen,[23] who suggested that the nonsymmorphic structure of tellurium could be approximated by one in which the chains were straightened out and there would be only a single atom per unit cell. Such a structure would have D_{4h} symmetry if the resulting straight lines of atoms were shifted so that the angles of 60° or 120° became 90°. The symmetry operations for the lattice are then C_4^z about the z-axis (the original c-axis), C_2^x normal to this axis, and J. The character system and the associated basis functions are shown in Table 5-25. Using arguments similar to those of Gáspár, we may show the justification of the energy band diagram, Fig. 5-45. It is easy to see, in fact, that the p_z orbitals (along the four-fold axis) will overlap so heavily that they do not even play a role at the true lattice spacing; instead, the pair (p_x, p_y) forms the valence band.

Turning next to the d functions, the d_{z^2} orbitals are similar to the p_z orbitals, but overlap by a greater amount. This band will therefore start at a greater lattice spacing and be wider than the p_z band; it is seen to define the lower edge of the conduction band. The remainder of the d functions will not overlap so much; as an example, the band due to the pair (d_{yz}, d_{xz}) is shown. It lies entirely within the d_{z^2} band. The other two d bands, omitted for clarity, will be similar to this one. Note that in this figure and the previous one we are using relative or reduced lattice spacings; that is, the horizontal scale is arbitrary.

These qualitative band structures have been used to explain the electrical and optical properties of selenium, tellurium, and their alloys. Let us consider briefly their application to the infrared properties of tellurium. We discussed in Chapter 3 the selection rules for the transition from an initial energy E_n to a final energy E_p for an oscillator in an applied electric field $\mathcal{E}$. We recall that light is a form of electromagnetic radiation and it is customary

[23] H. B. Callen, *J. Chem. Phys.*, **22**, 518 (1954).

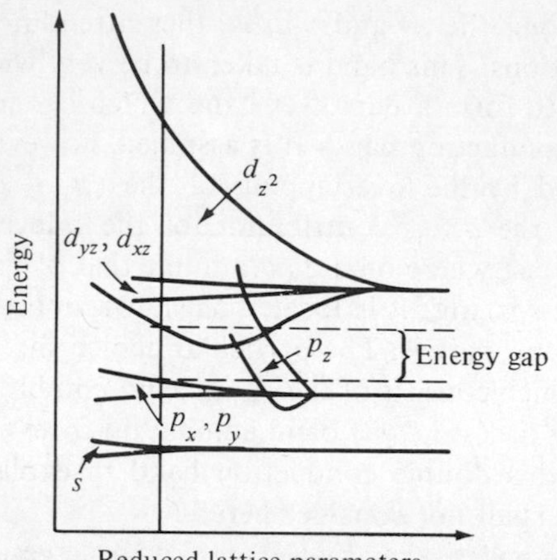

5-49 Energy band structure as a function of lattice space for tellurium, assuming D_{4h} symmetry

to specify its polarization in terms of the orientation of the vector $\mathscr{E}$. When we shine light on a tellurium crystal, we are furnishing energy to the electrons in a fashion analogous to the molecular processes of Chapter 3 and the restrictions found there carry over. In particular, we shall use Eq. (3-77). Suppose that the electric vector $\mathscr{E}$ of the light striking a crystal lies along the z-axis. Then the probability of a transition from an energy E_n to an energy E_p is zero if the integral

$$\int \psi_p z \psi_n \, dz$$

vanishes. This integral will be nonvanishing if the Kronecker product for the representations associated with z and ψ_n is identical to, or contains, the representation to which ψ_p belongs.

Before showing explicitly how this principle operates, let us consider the experimental data which needs explaining. Figure 5-50 shows the results of some very careful measurements by Nomura and Blakemore[24] of the absorption of infrared radiation for two different orientations (and at two temperatures). The wavelength of the radiation may be determined from the Planck relation

$$E = h\nu$$

which can also be written as[25]

$$E\,(\text{eV}) = \frac{1.24}{\lambda\,(\text{microns})}$$

[24]K. C. Nomura and J. S. Blakemore, *Phys. Rev.*, **127**, 1024 (1962).

[25]A. Nussbaum, *Semiconductor Device Physics*, Prentice-Hall, Inc., Englewood Cliffs, N. J., 1961.

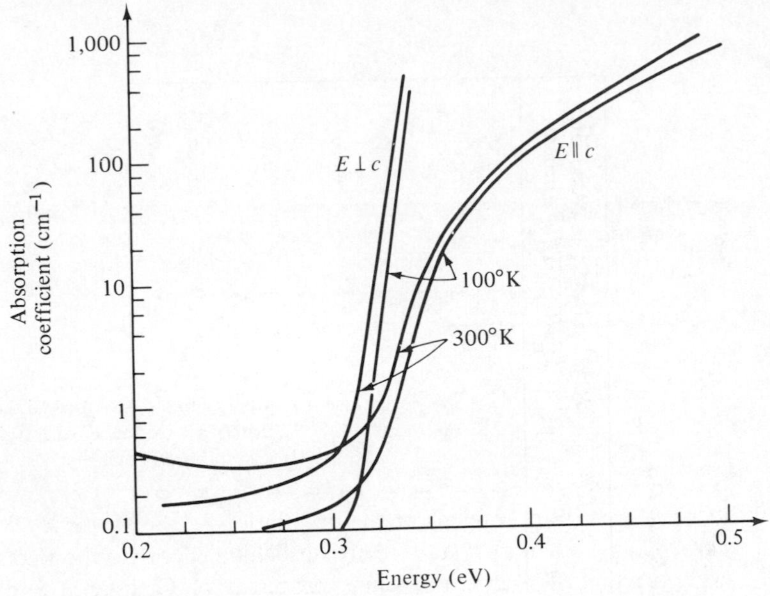

5-50 Absorption of infrared radiation by tellurium for the two principal polarization directions

The two curves for $\mathscr{E}$ in the xy-plane (that is, normal to the c-axis) indicate that an electron can absorb about 0.32 eV from the incident radiation. We interpret this as the amount of energy required to raise the electron to the conduction band from the valence band. On the other hand, for $\mathscr{E}$ along the symmetry axis, it takes about 0.35 eV for the same process. These energies are a measure of the width of the forbidden band, and the question which must be answered is why this quantity depends on the polarization of the light.

Callen's explanation can be given in connection with Table 5-25 and Fig. 5-49. His valence band, formed by the (p_x, p_y) pair, belongs to E_u. If $\mathscr{E}$ is parallel to the z-axis, then we are interested in the Kronecker product associated with the representations for z and for (p_x, p_y). But the former is A_{2u}, since z has the same symmetry as the p_z orbital. We then find that

$$A_{2u} \times E_u = E_g$$

By (3-77), the final function ψ_p must belong therefore to E_g. The associated functions are the pair (d_{yz}, d_{xz}) which form a band whose bottom edge lies slightly above the bottom of the conduction band.

On the other hand, light polarized in a plane normal to the c-axis is associated with x and y, or the representation E_u, and

TABLE 5-25 ***Character Table for the Group D_{4h}***

	E	$(C_4^z)^2$	C_4^z $(C_4^z)^3$	C_2^x $C_2^x(C_4^z)^2$	$C_2^xC_4^z$ $C_2^x(C_4^z)^3$	J	$J(C_4^z)^2$	JC_2^z $J(C_4^z)^3$	JC_2^x $JC_2^x(C_4^z)^2$	$JC_2^xC_4^z$ $JC_2^x(C_4^z)^3$	Associated orbitals
A_{1g}	1	1	1	1	1	1	1	1	1	1	s, d_{z^2}
A_{1u}	1	1	1	1	1	−1	−1	−1	−1	−1	
A_{2g}	1	1	1	−1	−1	1	1	1	−1	−1	
A_{2u}	1	1	1	−1	−1	−1	−1	−1	1	1	p_z
B_{1g}	1	1	−1	1	−1	1	1	−1	1	−1	$d_{x^2-y^2}$
B_{1u}	1	1	−1	1	−1	−1	−1	1	−1	1	
B_{2g}	1	1	−1	−1	1	1	1	−1	−1	1	d_{xy}
B_{2u}	1	1	−1	−1	1	−1	−1	1	1	−1	
E_g	2	−2	0	0	0	2	−2	0	0	0	(d_{yz}, d_{xz})
E_u	2	−2	0	0	0	−2	2	0	0	0	(p_x, p_y)

$$E_u \times E_u = A_{1g} + A_{2g} + B_{1g} + B_{2g}$$

The lowest band of the four possible ones is the one formed from the d_{z^2} orbitals, belonging to A_{1g}. Hence, the bottom of the conduction band is accessible to electrons excited by light in the xy-plane, and that is why this polarization corresponds to a smaller gap. It is interesting to note that the (d_{yz}, d_{xz}) band, belonging to E_g, probably lies too high to be observed and the p_z band, belonging to A_{2u}, is inaccessible no matter where it lies.

PROBLEM 5-36

Show how Fig. 5-49 provides an alternate explanation for the measurements of Blakemore and Nomura. Bear in mind that C_{2v} has only one-dimensional representations. Hence, the degenerate pair of coordinates (x, y) cannot go with any of these representations, so that there effectively are no selection rules which apply when $\mathscr{E}$ is perpendicular to the c-axis.

5-13 The close-packed hexagonal structure

When a large number of identical spheres are placed on a plane and pushed tightly together, they group themselves as shown in Fig. 5-51(a). Each sphere touches six others and we obtain the familiar hexagonal pattern. A second identical layer will fit into the interstices of the first, with each sphere resting on three below it and touching six spheres in its own layer [Fig. 5-51(b)]. The third layer can be placed in one of two ways; either over the interstices

5-51(a) One layer of a close-packed structure

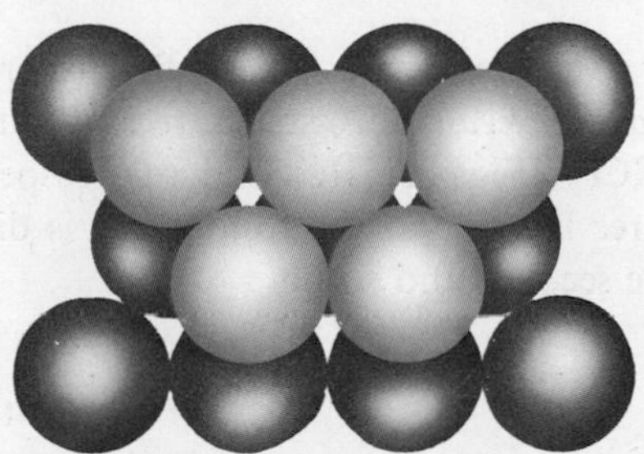

5-51(b) Top view of two layers of a close-packed structure

5-51(c) A three-layer cubic close-packed structure

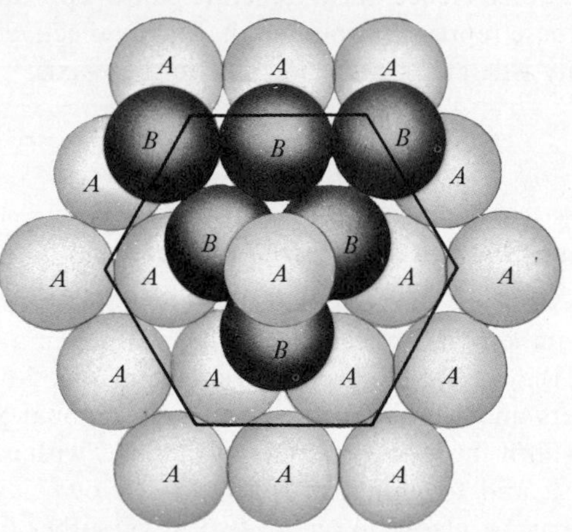

5-51(d) A three-layer hexagonal close-packed structure. Atoms labelled *A* are in the first, third, . . . , layers; those labelled *B* are in the second, fourth, . . . , layers.)

or over the spheres of the first layer. In the former case we obtain a *cubic close-packed* (CCP) structure and in the latter case, a *hexagonal close-packed* (HCP) [Figs. 5-51(c) and (d), respectively]. The cubic structure thus has three layers and the fourth one is directly over the first, the fifth one over the second, and so on.

PROBLEM 5-37

Demonstrate that Fig. 5-51(c) shows a FCC lattice.

The HCP structure is composed of two distinct layers, with the third one over the first layer, etc. For the HCP structure, all spheres in a given plane are equivalent, but only alternate planes are equivalent, so that we take two of the primitive lattice vectors as shown in Fig. 5-52 and the third one $\mathbf{a}_3$ extends for a distance corresponding to the separation of two equivalent planes. If we choose an origin at the center of a sphere, then another sphere in the next layer which touches this one will have its center at the point

$$\mathbf{r} = \tfrac{2}{3}\mathbf{a}_1 + \tfrac{1}{3}\mathbf{a}_2 + \tfrac{1}{2}\mathbf{a}_3 \tag{5-228}$$

and the rest of the structure can be generated from this basis by the symmetry operations as, we shall see.

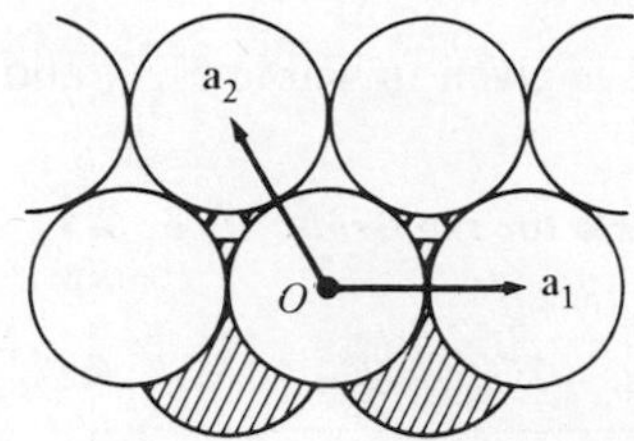

5-52 The HCP vectors

The HCP space group is D_{6h}^4 or $P\ 6_3/mmm$ (abbreviated), $P\ 6_3/m\ 2/m\ 2/m$ (full), which is nonsymmorphic. The point group D_{6h} or $6/m\ 2/m\ 2/m$ is (from the notation) a six-fold dihedral group. There are six each of two kinds of dihedral axes: one set through the corners and the other through the midpoints of the hexagon. Hence the symmetry operations are as listed in Table 5-26.

TABLE 5-26 ***Symmetry Operations for the Point Group D_{6h} or $6/m\ 2/m\ 2/m$***

E	The identity
C_6, C_3, C_2	Rotations of 60°, 120°, or 180° about the six-fold axis
C_2'	Dihedral axis through two opposite corners
C_2''	Dihedral axis through the midpoints of two opposite sides
σ_h	A reflection in the plane normal to the six-fold axis
$\sigma_h C_6, \sigma_h C_3, \sigma_h C_2$ $\sigma_h C_2'$ $\sigma_h C_2''$	Combination of σ_h with the other symmetry operations

PROBLEM 5-38

Eyring, Walker, and Kimball[6] describe the group D_{6h} as

$$D_{6h} = D_6 \times C_i$$

where D_6 is composed of the top half of Table 5-26. Hochstrasser,[26] on the other hand, lists the elements as

$$E, 2C_6, 2C_3, C_2, 3C_2', 3C_2'', J, 2S_3, 2S_6, \sigma_h, 3\sigma_d, 3\sigma_v$$

Show that these two descriptions are identical to that of Table 5-26.

The character system of D_6 or 622 is given in Table 5-27, and we obtain

TABLE 5-27 ***Character Table for the Group D_6 or 622***

	E	C_2	$2C_3$	$2C_6$	$3C_2'$	$3C_2''$
A_1	1	1	1	1	1	1
A_2	1	1	1	1	−1	−1
B_1	1	−1	1	−1	1	−1
B_2	1	−1	1	−1	−1	1
E_1	2	−2	−1	1	0	0
E_2	2	2	−1	−1	0	0

the characters for D_{6h} by taking the Kronecker product with those for C_i, as for Table 5-4.

To obtain the Cayley table, consider a point X_0 on the perimeter of the hexagon in Fig. 5-53, and an arbitrary vector **r**. In terms of the lattice vectors, we may write **r** as

$$\mathbf{r} = r_1\mathbf{a}_1 + r_2\mathbf{a}_2 + r_3\mathbf{a}_3 \tag{5-229}$$

Then the rotations of $\pm 60°$ about the six-fold z-axis produce the points X_1, X_{-1}, X_2, X_{-2}, and X_3; the dihedral operations produce the points $Y_0, Y_1, Y_{-1}, Y_2, Y_{-2}, Y_3$. These operations are most simply expressed in terms of the behavior of the $\mathbf{a}_i$. For example, X_1 converts $\mathbf{a}_1$ into $\mathbf{a}_1 + \mathbf{a}_2$, $\mathbf{a}_2$ into $-\mathbf{a}_1$, and $\mathbf{a}_3$ is unaltered. Thus, by (5-229)

[26] R. H. Hochstrasser, *Molecular Aspects of Symmetry*, W. A. Benjamin, New York, 1966.

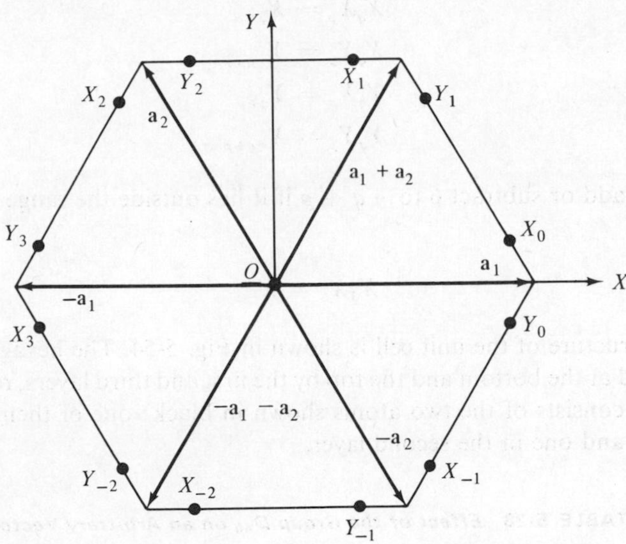

5-53 Symmetry operations for a hexagonal structure

$$X_1\mathbf{r} = r_1(\mathbf{a}_1 + \mathbf{a}_2) + r_2(-\mathbf{a}_1) + r_3\mathbf{a}_3$$
$$= (r_1 - r_2)\mathbf{a}_1 + r_1\mathbf{a}_2 + r_3\mathbf{a}_3$$

and as a further example

$$X_2\mathbf{r} = r_1(\mathbf{a}_2) + r_2(-\mathbf{a}_1 - \mathbf{a}_2) + r_3\mathbf{a}_3$$
$$= -r_2\mathbf{a}_1 + (r_1 - r_2)\mathbf{a}_2 + r_3\mathbf{a}_3$$

where, using the notation of Slater, we shall regard the X_i and Y_j as symmetry operations. The effect of each member of D_6 on r_1, r_2, r_3 is listed in Table 5-28. The other twelve operations of D_{6h}, which are denoted X_i' and Y_j' in Slater's notation, simply reverse the sign of r_3. Both the Schoenflies notation and the Slater notation are given.

For two successive operations, we see that $X_2X_1\mathbf{r}$ may be determined from the fact that X_2 converts the coefficient of $\mathbf{a}_1$ into the negative of the coefficient of $\mathbf{a}_2$, and the coefficient of $\mathbf{a}_2$ becomes the difference of the coefficients of $\mathbf{a}_1$ and $\mathbf{a}_2$. Using this rule on $X_1\mathbf{r}$ gives

$$X_2(X_1\mathbf{r}) = -r_1\mathbf{a}_1 - r_2\mathbf{a}_2 + r_3\mathbf{a}_3 = X_3\mathbf{r}$$

or
$$X_2X_1 = X_3$$

All relations of this type may be summarized by the rules

$$\begin{aligned} X_q X_p &= X_{q+p} \\ X_q Y_p &= Y_{-q+p} \\ Y_q X_p &= Y_{q+p} \\ Y_q Y_p &= X_{-q+p} \end{aligned} \tag{5-230}$$

where we add or subtract 6 to $\pm q + p$ if it lies outside the range $-2, 3$. For example,

$$X_3 X_1 = X_{-2}$$

The structure of the unit cell is shown in Fig. 5-54. The hexagonal prism is bounded at the bottom and the top by the first and third layers, respectively. The basis consists of the two atoms shown in black; one of them lies in the first layer and one in the second layer.

TABLE 5-28 ***Effect of the Group D_{6h} on an Arbitrary Vector***

E	X_0	r_1	r_2	r_3
C_6	X_1	$r_1 - r_2$	r_1	r_3
C_6^{-1}	X_{-1}	r_2	$-r_1 + r_2$	r_3
C_3	X_2	$-r_2$	$r_1 - r_2$	r_3
C_3^{-1}	X_{-2}	$-r_1 + r_2$	$-r_1$	r_3
C_2	X_3	$-r_1$	$-r_2$	r_3
C_2'	Y_0	$r_1 - r_2$	$-r_2$	$-r_3$
C_2''	Y_1	r_1	$r_1 - r_2$	$-r_3$
C_2''	Y_{-1}	$-r_2$	$-r_1$	$-r_3$
C_2'	Y_2	r_2	r_1	$-r_3$
C_2'	Y_{-2}	$-r_1$	$-r_1 + r_2$	$-r_3$
C_2''	Y_3	$-r_1 + r_2$	r_1	$-r_3$

PROBLEM 5-39

Explain why this unit cell differs from those of Fig. 1-32(n) and Fig. 1-32(o) and why it is not a new Bravais lattice.

Returning to the HCP structure, Fig. 5-55 shows two layers indicated by black and white circles, with a separation of $a_3/2$ between them. We have chosen the origin so that the z-axis is a six-fold screw axis, for a 60° rotation

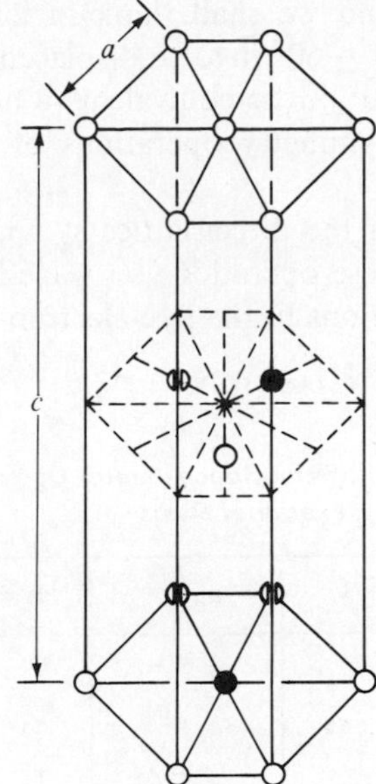

5-54 Unit cell for the hexagonal close-packed structure

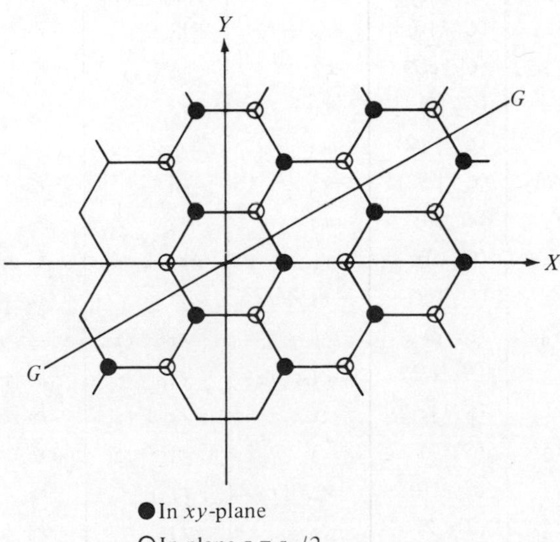

5-55 Top view of the HCP structure

followed by a displacement of $a_3/2$ is a covering operation. We also note that the line GG indicates the position of a glide-plane passing through the z-axis, since a displacement of any atom by $\pm a_3/2$ in the z-direction followed by a reflection in the glide-plane is also a covering operation. These two compound operations are equivalent, and we shall think in terms of a screw axis. With this in mind, rotations of $\pm 60°$ involve displacements of $\pm a_3/2$, but rotations of $\pm 120°$ do not, since $\pm a_3$ is equivalent to no displacement. These considerations lead to the symmetry operations of the group D^4_{6h} listed in Table 5-29.

Having established the nature of the direct-lattice symmetry, we would next like to consider the effect of these operations on wave-functions in the Brillouin zone. The plane-wave solutions to the free-electron equation are

$$\psi = e^{i(\mathbf{k}-2\pi m\mathbf{B})\cdot\mathbf{r}} \tag{5-76}$$

TABLE 5-29 ***Symmetry Operations of the Space Group D^4_{6h} or $P\,6_3/mmm$ and Their Effect on a Hexagonal Basis***

		$\mathbf{a}_1$	$\mathbf{a}_2$	$\mathbf{a}_3$
X_0	$(E\|0)$	r_1	r_2	r_3
X_1	$(C_6\|c/2)$	$r_1 - r_2$	r_1	$r_3 + \frac{1}{2}$
X_{-1}	$(C_6^{-1}\|c/2)$	r_2	$-r_1 + r_2$	$r_3 + \frac{1}{2}$
X_2	$(C_3\|0)$	$-r_2$	$r_1 - r_2$	r_3
X_{-2}	$(C_3^{-1}\|0)$	$-r_1 + r_2$	$-r_1$	r_3
X_3	$(C_2\|c/2)$	$-r_1$	$-r_2$	$r_3 + \frac{1}{2}$
Y_0	$(C'_2\|0)$	$r_1 - r_2$	$-r_2$	$-r_3$
Y_1	$(C''_2\|c/2)$	r_1	$r_1 - r_2$	$-r_3 + \frac{1}{2}$
Y_{-1}	$(C''_2\|c/2)$	$-r_2$	$-r_1$	$-r_3 + \frac{1}{2}$
Y_2	$(C'_2\|0)$	r_2	r_1	$-r_3$
Y_{-2}	$(C'_2\|0)$	$-r_1$	$-r_1 + r_2$	$-r_3$
Y_3	$(C''_2\|c/2)$	$-r_1 + r_2$	r_1	$-r_3 + \frac{1}{2}$
X'_0	$(\sigma_h\|0)$	r_1	$r_1 - r_2$	r_3
X'_1	$(S_6\|c/2)$	$r_1 - r_2$	r_1	$-r_3 + \frac{1}{2}$
X'_{-1}	$(S_6^{-1}\|c/2)$	r_2	$-r_1 + r_2$	$-r_3 + \frac{1}{2}$
X'_2	$(S_3\|0)$	$-r_2$	$r_1 - r_2$	$-r_3$
X'_{-2}	$(S_3^{-1}\|0)$	$-r_1 + r_2$	$-r_1$	$-r_3$
X'_3	$(J\|c/2)$	$-r_2$	$-r_1$	$-r_3 + \frac{1}{2}$
Y'_0	$(\sigma_v\|0)$	$r_1 - r_2$	$-r_2$	r_3
Y'_1	$(\sigma_d\|c/2)$	r_1	$r_1 - r_2$	$r_3 + \frac{1}{2}$
Y'_{-1}	$(\sigma_d\|c/2)$	$-r_2$	$-r_1$	$r_3 + \frac{1}{2}$
Y'_2	$(\sigma_v\|0)$	r_2	r_1	r_3
Y'_{-2}	$(\sigma_v\|0)$	$-r_1$	$-r_1 + r_2$	r_3
Y'_3	$(\sigma_d\|c/2)$	$-r_1 + r_2$	r_1	$r_3 + \frac{1}{2}$

and we shall use them as the basis for a representation of the group D_{6h}^4 in reciprocal or k-space.

PROBLEM 5-40

Orienting $\mathbf{a}_1$ and $\mathbf{a}_2$ as shown in Fig. 5-52, prove that

$$\mathbf{B} = \begin{pmatrix} 1/a & 1/\sqrt{3}\,a & 0 \\ 0 & 2/\sqrt{3}\,a & 0 \\ 0 & 0 & 1/c \end{pmatrix}$$

Equation (5-71) is equivalent to

$$\mathbf{mB} = m_1\mathbf{b}_1 + m_2\mathbf{b}_2 + m_3\mathbf{b}_3 \tag{5-231}$$

To agree with Slater's notation, we introduce three new integers h_1, h_2, h_3 by

$$h_i = -m_i \tag{5-232}$$

and we write $\mathbf{k}$ in terms of its components as

$$\mathbf{k} = 2\pi(p_1\mathbf{b}_1 + p_2\mathbf{b}_2 + p_3\mathbf{b}_3) \tag{5-233}$$

PROBLEM 5-41

Show from (5-77) that

$$\xi = 2\pi p_1$$

or

$$\xi = p_1$$

with similar relations for h and ξ.

Using these relations, plus the fact that $\mathbf{a}_i \cdot \mathbf{b}_j = \delta_{ij}$, Eq. (5-76) for the free electron wave functions becomes

$$\psi = \exp\,[2\pi i\{(p_1 + h_1)r_1 + (p_2 + h_2)r_2 + (p_3 + h_3)r_3\}] \tag{5-234}$$

Then the effect of a typical operation X_1 of D_{6h}^4 on ψ is

$$
\begin{aligned}
X_1\psi &= X_1 \exp\,[2\pi i\{(p_1 + h_1)r_1 + (p_2 + h_2)r_2 + (p_3 + h_3)r_3\}] \\
&= \exp\,[2\pi i\{(p_1 + h_1)(r_1 - r_2) + (p_2 + h_2)r_1 + (p_3 + h_3)(r_3 + \tfrac{1}{2})\}] \\
&= \exp\,[2\pi i(p_1 + p_2 + h_1 + h_2)r_1 - (p_1 + h_1)r_2 + (p_3 + h_3)r_3]\cdot \\
&\qquad \exp\,[\pi i(p_3 + h_3)]
\end{aligned}
$$

In a similar way, we obtain the transformed wave functions for all the unprimed operators listed in Table 5-30.

The effect of the product of two of the operations—for example, X_1 followed by X_2—on the function ψ may be determined in the same way that we obtained the point group rules of Eq. (5-166); X_1 takes the coefficients of r_1 and r_2 in the original function and adds them to form the coefficient of r_1 in the function $X_1\psi$. Expressing the effect of X_1 on the rest of ψ in this way, as well as the effect of X_2 on the function, we have that

$$
\begin{aligned}
X_2(X_1\psi) = \exp\,[2\pi i\{-(p_1 + h_1)r_1 - (p_2 + h_2)r_2 + (p_3 + h_3)r_3\}]\cdot \\
\exp\,[\pi i(p_3 + h_3)] = x_3\psi_k
\end{aligned}
$$

or

$$X_2X_1 = X_3 \tag{5-235}$$

Similarly,

$$X'_2X_1 = X_3 \tag{5-236}$$

TABLE 5-30 ***Basis Functions Under the Operations of the Group D_{6h}^4***

Operator	Coefficient of $2\pi i$	Coefficient of πi
X_0	$(p_1 + h_1)r_1 + (p_2 + h_2)r_2 + (p_3 + h_3)r_3$	0
X_1	$(p_1 + h_1 + p_2 + h_2)r_1 - (p_1 + h_1)r_2 + (p_3 + h_3)r_3$	$(p_3 + h_3)$
X_{-1}	$-(p_2 + h_2)r_1 + (p_1 + h_1 + p_2 + h_2)r_2 + (p_3 + h_3)r_3$	$(p_3 + h_3)$
X_2	$(p_2 + h_2)t_1 - (p_1 + h_1 + p_2 + h_2)r_2 + (p_3 + h_3)r_3$	0
X_{-2}	$-(p_1 + h_1 + p_2 + h_2)r_1 + (p_1 + h_1)r_2 + (p_3 + h_3)r_3$	0
X_3	$-(p_1 + h_1)r_1 - (p_2 + h_2)r_2 + (p_3 + h_3)r_3$	$(p_3 + h_3)$
Y_0	$(p_1 + h_1)r_1 - (p_1 + h_1 + p_2 + h_2)r_2 - (p_3 + h_3)r_3$	0
Y_1	$(p_1 + h_1 + p_2 + h_2)r_1 - (p_2 + h_2)r_2 - (p_3 + h_3)r_3$	$(p_3 + h_3)$
Y_{-1}	$-(p_2 + h_2)r_1 - (p_1 + h_1)r_1 - (p_3 + h_3)r_3$	$(p_3 + h_3)$
Y_2	$(p_2 + h_2)r_2 + (p_1 + h_1)r_1 - (p_3 + h_3)r_3$	0
Y_{-2}	$-(p_1 + h_1 + p_2 + h_2)r_1 + (p_2 + h_2)r_2 - p_3 + h_3)r_3$	0
Y_3	$(-p_1 - h_1 + p_2 + h_2)r_1 + (p_1 + h_1)r_2 - (p_3 + h_3)r_3$	$(p_3 + h_3)$

However, for the product $X_2'X_1$, we obtain

$$\begin{aligned} X_2' \exp [2\pi i\{-(p_1 + h_1)r_1 - (p_2 + h_2)r_2 + (p_3 + h_3)r_3]\cdot \\ \exp [\pi i(p_3 + h_3)] = \exp [2\pi i\{-(p_2 + h_2)r_1 \\ + (+p_1 + h_1 + p_2 + h_2)r_2 - (p_3 + h_3)r_3]\cdot \\ \exp [\pi i(p_3 + h_3)] \exp [\pi i(p_3 + h_3)] = X'_1\psi e^{2\pi i p_3} \end{aligned} \tag{5-237}$$

since $e^{2\pi i h_3} = 1$ because the h_i are integers.

Thus, (5-237) shows that the elements of Table 5-30 do *not* form a group because of the factor $e^{2\pi i p_3}$ which appears in some of the products. We may verify as above that the combining rules for the operations of Table 5-30 are:

1. The subscripts of the products obey the same rules as for the point group; i.e., Eqs. (5-230).
2. The extra factor appears only in the following cases:

$$\begin{aligned} &\text{unprimed odd} \times \text{unprimed odd: } e^{2\pi i p_3} \\ &\text{unprimed odd} \times \text{primed odd: } e^{-2\pi i p_3} \\ &\text{primed even} \times \text{unprimed odd: } e^{2\pi i p_3} \\ &\text{primed even} \times \text{primed odd: } e^{-2\pi i p_3} \end{aligned} \tag{5-238}$$

Thus, when the symmetry elements of the HCP lattice are used as basis, D_{6h}^4 is a space group, but when the plane waves of (5-234) are used as a basis, the group requirements are no longer obeyed. We note, however, that at $\mathbf{k} = 0$, Eq. (5-232) shows that $p_1, p_2, p_3 = 0$, so that $e^{2\pi i p_3}$ does not enter. Hence, the space group multiplication table is identical to that of the point group.

Constructing the Brillouin zone of the HCP structure by drawing perpendicular-bisectors, we obtain the hexagonal prism of Fig. 5-56. The point Γ corresponds to $\mathbf{k} = 0$ in reciprocal space, so that it is a straightforward matter to obtain the character system from the Cayley table and this is presented in Table 5-31. This table also gives the matrices for the degenerate irreducible representations, since these are needed to compute projection operators. For two-dimensional representations, we may use matrices of the form

$$C_3 = \begin{pmatrix} -1/2 & -\sqrt{3}/2 \\ \sqrt{3}/2 & -1/2 \end{pmatrix}$$

already encountered in Chapter 2, or we may use the complex form involving the factor

$$\omega = e^{2\pi i/3}$$

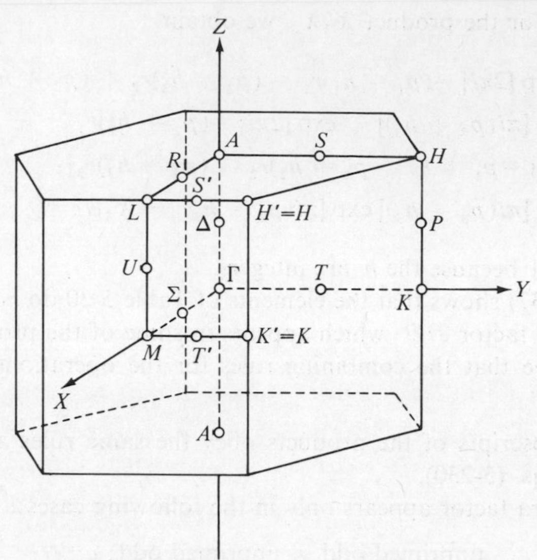

5-56 Brillouin zone for a hexagonal structure

TABLE 5-31 ***Irreducible Representations at Γ for the Group D_{6h}^4***

	X_0	X_1	X_{-1}	X_2	X_{-2}	X_3	Y_0	Y_1	Y_{-1}	Y_2	Y_{-2}	Y_3
Γ_1^+, Γ_2^-	1	1	1	1	1	1	1	1	1	1	1	1
Γ_2^+, Γ_1^-	1	1	1	1	1	1	−1	−1	−1	−1	−1	−1
Γ_3^-, Γ_4^+	1	−1	−1	1	1	−1	−1	1	1	−1	−1	1
Γ_4^-, Γ_3^+	1	−1	−1	1	1	−1	1	−1	−1	1	1	−1
$(\Gamma_5^+, \Gamma_5^-)_{11}$	1	ω	ω^2	ω^2	ω	1	0	0	0	0	0	0
$(\Gamma_5^+, \Gamma_5^-)_{21}$	0	0	0	0	0	0	1	ω	ω^2	ω^2	ω	1
$(\Gamma_5^+, \Gamma_5^-)_{12}$	0	0	0	0	0	0	1	ω^2	ω	ω	ω^2	1
$(\Gamma_5^+, \Gamma_5^-)_{22}$	1	ω^2	ω	ω	ω^2	1	0	0	0	0	0	0
$\chi(\Gamma_5^+, \Gamma_5^-)$	2	−1	−1	−1	−1	2	0	0	0	0	0	0
$(\Gamma_6^-, \Gamma_6^+)_{11}$	1	$-\omega^2$	$-\omega$	ω	ω^2	−1	0	0	0	0	0	0
$(\Gamma_6^-, \Gamma_6^+)_{21}$	0	0	0	0	0	0	1	$-\omega^2$	$-\omega$	ω	ω^2	−1
$(\Gamma_6^-, \Gamma_6^+)_{12}$	0	0	0	0	0	0	1	$-\omega$	$-\omega^2$	ω^2	ω	−1
$(\Gamma_6^-, \Gamma_6^+)_{22}$	1	$-\omega$	$-\omega^2$	ω^2	ω	−1	0	0	0	0	0	0
$\chi(\Gamma_6^-, \Gamma_6^+)$	2	1	1	−1	−1	−2	0	0	0	0	0	0

$$\omega = e^{2\pi i/3}$$

introduced in the previous section; Slater prefers the complex form. We note that this table is compressed; only 12 of the 24 operations are indicated. For the primed operations, the representations for Γ_1^+ are identical to the unprimed entries and for Γ_2^-, they are the negative of those shown; a similar system is used for the remaining five pairs.

As we have indicated, the factor

$$\alpha = e^{i\pi p_3} \tag{5-239}$$

which appears in the plane wave expressions at every point in the Brillouin zone except Γ, causes complications. The way out of this difficulty is through the properties of the projection operator. These operators convert a given basis function into a symmetrized linear combination of functions. Let P_1 be that part of a projection operator containing only odd-subscript symmetry operations; by (5-238) these are the only ones which can produce the factor α of (5-239). If P_1 operates on some basis function ψ to produce a new function ψ_1 and this is followed by an operator R satisfying (5-238), the result will be

$$RP_1\psi = \alpha\psi_1$$

To avoid this situation, we can incorporate a factor α^* into the projection operators, so that

$$R(\alpha^* P_1)\psi = \alpha^*\alpha\psi_1 = \psi_1$$

Actually, the definition for a projection operator (2-138) for complex functions should be altered to

$$P_{ij}^{(k)} = \Sigma\, \Gamma_{ij}^{(k)}(G_l)^* G_l \tag{5-240}$$

If the factor α is incorporated into the representation of every odd-subscript symmetry operator, that is, every one which incorporates a translation of $\mathbf{a}_3/2$, then the α^* appearing in (5-240) will have the desired effect. Combining this idea with what we already know about the properties of character tables leads to the system of Table 5-32, which applies to any point on the Δ-axis. We note that the little group contains *no* primed operations, since σ_h does not leave any point invariant on the Δ-axis.

The situation is different for the point A at the top of the Δ-axis. The point at the bottom of this axis is also labelled A, since they have the same geometrical symmetry. However, we note that p_3 in (5-234) is equal to 0.5 at the top of the zone and to -0.5 at the bottom. Hence, the factor α becomes

$$e^{+i\pi/2} = i \quad \text{or} \quad e^{-i\pi/2} = -i$$

TABLE 5-32 ***Irreducible Representations Along the Δ-axis for the Group*** D_{6h}^4

	X_0	X_1	X_{-1}	X_2	X_{-2}	X_3	Y_0	Y_1	Y_{-1}	Y_2	Y_{-2}	Y_3
Δ_1	1	α	α	1	1	α	1	α	α	1	1	α
Δ_2	1	$-\alpha$	$-\alpha$	1	1	$-\alpha$	1	$-\alpha$	$-\alpha$	1	1	$-\alpha$
Δ_3	1	α	α	1	1	α	-1	$-\alpha$	$-\alpha$	-1	-1	$-\alpha$
Δ_4	1	$-\alpha$	$-\alpha$	1	1	$-\alpha$	-1	α	α	-1	-1	α
$(\Delta_5)_{11}$	1	$\omega\alpha$	$\omega^2\alpha$	ω^2	ω	α	0	0	0	0	0	0
$(\Delta_5)_{21}$	0	0	0	0	0	0	1	$\omega\alpha$	$\omega^2\alpha$	ω^2	ω	α
$(\Delta_5)_{12}$	0	0	0	0	0	0	1	$\omega^2\alpha$	$\omega\alpha$	ω	ω^2	α
$(\Delta_5)_{22}$	1	$\omega^2\alpha$	$\omega\alpha$	ω	ω^2	α	0	0	0	0	0	0
$\chi(\Delta_5)$	2	$-\alpha$	$-\alpha$	-1	-1	2α	0	0	0	0	0	0
$(\Delta_6)_{11}$	1	$-\omega^2\alpha$	$-\omega\alpha$	ω	ω^2	$-\alpha$	0	0	0	0	0	0
$(\Delta_6)_{21}$	0	0	0	0	0	0	1	$-\omega^2\alpha$	$-\omega\alpha$	ω	ω^2	$-\alpha$
$(\Delta_6)_{12}$	0	0	0	0	0	0	1	$-\omega\alpha$	$-\omega^2\alpha$	ω^2	ω	$-\alpha$
$(\Delta_6)_{22}$	1	$-\omega\alpha$	$-\omega^2\alpha$	ω^2	ω	$-\alpha$	0	0	0	0	0	0
$\chi(\Delta_6)$	2	α	α	-1	-1	-2α	0	0	0	0	0	0

$\omega = e^{2\pi i/3}$ $\alpha = e^{i\pi p_3}$

This result means that σ_h is a member of the little group associated with A, since the term $e^{-i\pi/2}$ is a constant times $e^{i\pi/2}$. There is an additional complexity, however. If we write the free electron wave functions in the original Bloch form

$$\psi = u(z)e^{ikz}$$

and if we operate twice on ψ with any compound operation $(R\,|\,c/2)$ containing a displacement, then

$$(R\,|\,c/2)^2\psi = u(z + c)e^{ik(z+c)} = u(z)e^{ikz}e^{ikc}$$

since $u(z)$ is periodic with period c. At the point A, we see that

$$e^{ikz}e^{ikc} = e^{ikz}e^{i(\pi/c)c} = -e^{ikz}$$

and

$$(R\,|\,c/2)^2\psi = -\psi$$

Hence, the translation c on the basis functions at A is *not* a symmetry operation of the group for Γ; *it is a new operation which we denote by T.*

PROBLEM 5-42

Prove that

$$T^2 = E \tag{5-241}$$

The addition of T to the group at A enormously complicates the relations among the symmetry elements. For one thing, we see that all 24 members of the group belonging to $\mathbf{\Gamma}$ also belong to A, for any operation of Table 5-29 will convert the function

$$\psi = u(z)e^{i\pi z/c}$$

into $\pm\psi$ or $\pm i\psi$, each of which is a constant times ψ.

PROBLEM 5-43

Prove this statement.

In addition, the combination of T, or $(E\,|\,c)$, with the elements of $\mathbf{\Gamma}$ generates new elements in the group at A. For example,

TABLE 5-33 ***Symmetry Operations of the Point A for the Group*** D_{6h}^4

K_1	$(E\,|\,0)$
K_2	$(C_2\,|\,c/2)$, $(C_2\,|\,-c/2)$
K_3	$(C_3\,|\,0)$, $(C_3^{-1}\,|\,0)$
K_4	$(C_6\,|\,c/2)$, $(C_6^{-1}\,|\,c/2)$, $(C_6\,|\,-c/2)$, $(C_6^{-1}\,|\,-c/2)$
K_5	$3(C_2'\,|\,0)$, $3(C_2'\,|\,c)$
K_6	$3(C_2''\,|\,0)$, $3(C_2''\,|\,c)$
K_1'	$(J\,|\,c/2)$, $(J\,|\,-c/2)$
K_2'	$(\sigma_h\,|\,0)$, $(\sigma_h\,|\,c)$
K_3'	$(S_6\,|\,c/2)$, $(S_6^{-1}\,|\,c/2)$, $(S_6\,|\,-c/2)$, $(S_6^{-1}\,|\,-c/2)$
K_4'	$(S_3\,|\,0)$, $(S_3^{-1}\,|\,0)$, $(S_3\,|\,c)$, $(S_3^{-1}\,|\,c)$
K_5'	$(\sigma_c\,|\,c/2)$, $3(\sigma_d\,|\,-c/2)$
K_6'	$3(\sigma_v\,|\,0)$
K_7	$(E\,|\,c)$
K_8	$(C_3\,|\,c)$,$(C_3^{-1}\,|\,c)$
K_9	$3(C_v\,|\,c)$

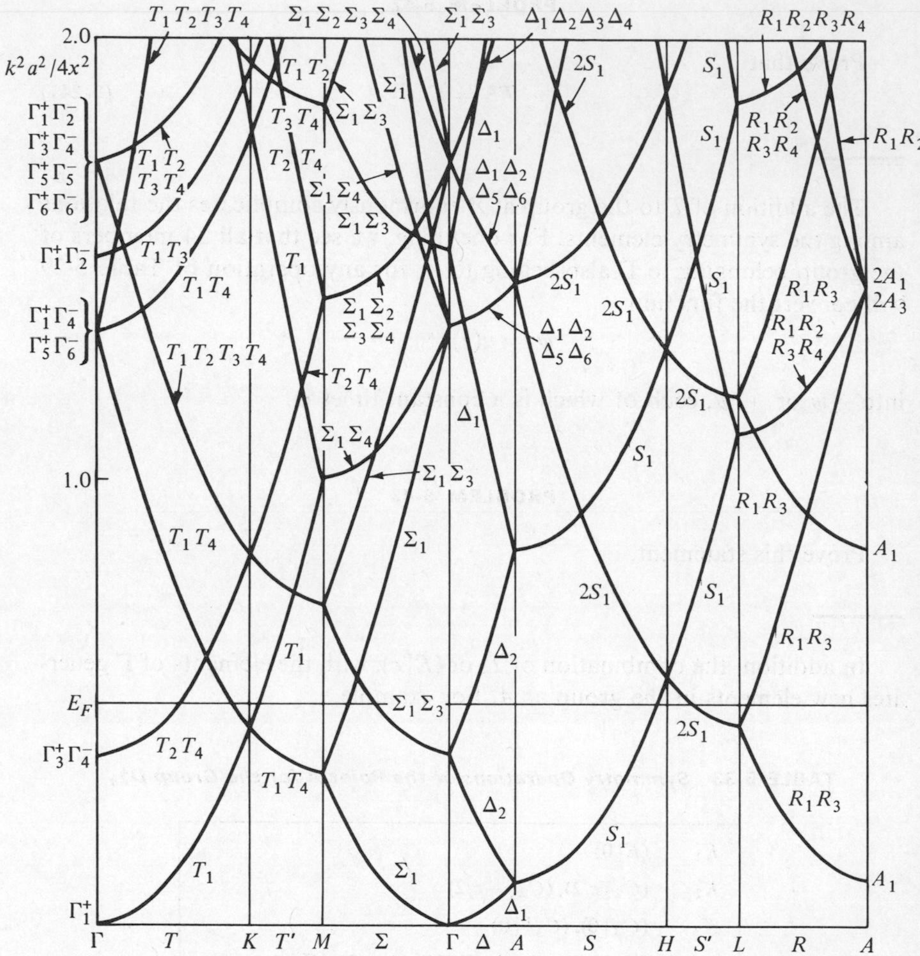

5-57 Free electron bands for the HCP structure

$$(E\,|\,c)(C_3\,|\,0) = (C_3\,|\,c) \quad \text{and} \quad (E\,|\,c)(C_2\,|\,c/2) = (C_2\,|\,-c/2)$$

since $3c/2$ is equivalent to $-c/2$. We find, in fact, that all elements of Table 5-29 of the form $(R\,|\,0)$ generate a new element of the form $(R\,|\,c)$ and all those of the form $(R\,|\,c/2)$ produce an element of the form $(R\,|\,-c/2)$. Hence; the "little" group at A actually contains 48 members divided into 15 classes, as given in Table 5-33.

Table 5-33 will reduce to the 24 elements of Γ if we eliminate every element containing a displacement $-c/2$ or $-c$, which is the case when we go from A to Γ. Hence, the group of Γ contains the 12 classes K_1 through K_6

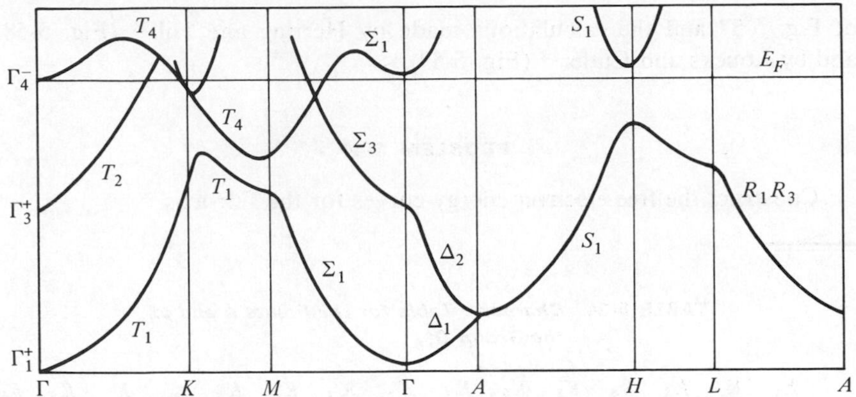

5-58 Band structure for the HCP beryllium crystal as calculated by Herring and Hill[27]

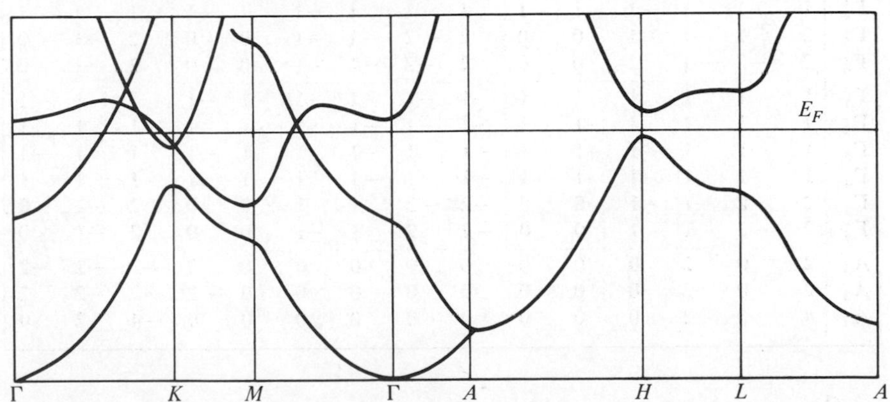

5-59 Band structure for the HCP beryllium crystal as calculated by Loucks and Cutler[28]

and K'_1 through K'_6; classes K_7, K_8, and K_9 become identical to K_1, K_3, and K'_6, respectively. Table 5-34 shows the character system for *both* **Γ** and A; for **Γ**, only the first 12 rows and the first 12 columns need be considered. Note that this portion of the table has the same 2×2 structure as Table 5-4. In this case, however, the antisymmetry of the fourth block in the **Γ** table is with respect to the compound operation $(J|-c/2)$. The three new A representations for 24 of the 48 operations are shown in Table 5-35, as taken from Slater.[5] Although Slater does not explicitly indicate that his table is incomplete, the other half may be easily obtained by simple matrix multiplication.

The element beryllium crystallizes in the HCP configuration; Slater points out that there is a remarkable resemblance between the free electron bands of Fig. 5-57 and the calculations made by Herring and Hill[27] (Fig. 5-58) and by Loucks and Cutler[28] (Fig. 5-59).

PROBLEM 5-44

Construct the free electron energy curves for the ΓM-axis.

TABLE 5-34 ***Character Table for the Points A and of the Group D_{6h}^4***

	K_1	K_2	K_3	K_4	K_5	K_6	K'_1	K'_2	K'_3	K'_4	K'_5	K'_6	K_7	K_8	K_9
Γ_1^+	1	1	1	1	1	1	1	1	1	1	1	1	1	1	1
Γ_2^+	1	1	1	1	−1	−1	1	1	1	1	−1	−1	1	−1	−1
Γ_3^+	1	−1	1	−1	−1	1	1	−1	1	−1	−1	1	1	1	1
Γ_4^+	1	−1	1	−1	1	−1	1	−1	1	−1	1	−1	1	1	−1
Γ_5^+	2	2	−1	−1	0	0	2	2	−1	−1	0	0	2	−1	0
Γ_6^+	2	−2	−1	1	0	0	2	−2	−1	1	0	0	2	−1	0
Γ_1^-	1	1	1	1	1	1	−1	−1	−1	−1	−1	−1	1	1	−1
Γ_2^-	1	1	1	1	−1	−1	−1	−1	−1	−1	1	1	1	1	1
Γ_3^-	1	−1	1	−1	−1	1	−1	1	−1	1	1	−1	1	1	−1
Γ_4^-	1	−1	1	−1	1	−1	−1	1	−1	1	−1	1	1	1	1
Γ_5^-	2	2	−1	−1	0	0	−2	−2	1	1	0	0	2	−1	0
Γ_6^-	2	−2	−1	1	0	0	−2	2	1	−1	0	0	2	−1	0
A_1	2	0	2	0	0	0	0	0	0	0	0	2	−2	−2	−2
A_2	2	0	2	0	0	0	0	0	0	0	0	−2	−2	−2	2
A_3	4	0	−2	0	0	0	0	0	0	0	0	0	−4	2	0

Our approach to the problem of generating irreducible representations has been semiempirical; that is, we have combined group theoretical principles with educated guesses. What we would prefer would be a systematic method applicable to any nonsymmorphic space group. Such a method has been published by Zak,[29] improved by Klauder and Gay,[30] and applied by Gay *et al.*[31] Another approach is that of Raghavacharyulu,[32] whose technique was used by Miller and Love[33] to obtain the representations for all the

[27]C. Herring and A. G. Hill, *Phys. Rev.*, **58**, 132 (1940).

[28]T. L. Loucks and P. H. Cutler, *Phys. Rev.*, **133**, A819 (1964).

[29]J. Zak, *J. Math. Phys.*, **1**, 165 (1960).

[30]L. T. Klauder, Jr. and J. G. Gay, *J. Math Phys.*, **9**, 1488 (1968).

[31]J. G. Gay, W. A. Albers, Jr. and F. J. Arlinghaus, *J. Phys. Chem. Soleds*, **29**, 1449 (1968).

[32]I. V. V. Raghavacharyulu, *Can. J. Phys.*, **39**, 830 (1961).

[33]S. C. Miller and W. F. Love, *Tables of Irreducible Representations of Space and Co-Representations of Magnetic Space Groups*, Pruitt Press, Boulder, Colo., 1967.

TABLE 5-35 *Irreducible Representations at the Point A for the Group D_{6h}^4*

	X_0	X_1	X_{-1}	X_2	X_{-2}	X_3	Y_0	Y_1	Y_{-1}	Y_2	Y_{-2}	Y_3	X'_0	X'_1	X'_{-1}	X'_2	X'_{-2}	X'_3	Y'_0	Y'_1	Y'_{-1}	Y'_2	Y'_{-2}	Y'_3
$(A_1)_{11}$	1	i	i	1	1	i	1	i	i	1	1	i	0	0	0	0	0	0	0	0	0	0	0	0
$(A_1)_{21}$	0	0	0	0	0	0	0	0	0	0	0	0	1	$-i$	$-i$	1	1	$-i$	1	$-i$	$-i$	1	1	$-i$
$(A_1)_{12}$	0	0	0	0	0	0	0	0	0	0	0	0	1	i	i	1	1	i	1	i	i	1	1	i
$(A_1)_{22}$	1	$-i$	$-i$	1	1	$-i$	1	$-i$	$-i$	1	1	$-i$	0	0	0	0	0	0	0	0	0	0	0	0
$\chi(A_1)$	2	0	0	2	2	0	2	0	0	2	2	0	0	0	0	0	0	0	0	0	0	0	0	0
$(A_2)_{11}$	1	i	i	1	1	i	-1	$-i$	$-i$	-1	-1	$-i$	0	0	0	0	0	0	0	0	0	0	0	0
$(A_2)_{21}$	0	0	0	0	0	0	0	0	0	0	0	0	1	$-i$	$-i$	1	1	$-i$	-1	i	i	-1	-1	i
$(A_2)_{12}$	0	0	0	0	0	0	0	0	0	0	0	0	1	i	i	1	1	i	-1	$-i$	$-i$	-1	-1	$-i$
$(A_2)_{22}$	1	$-i$	$-i$	1	1	$-i$	-1	i	i	-1	-1	i	0	0	0	0	0	0	0	0	0	0	0	0
$\chi(A_2)$	2	0	0	2	2	0	-2	0	0	-2	-2	0	0	0	0	0	0	0	0	0	0	0	0	0
$(A_3)_{11}$	1	$i\omega$	$i\omega^2$	ω^2	ω	i	0	0	0	0	0	0	0	0	0	0	0	0	0	0	0	0	0	0
$(A_3)_{21}$	0	0	0	0	0	0	1	$i\omega$	$i\omega^2$	ω^2	ω	i	0	0	0	0	0	0	0	0	0	0	0	0
$(A_3)_{31}$	0	0	0	0	0	0	0	0	0	0	0	0	1	$-i\omega$	$-i\omega^2$	ω^2	ω	$-i$	0	0	0	0	0	0
$(A_3)_{41}$	0	0	0	0	0	0	0	0	0	0	0	0	0	0	0	0	0	0	1	$-i\omega$	$-i\omega^2$	ω^2	ω	$-i$
$(A_3)_{12}$	0	0	0	0	0	0	1	$i\omega^2$	$i\omega$	ω	ω^2	i	0	0	0	0	0	0	0	0	0	0	0	0
$(A_3)_{22}$	1	$i\omega^2$	$i\omega$	ω	ω^2	i	0	0	0	0	0	0	0	0	0	0	0	0	0	0	0	0	0	0
$(A_3)_{32}$	0	0	0	0	0	0	0	0	0	0	0	0	0	0	0	0	0	0	1	$-i\omega^2$	$-i\omega$	ω	ω^2	$-i$
$(A_3)_{42}$	0	0	0	0	0	0	0	0	0	0	0	0	1	$-i\omega^2$	$-i\omega$	ω	ω^2	$-i$	0	0	0	0	0	0
$(A_3)_{13}$	0	0	0	0	0	0	0	0	0	0	0	0	1	$i\omega$	$i\omega^2$	ω^2	ω	i	0	0	0	0	0	0
$(A_3)_{23}$	0	0	0	0	0	0	0	0	0	0	0	0	0	0	0	0	0	0	1	$i\omega$	$i\omega^2$	ω^2	ω	i
$(A_3)_{33}$	1	$-i\omega$	$-i\omega^2$	ω^2	ω	$-i$	0	0	0	0	0	0	0	0	0	0	0	0	0	0	0	0	0	0
$(A_3)_{43}$	0	0	0	0	0	0	1	$-i\omega$	$-i\omega^2$	ω^2	ω	$-i$	0	0	0	0	0	0	0	0	0	0	0	0
$(A_3)_{14}$	0	0	0	0	0	0	0	0	0	0	0	0	0	0	0	0	0	0	1	$i\omega^2$	$i\omega$	ω	ω^2	i
$(A_3)_{24}$	0	0	0	0	0	0	0	0	0	0	0	0	1	$i\omega^2$	$i\omega$	ω	ω^2	i	0	0	0	0	0	0
$(A_3)_{34}$	0	0	0	0	0	0	i	$-i\omega^2$	$-i\omega$	ω	ω^2	$-i$	0	0	0	0	0	0	0	0	0	0	0	0
$(A_3)_{44}$	1	$-i\omega^2$	$-i\omega$	ω	ω^2	$-i$	0	0	0	0	0	0	0	0	0	0	0	0	0	0	0	0	0	0
$\chi(A_3)$	4	0	0	-2	-2	0	0	0	0	0	0	0	0	0	0	0	0	0	0	0	0	0	0	0

groups. Finally, a thorough treatment of the symmorphic groups, including the coefficients used in forming the symmetrized linear combinations of wave functions, has been published by Luehrmann.[34] It is a matter of individual taste as to whether to compute the representations separately for each group or to go to the trouble of mastering a more general approach.

PROBLEM 5-45

Use references 29, 30, and 31 to show how the representations for the space group D_{4h}^{14} are obtained. Explain how and why these tables differ from those of Slater.

PROBLEM 5-46

Use references 32 and 33 to obtain the same information considered in the previous problem. Compare the two sets of tables.

5-14 The band structure of silicon and germanium

The diamond structure consists of two identical, interlocking face-centered cubic lattices, as may be seen by examining Fig. 5-60. One of the lattices is displaced from the other by a distance $(a/4, a/4, a/4)$ where a is the lattice constant, and this feature is illustrated in Fig. 5-61. A face-centered cubic direct lattice has a body-centered cubic reciprocal lattice, and the first Brillouin zone is shown in Fig. 5-62.

The space group O_h^7 for diamond can be described with reference to Fig. 5-61. We shall choose the atom in the lower left-hand corner of the diagram as the origin O, and place a set of axes along the cube edges. The nearest neighbor of this atom lies at $(a/4, a/4, a/4)$, and the inversion J through the corner does not represent a symmetry operation for this atom, but a 180° rotation J_4^2 about an edge is such an operation. The effect of J is to place the nearest neighbor at $(-a/4, -a/4, -a/4)$, and we then return it to a lattice position by a translation **T**, where

$$\mathbf{T} = \left(\frac{a}{4}, \frac{a}{4}, \frac{a}{4}\right) \tag{5-242}$$

[34] A. W. Luehrmann, *Adv. in Phys.*, **17**, 65 (1968).

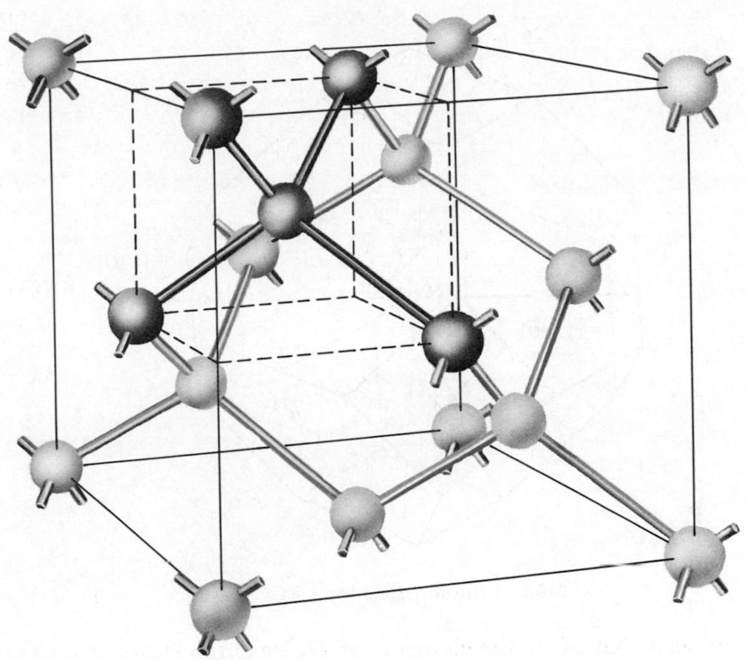

5-60 The diamond lattice

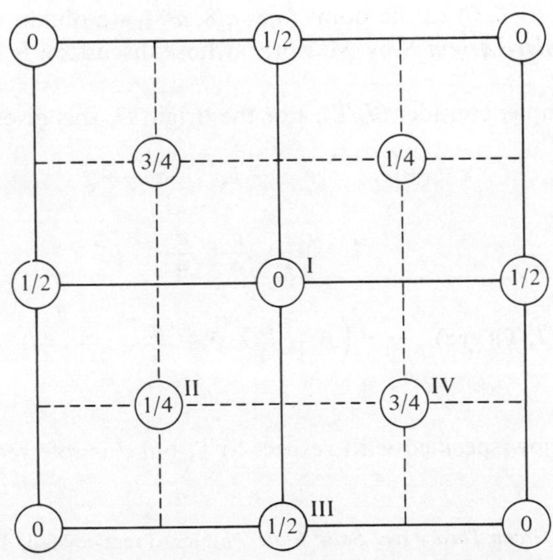

5-61 Unit cell of the diamond lattice (screw axis indicated by Roman numerals)

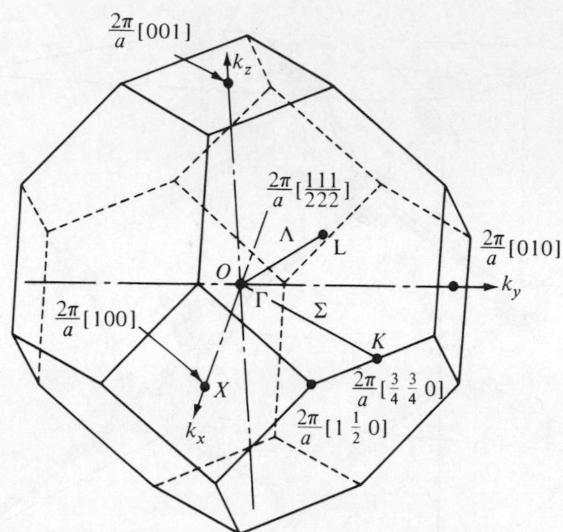

5-62 Brillouin zone for a FCC crystal

It can be seen that 24 of the elements of O_h leave the lattice invariant, and these are given in the top half of Table 5-36. The other 24 must be combined with the translation **T**, and the table shows the effect on a general point x, y, z expressed with respect to axes along the cube edges. If we shift the origin O from (0, 0, 0) to the point ($a/8$, $a/8$, $a/8$), we obtain what has been called the *standard origin* S by Mariot,[35] whose discussion is based on Herman's thesis.[36]

As an example, consider (J, **T**). For the origin O, this gives

$$x \longrightarrow \bar{x}, \qquad y \longrightarrow \bar{y}, \qquad z \longrightarrow \bar{z}$$

followed by

$$\mathbf{T} = \frac{a}{4}, \frac{a}{4}, \frac{a}{4}$$

so that

$$(J, \mathbf{T})(xyz) \longrightarrow \left(\bar{x} + \frac{a}{4}, \bar{y} + \frac{a}{4}, \bar{z} + \frac{a}{4}\right)$$

as shown.

If (xyz) is now specified with respect to S, but J is referred to O, then J has the effect

[35]L. Mariot, *Group Theory and Solid State Physics*, Prentice-Hall, Inc., Englewood Cliffs, N.J., 1960.

[36]F. Herman, "Electronic Structure of Diamond-Type Crystals," Ph.D. thesis, Columbia University, New York, 1953.

TABLE 5-36 ***Symmetry Operations with Respect to Coordinate Origin O and Standard Origin S for the Diamond Lattice***

Operation	O			S		
$(E_2, 0)$	xyz			x	y	z
$(C_4^2, 0)$	$\bar{x}\bar{y}z$			$\bar{x} - \frac{1}{4}a$	$\bar{y} - \frac{1}{4}a$	z
	$x\bar{y}\bar{z}$			x	$\bar{y} - \frac{1}{4}a$	$\bar{z} - \frac{1}{4}a$
	$\bar{x}y\bar{z}$			$\bar{x} - \frac{1}{4}a$	y	$\bar{z} - \frac{1}{4}a$
$(JC_4, 0)$	$y\bar{x}\bar{z}$			y	$\bar{x} - \frac{1}{4}a$	$\bar{z} - \frac{1}{4}a$
	$\bar{y}x\bar{z}$			$\bar{y} - \frac{1}{4}a$	x	$\bar{z} - \frac{1}{4}a$
	$\bar{x}z\bar{y}$			$\bar{x} - \frac{1}{4}a$	z	$\bar{y} - \frac{1}{4}a$
	$\bar{x}\bar{z}y$			$\bar{x} - \frac{1}{4}a$	$\bar{z} - \frac{1}{4}a$	y
	$\bar{z}\bar{y}x$			$\bar{z} - \frac{1}{4}a$	$\bar{y} - \frac{1}{4}a_v$	x
	$z\bar{y}\bar{x}$			z	$\bar{y} - \frac{1}{4}a$	$\bar{x} - \frac{1}{4}a$
$(JC_2, 0)$	$\bar{y}\bar{x}z$			$\bar{y} - \frac{1}{4}a$	$\bar{x} - \frac{1}{4}a$	z
	$\bar{z}y\bar{x}$			$\bar{z} - \frac{1}{4}a$	y	$\bar{x} - \frac{1}{4}a$
	$x\bar{z}\bar{y}$			x	$\bar{z} - \frac{1}{4}a$	$\bar{y} - \frac{1}{4}a$
	yxz			y	x	z
	zyx			z	y	x
	xzy			x	z	y
(C_3, O)	zxy			z	x	y
	yzx			y	z	x
	$z\bar{x}\bar{y}$			z	$\bar{x} - \frac{1}{4}a$	$\bar{y} - \frac{1}{4}a$
	$\bar{y}\bar{z}x$			$\bar{y} - \frac{1}{4}a$	$\bar{z} - \frac{1}{4}a$	x
	$\bar{z}\bar{x}y$			$\bar{z} - \frac{1}{4}a$	$\bar{x} - \frac{1}{4}a$	y
	$\bar{y}z\bar{x}$			$\bar{y} - \frac{1}{4}a$	z	$\bar{x} - \frac{1}{4}a$
	$\bar{z}x\bar{y}$			$\bar{z} - \frac{1}{4}a$	x	$\bar{y} - \frac{1}{4}a$
	$y\bar{z}\bar{x}$			y	$\bar{z} - \frac{1}{4}a$	$\bar{x} - \frac{1}{4}a$
(J, T)	$\bar{x} + \frac{1}{4}a$	$\bar{y} + \frac{1}{4}a$	$\bar{z} + \frac{1}{4}a$	$\bar{x}$	$\bar{y}$	$\bar{z}$
(JC_4^2, T)	$x + \frac{1}{4}a$	$y + \frac{1}{4}a$	$\bar{z} + \frac{1}{4}a$	$x + \frac{1}{4}a$	$y + \frac{1}{4}a$	$\bar{z}$
	$\bar{x} + \frac{1}{4}a$	$y + \frac{1}{4}a$	$z + \frac{1}{4}a$	$\bar{x}$	$y + \frac{1}{4}a$	$z + \frac{1}{4}a$
	$x + \frac{1}{4}a$	$\bar{y} + \frac{1}{4}a$	$z + \frac{1}{4}a$	$x + \frac{1}{4}a$	$\bar{y}$	$z + \frac{1}{4}a$
(C_4, T)	$\bar{y} + \frac{1}{4}a$	$x + \frac{1}{4}a$	$z + \frac{1}{4}a$	$\bar{y}$	$x + \frac{1}{4}a$	$z + \frac{1}{4}a$
	$y + \frac{1}{4}a$	$\bar{x} + \frac{1}{4}a$	$z + \frac{1}{4}a$	$y + \frac{1}{4}a$	$\bar{x}$	$z + \frac{1}{4}a$
	$x + \frac{1}{4}a$	$\bar{z} + \frac{1}{4}a$	$y + \frac{1}{4}a$	$x + \frac{1}{4}a$	$\bar{z}$	$y + \frac{1}{4}a$
	$x + \frac{1}{4}a$	$z + \frac{1}{4}a$	$\bar{y} + \frac{1}{4}a$	$x + \frac{1}{4}a$	$z + \frac{1}{4}a$	$\bar{y}$
	$z + \frac{1}{4}a$	$y + \frac{1}{4}a$	$\bar{x} + \frac{1}{4}a$	$z + \frac{1}{4}a$	$y + \frac{1}{4}a$	$\bar{x}$
	$\bar{z} + \frac{1}{4}a$	$y + \frac{1}{4}a$	$x + \frac{1}{4}a$	$\bar{z}$	$y + \frac{1}{4}a$	$x + \frac{1}{4}a$
(C_2, T)	$y + \frac{1}{4}a$	$x + \frac{1}{4}a$	$\bar{z} + \frac{1}{4}a$	$y + \frac{1}{4}a$	$x + \frac{1}{4}a$	$\bar{z}$
	$z + \frac{1}{4}a$	$\bar{y} + \frac{1}{4}a$	$x + \frac{1}{4}a$	$z + \frac{1}{4}a$	$\bar{y}$	$x + \frac{1}{4}a$
	$\bar{x} + \frac{1}{4}a$	$z + \frac{1}{4}a$	$y + \frac{1}{4}a$	$\bar{x}$	$z + \frac{1}{4}a$	$y + \frac{1}{4}a$
	$\bar{y} + \frac{1}{4}a$	$\bar{x} + \frac{1}{4}a$	$\bar{z} + \frac{1}{4}a$	$\bar{y}$	$\bar{x}$	$\bar{z}$
	$\bar{z} + \frac{1}{4}a$	$\bar{y} + \frac{1}{4}a$	$\bar{x} + \frac{1}{4}a$	$\bar{z}$	$\bar{y}$	$\bar{x}$
	$\bar{x} + \frac{1}{4}a$	$\bar{z} + \frac{1}{4}a$	$\bar{y} + \frac{1}{4}a$	$\bar{x}$	$\bar{z}$	$\bar{y}$
(JC_3, T)	$\bar{z} + \frac{1}{4}a$	$\bar{x} + \frac{1}{4}a$	$\bar{y} + \frac{1}{4}a$	$\bar{z}$	$\bar{x}$	$\bar{y}$
	$\bar{y} + \frac{1}{4}a$	$\bar{z} + \frac{1}{4}a$	$\bar{x} + \frac{1}{4}a$	$\bar{y}$	$\bar{z}$	$\bar{x}$
	$\bar{z} + \frac{1}{4}a$	$x + \frac{1}{4}a$	$y + \frac{1}{4}a$	$\bar{z}$	$x + \frac{1}{4}a$	$y + \frac{1}{4}a$
	$y + \frac{1}{4}a$	$z + \frac{1}{4}a$	$\bar{x} + \frac{1}{4}a$	$y + \frac{1}{4}a$	$z + \frac{1}{4}a$	$\bar{x}$
	$z + \frac{1}{4}a$	$x + \frac{1}{4}a$	$\bar{y} + \frac{1}{4}a$	$z + \frac{1}{4}a$	$x + \frac{1}{4}a$	$\bar{y}$
	$y + \frac{1}{4}a$	$\bar{z} + \frac{1}{4}a$	$x + \frac{1}{4}a$	$y + \frac{1}{4}a$	$\bar{z}$	$x + \frac{1}{4}a$
	$z + \frac{1}{4}a$	$\bar{x} + \frac{1}{4}a$	$y + \frac{1}{4}a$	$z + \frac{1}{4}a$	$\bar{x}$	$y + \frac{1}{4}a$
	$\bar{y} + \frac{1}{4}a$	$z + \frac{1}{4}a$	$x + \frac{1}{4}a$	$\bar{y}$	$z + \frac{1}{4}a$	$x + \frac{1}{4}a$

$$x \longrightarrow \bar{x} - \frac{a}{4}, \qquad y \longrightarrow \bar{y} - \frac{a}{4}, \qquad z \longrightarrow \bar{z} - \frac{a}{4}$$

and **T** again produces the entry in the table.

The ten operations of Table 5-36 can be regarded as a combination of the point group elements and a compound operation known as a *diagonal glide*. Jones[12] has shown how to define this glide with respect to the origin S (Fig. 5-63). Using the coordinate systems shown, a typical glide operation G consists of a displacement $a/4$ along the y'-axis, a displacement $a/4$ along the z'-axis, and a reflection in the plane $x' = 0$. For a point (xyz) in the unprimed system, this operation can be expressed as

$$G(x, y, z) \longrightarrow \left(-x + \frac{a}{4}, y + \frac{a}{4}, z + \frac{a}{4}\right)$$

This operation is equivalent to $\mathbf{T}m$, where $\mathbf{T}$ is defined by (5-242) and m is a reflection in the plane $x = 0$ (rather than $x' = 0$). We note that cubic symmetry implies that m could just as well be a reflection in the planes $y = 0$ or $z = 0$, and this symbol will denote any of the three possibilities.

The 48 operations of Table 5-36 can be expressed as combinations of the 48 point group operations with $\mathbf{T}m$ by the rules

$$(R, 0) \longrightarrow R, \qquad (R, \mathbf{T}) \longrightarrow \mathbf{T}mJR$$
$$(JR, 0) \longrightarrow JR, \qquad (JR, \mathbf{T}) \longrightarrow \mathbf{T}mR$$

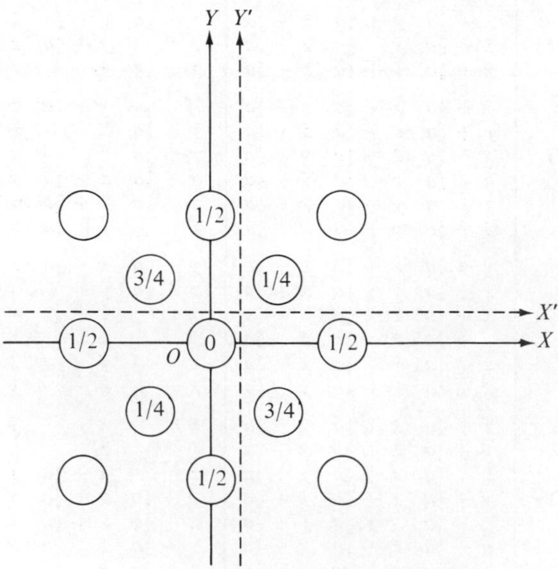

5-63 Glide planes of the diamond lattice

Thus, we have the correspondences

$$\begin{aligned}
(E, 0) &\longrightarrow E, & (J, \mathbf{T}) &\longrightarrow \mathbf{T}m \\
(C_4^2, 0) &\longrightarrow C_4, & (JC_4^2, \mathbf{T}) &\longrightarrow \mathbf{T}mC_4^2 \\
(JC_4, 0) &\longrightarrow JC_4, & (C_4, \mathbf{T}) &\longrightarrow \mathbf{T}mJC_2 \\
(JC_2, 0) &\longrightarrow JC_2, & (C_2, \mathbf{T}) &\longrightarrow \mathbf{T}mJC_2 \\
(C_3, 0) &\longrightarrow C_3, & (JC_3, \mathbf{T}) &\longrightarrow \mathbf{T}mC_3
\end{aligned}$$

We may remark here that the diagonal glide is not a unique compound operation, since it is equivalent to a four-fold screw axis. Such an axis is indicated in Fig. 5-61 by the Roman numerals I–IV. It should also be noted that the operations listed above do not form a group, the situation being analogous to that of tellurium. For example,

$$\mathbf{T}m\mathbf{T}m(x, y, z) \longrightarrow \left(x, y + \frac{a}{2}, z + \frac{a}{2}\right)$$

However, we see that $(\mathbf{T}m)^2$ corresponds to a displacement $(0, \frac{1}{2}, \frac{1}{2})$, so that any atom in the lattice is moved to the site of its nearest neighbor. Hence we can make the identification

$$(\mathbf{T}m)^2 = E$$

which is similar to Eq. (5-222) for tellurium. A Cayley table may then be constructed and the 48 operations are isomorphic to those of the point group, so that the character table applicable to Γ is Table 5-4. It has been shown by Herman[36] that the first half of the table corresponds to free electron functions which are symmetrical with respect to $\mathbf{T}m$ and the second half to functions which are antisymmetrical. As we might expect, the tables needed to determine the projection operators associated with points of high symmetry in the Brillouin zone of the diamond lattice are rather extensive, and

TABLE 5-37 ***Character Table for the Δ-Axis of the Diamond Lattice***

	E	C_4^2	JC_2	JC_4^2	C_4
Δ_1	1	1	1	α	α
Δ_2	1	1	-1	α	$-\alpha$
Δ_2'	1	1	-1	$-\alpha$	$-\alpha$
Δ_1'	1	1	-1	$-\alpha$	α
Δ_5	2	-2	0	0	0

$\alpha = \exp(ika)$

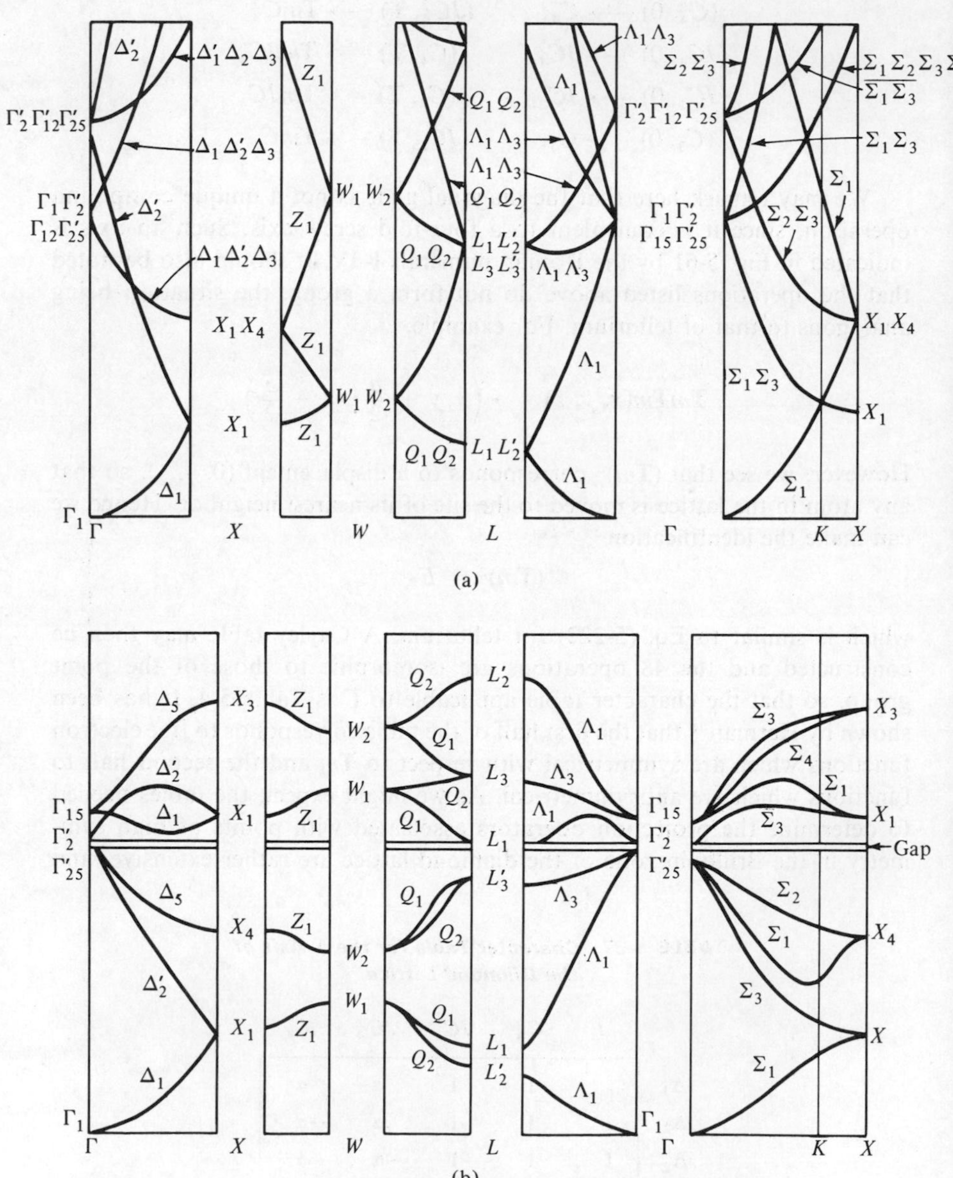

5-64 (a) Free electron energy bands for the diamond lattice; (b) energy bands for germanium

we refer the reader to Appendix 3 of Slater's book.[5] As a simple example, the character system along the Δ-axis, Table 5-37, contains the additional factor $\alpha = \exp(ika)$.

The free electron energy curves for germanium as given by Herman[37] or Slater,[5] are shown in Fig. 5-64(a). The calculations are analogous to those for tellurium, and the position of the gap is determined by the location of the first four levels, so that the division is a curve rather than the shaded regions of Fig. 5-45. In Fig. 5-64(b), the energy bands for germanium, obtained partly by use of the OPW method and partly by symmetry arguments, is shown. This diagram is the result of a computer calculation by Herman, and another version,[38] showing the way the band edges vary as a function of distance in the Brillouin zone, is given in Fig. 5-65(b). A similar diagram for silicon is given in Fig. 5-65(a). We shall consider the two lower diagrams, showing the effect of spin-orbit splitting, in a later section.

As we have seen in connection with beta-brass, the OPW method involves two separate procedures. One is the use of group theory to simplify the secular determinants and the other is the evaluation of integrals involving the crystal potential $V(r)$. Let us consider as an example the plane waves associated with $\mathbf{m} = (111)$ at the center Γ of the zone. Since $\xi = \eta = \zeta = 0$, Eq. (5-79) shows that this point corresponds to an eight-fold degeneracy. That is, $\mathbf{m}$ may take on the values $(\pm 1, \pm 1, \pm 1)$. The secular determinant involving these eight plane waves is

	(111)	$(1\bar{1}\bar{1})$	$(\bar{1}1\bar{1})$	$(\bar{1}\bar{1}1)$	$(\bar{1}\bar{1}\bar{1})$	$(\bar{1}11)$	$(1\bar{1}1)$	$(11\bar{1})$
$(111)^*$	$V_{000} + E_0 - E$	$V_{0\bar{2}\bar{2}}$	$V_{\bar{2}0\bar{2}}$	$V_{\bar{2}\bar{2}0}$	$V_{\bar{2}\bar{2}\bar{2}}$	$V_{\bar{2}00}$	$V_{0\bar{2}0}$	$V_{00\bar{2}}$
$(1\bar{1}\bar{1})^*$	V_{022}	$V_{000} + E_0 - E$	$V_{\bar{2}20}$	$V_{\bar{2}02}$	$V_{\bar{2}00}$	$V_{\bar{2}22}$	V_{002}	V_{020}
$(\bar{1}1\bar{1})^*$	V_{202}	$V_{2\bar{2}0}$	$V_{000} + E_0 - E$	$V_{0\bar{2}2}$	$V_{0\bar{2}0}$	V_{002}	$V_{2\bar{2}2}$	V_{200}
$(\bar{1}\bar{1}1)^*$	V_{220}	$V_{20\bar{2}}$	$V_{02\bar{2}}$	$V_{000} + E_0 - E$	$V_{00\bar{2}}$	V_{020}	V_{200}	$V_{22\bar{2}}$
$(\bar{1}\bar{1}\bar{1})^*$	V_{222}	V_{200}	V_{020}	V_{002}	$V_{000} + E_0 - E$	V_{022}	V_{202}	V_{220}
$(\bar{1}11)^*$	V_{200}	$V_{2\bar{2}\bar{2}}$	$V_{00\bar{2}}$	$V_{0\bar{2}0}$	$V_{0\bar{2}\bar{2}}$	$V_{000} + E_0 - E$	$V_{2\bar{2}0}$	$V_{20\bar{2}}$
$(1\bar{1}1)^*$	V_{020}	$V_{00\bar{2}}$	$V_{\bar{2}2\bar{2}}$	$V_{\bar{2}00}$	$V_{\bar{2}0\bar{2}}$	$V_{\bar{2}20}$	$V_{000} + E_0 - E$	$V_{02\bar{2}}$
$(11\bar{1})^*$	V_{002}	$V_{0\bar{2}0}$	$V_{\bar{2}00}$	$V_{\bar{2}\bar{2}2}$	$V_{\bar{2}\bar{2}0}$	$V_{\bar{2}02}$	$V_{0\bar{2}2}$	$V_{000} + E_0 - E$

[37]F. Herman, *Rev. Mod. Phys.*, **30**, 102 (1958).

[38]F. Herman, *Proc. IRE*, **43**, 1703 (1955).

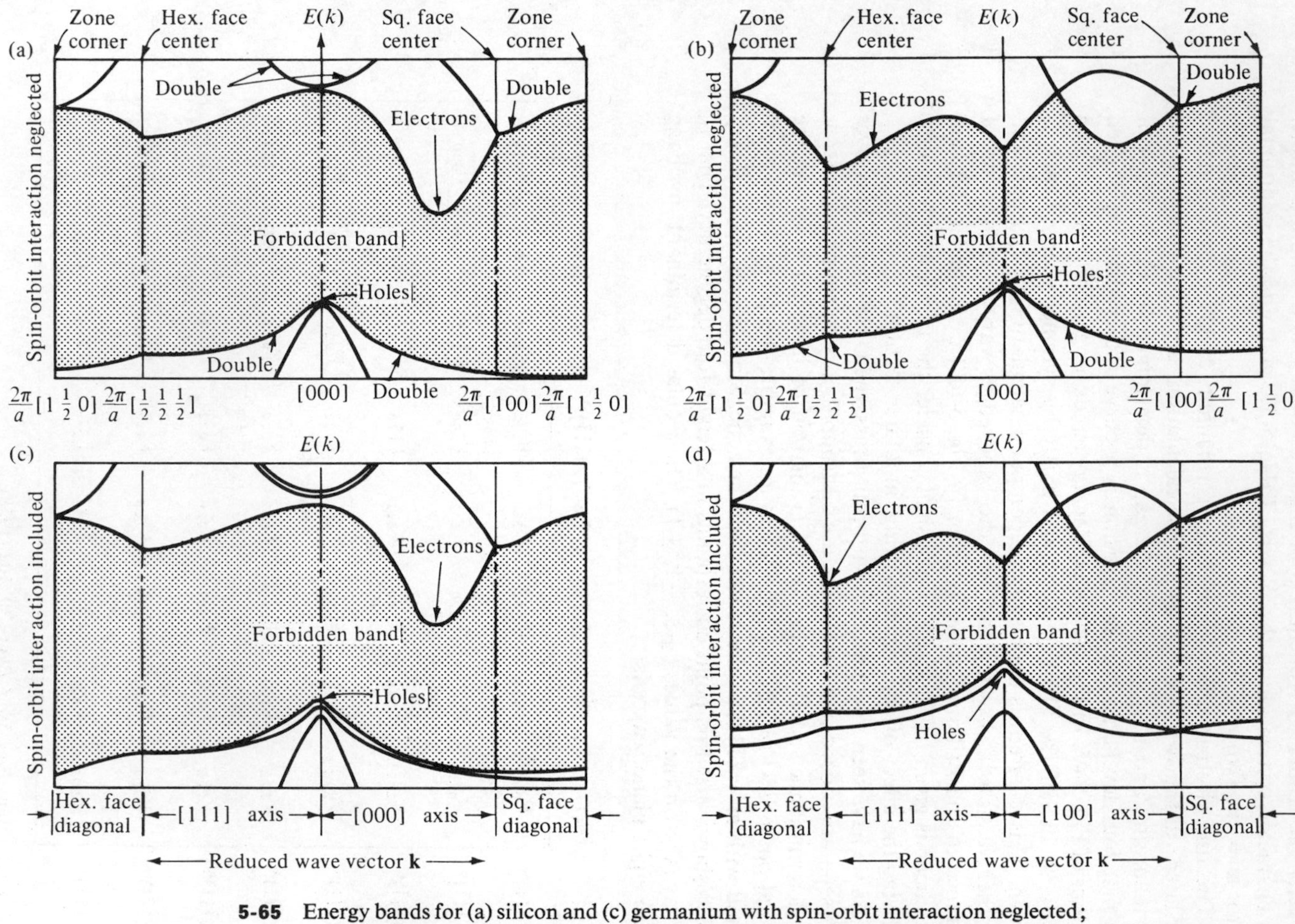

5-65 Energy bands for (a) silicon and (c) germanium with spin-orbit interaction neglected; energy bands for (b) silicon and (d) germanium with spin-orbit interaction taken into account

TABLE 5-38 ***Effect of Diamond Lattice Operations on (111) Plane Waves***

Operation index	Function index 1	2	3	4	5	6	7	8
1	1	2	3	4	5	6	7	8
2	−4	3	2	−1	−8	7	6	−5
3	−2	−1	4	3	−6	−5	8	7
4	−3	4	−1	2	−7	8	−5	6
5	−3	−1	4	2	−7	−5	8	6
6	−2	4	−1	3	−6	8	−5	7
7	−4	3	−1	2	−8	7	−5	6
8	−3	4	2	−1	−7	8	6	−5
9	−2	3	4	−1	−6	7	8	−5
10	−4	−1	2	3	−8	−5	6	7
11	−4	2	3	−1	−8	6	7	−5
12	−3	2	−1	4	−7	6	−5	8
13	−2	−1	3	4	−6	−5	7	8
14	1	3	2	4	5	7	6	8
15	1	4	3	2	5	8	7	6
16	1	2	4	3	5	6	8	7
17	1	4	2	3	5	8	6	7
18	1	3	4	2	5	7	8	6
19	−4	−1	3	2	−8	−5	7	6
20	−2	4	3	−1	−6	8	7	−5
21	−3	2	4	−1	−7	6	8	−5
22	−4	2	−1	3	−8	6	−5	7
23	−2	3	−1	4	−6	7	−5	8
24	−3	−1	2	4	−7	−5	6	8
25	5	6	7	8	1	2	3	4
26	−8	7	6	−5	−4	3	2	−1
27	−6	−5	8	7	−2	−1	4	3
28	−7	8	−5	6	−3	4	−1	2
29	−7	−5	8	6	−3	−1	4	2
30	−6	8	−5	7	−2	4	−1	3
31	−8	7	−5	6	−4	3	−1	2
32	−7	8	7	−5	−3	4	2	−1
33	−6	7	8	−5	−2	3	4	−1
34	−8	−5	6	7	−4	−1	2	3
35	−8	6	7	−5	−4	2	3	−1
36	−7	6	−5	8	−3	2	−1	4
37	−6	−5	7	8	−2	−1	3	4
38	5	7	6	8	1	3	2	4
39	5	8	6	7	1	4	3	2
40	5	6	8	7	1	2	4	3
41	5	8	6	7	1	4	2	3
42	5	7	8	6	1	3	4	2
43	−8	−5	7	6	−4	−1	3	2
44	−6	8	7	−5	−2	4	3	−1
45	−7	6	8	−5	−3	2	4	−1
46	−8	6	−5	7	−4	2	−1	3
47	−6	7	−5	8	−2	3	−1	4
48	−7	−5	6	8	−3	−1	2	4

where the subscripts on the V_{ijk} are the differences between the column and row subscripts.

In order to work with the projection operators, as we did for beta-brass, it is convenient to number the operations in Table 5-36 from 1 to 48, in the order given, and to label the eight plane waves as

$$\begin{aligned} \psi_1 &= [111], & \psi_2 &= [1\bar{1}\bar{1}] \\ \psi_3 &= [\bar{1}1\bar{1}], & \psi_4 &= [\bar{1}\bar{1}1] \\ \psi_5 &= [\bar{1}\bar{1}\bar{1}], & \psi_6 &= [\bar{1}11] \\ \psi_7 &= [1\bar{1}1], & \psi_8 &= [11\bar{1}] \end{aligned}$$

Using the origin S, the effect of each operation can be determined by starting with the plane wave solution in the form

$$\psi_1 = [111] = \exp\left\{\frac{-2\pi i(m_1 x + m_2 y + m_3 z)}{a}\right\} \tag{5-243}$$

as obtained from (5-80). Then

$$\begin{aligned} (JC_4^2, \mathbf{T})[111] &= \exp\left[\frac{-2\pi i\{m_1(x + a/4) + m_2(y + a/4) - m_3 z\}}{a}\right] \\ &= [11\bar{1}] \exp\left(\frac{-\pi i}{2}\right) \exp\left(\frac{-\pi i}{2}\right) = -[11\bar{1}] \end{aligned} \tag{5-244}$$

All operations of this type are given in Table 5-38, taken from Mariot.[35]

Combining the information in this table with Table 5-4 the reducible characters are

	E	C_4^2	JC_4	JC_2	C_3	J	JC_4^2	C_4	C_2	JC_3
χ	8	0	0	4	2	0	0	0	0	0

By (2-63), we have

$$n_{\Gamma_1} = 1, \qquad n_{\Gamma'_2} = 1, \qquad n_{\Gamma'_{25}} = 1, \qquad n_{\Gamma_{15}} = 1$$

with all other n_i vanishing. Thus, this representation can be decomposed into two nondegenerate irreducible representations and two triply-degenerate ones.

As we showed in connection with Eq. (5-157), cubic symmetry imposes some relations among the Fourier coefficients of the secular determinant. In addition, arguments like that associated with Eq. (5-244) produce other relations of the form

$$V_{200} = V_{\bar{2}00}$$

and the secular determinant simplifies to

	(111)	$(1\bar{1}\bar{1})$	$(\bar{1}1\bar{1})$	$(\bar{1}\bar{1}1)$	$(\bar{1}\bar{1}\bar{1})$	$(\bar{1}11)$	$(1\bar{1}1)$	$(11\bar{1})$
(111)*	$V_{000}+E_0-E$	V_{220}	V_{220}	V_{220}	V_{222}	$-V_{200}$	$-V_{200}$	$-V_{200}$
$(1\bar{1}\bar{1})$*	V_{220}	$V_{000}+E_0-E$	$-V_{220}$	$-V_{220}$	$-V_{200}$	V_{222}	V_{200}	V_{200}
$(\bar{1}1\bar{1})$*	V_{220}	$-V_{220}$	$V_{000}+E_0-E$	$-V_{220}$	$-V_{200}$	V_{200}	V_{222}	V_{200}
$(\bar{1}\bar{1}1)$*	V_{220}	$-V_{220}$	$-V_{220}$	$V_{000}+E_0-E$	$-V_{200}$	V_{200}	V_{200}	V_{222}
$(\bar{1}\bar{1}\bar{1})$*	V_{222}	V_{200}	V_{200}	V_{200}	$V_{000}+E_0-E$	V_{220}	V_{220}	V_{220}
$(\bar{1}11)$*	V_{200}	V_{222}	$-V_{200}$	$-V_{200}$	V_{220}	$V_{000}+E_0-E$	$-V_{220}$	$-V_{220}$
$(1\bar{1}1)$*	V_{200}	$-V_{200}$	V_{222}	$-V_{200}$	V_{220}	$-V_{220}$	$V_{000}+E_0-E$	$-V_{220}$
$(11\bar{1})$*	V_{200}	$-V_{200}$	$-V_{200}$	V_{222}	V_{220}	$-V_{220}$	$-V_{220}$	$V_{000}+E_0-E$

It is quite evident that computing eight projection operators each involving 48 terms is a long and tedious process, so that we shall merely state the results. It is found that

$$\Gamma_1: \quad E - E_0 = V_{000} - 3V_{220} + V_{222}$$
$$\Gamma'_2: \quad E - E_0 = V_{000} - 3V_{220} - V_{222}$$
$$\Gamma'_{25}: \quad E - E_0 = V_{000} + V_{220} + V_{222}$$
$$\Gamma_{15}: \quad E - E_0 = V_{000} + V_{220} - V_{222}$$

Herman[36] has given an explicit example of a factorized secular determinant involving [000], [111], [200], and [220] type plane waves; this is given as Table VII in his thesis.

The remainder of the procedure for determining the energy band structure is essentially numerical. We refer the reader to Herman *et al.*[39] for the latest results along these lines.

[39]F. Herman, R. L. Kortum, C. D. Kuglin, and R. A. Short, "New Studies of the Band Structure of Silicon, Germanium, and Grey Tin" in *Quantum Theory of Atoms, Molecules, and the Solid-State: A Tribute to J. C. Slater*, ed. Per-Olov Löwdin, Academic Press, Inc., New York, 1966.

5-15 Spin and double groups

One of the refinements in energy band theory which has been found to be significant in the case of silicon and germanium is the effect of spin. Spin introduces a new symmetry operation into any space group, because the arrangement of the spins must be consistent with the other symmetries.

To consider quantitatively the effect of spin, we realize that the solutions to the Schroedinger equation for a spherically symmetric potential $V(r)$ are of the form

$$\psi(r, \theta, \phi) = R(r)\Theta(\theta)\Phi(\phi)$$

where

$$\Phi(\phi) = \exp(\pm i m_l \phi)$$

A rotation of coordinate axes about the z-axis through an angle α can be expressed as

$$(\alpha, E)\psi = R\Theta \exp[\pm i m_l(\phi - \alpha)] = \exp[\mp i m_l \alpha]$$

For a given value of l, there are $(2l + 1)$ values of m_l, and the effect of this rotation on the associated atomic orbitals can be expressed in terms of the matrix

$$\Gamma^{(l)}(\alpha) = \begin{bmatrix} \exp(-il\alpha) & 0 & 0 & \cdots & 0 \\ 0 & \exp[-i(l-1)\alpha] & 0 & \cdots & 0 \\ 0 & 0 & \exp[-i(l-2)\alpha] & \cdots & 0 \\ \vdots & \vdots & \vdots & & \vdots \\ 0 & 0 & 0 & \cdots & \exp(il\alpha) \end{bmatrix}$$

The character is then

$$\begin{aligned} \chi^{(l)}(\alpha) &= \exp(-il\alpha) + \cdots + \exp(il\alpha) = \exp(-il\alpha) \sum_{k=0}^{2l} [\exp(i\alpha)]k \\ &= \exp \frac{(-il\alpha)(\{\exp[i(2l+1)\alpha] - 1\}}{[\exp(i\alpha) - 1]} \qquad \text{(5-245)} \\ &= \frac{\exp[i(l+\frac{1}{2})\alpha] - \exp[i(l+\frac{1}{2})\alpha]}{\exp(i\alpha/2) - \exp(-i\alpha/2)} = \frac{\sin(l+\frac{1}{2})\alpha}{\sin(\alpha/2)} \end{aligned}$$

For a rotation $(\alpha + 2\pi)$

$$\chi^{(l)}(\alpha + 2\pi) = \frac{\sin[(l+\frac{1}{2})(\alpha + 2\pi)]}{\sin(\alpha/2 + \pi)} = \frac{\sin(l+\frac{1}{2})\alpha}{-\sin\alpha/2} = -\chi^{(l)}(\alpha)$$

For $\alpha + 4\pi$, we find that

$$\chi^{(l)}(\alpha + 4\pi) = \chi^{(l)}(\alpha)$$

Hence, the characters change sign after a rotation of 2π, but come back to their original value for a second such rotation. We make the artificial stipulation that a rotation of 2π does not map the lattice into itself, but that a rotation of 4π does. The 2π rotation can be considered as a new symmetry element $\bar{E}$, called the *anti-identity*, which reverses the signs of the characters. We then have the relation

$$\bar{E}^2 = E \tag{5-246}$$

so that $\bar{E}$ is analogous to the operator T introduced in connection with Table 1-19. The effect of T on the character table is to double the number of operations at the top or bottom of the Brillouin zone. A similar result will hold for $\bar{E}$, leading to what are known as *double groups*. Double groups have twice as many operations as ordinary groups and more classes (although not necessarily twice the number).

Our discussion above has been in terms of orbital angular momentum, but it could equally well apply to total (orbital plus spin) angular momentum, or to spin alone. If we associate the possibilities of spin up or spin down with each lattice point, then we are led to double groups, because an additional symmetry element is introduced. Such groups are also known as Shubnikov[40] or *black-and-white* groups and are of importance in the theory of magnetism. Suppose that the operation $\bar{E}$, instead of being a rotation, has the effect of changing a lattice point from black to white, or vice versa. This operation could not be combined with a three-fold rotation, C_3, for example, for we cannot construct an equilateral triangle with alternating black and white corners. However, a hexagon can be made in this fashion, so that in the group D_{6h} (the symmetry group of the hexagonal prism), $\bar{E}$ would combine with some operations but not with others.

Let us consider the tetrahedral group T_d as an example. The characters of the single group are given in Table 5-39 (or 3-4). The double group will then have 48 operations, but only eight classes, because C_4^2 belong to one class, and so do JC_2 and RJC_2. Jones[12] proves this analytically by considering the properties of the transformation matrix for an arbitrary displacement which leaves one point invariant (i.e., one involving the Euler angles). However, we can intuitively anticipate this, since C_4^2 and JC_2 are both equivalent to a rotation of π. Now Eq. (5-245) shows that for such an operation, $\chi = 0$, since this corresponds to $\sin (l + \frac{1}{2})\pi$, where l is an integer. It may be surprising to find a color reversal associated with a three-fold operation C_3,

[40]A. V. Shubnikov, N. V. Belov, and others, *Colored Symmetry*, The Macmillan Company, New York, 1964.

TABLE 5-39 ***Character Table for Point Group T_d***

	E	$3C_4^2$	$8C_3$	$6JC_4$	$6JC_2$
Γ_1	1	1	1	1	1
Γ_2	1	1	1	-1	-1
Γ_3	2	2	-1	0	0
Γ_4	3	-1	0	-1	1
Γ_5	3	-1	0	1	-1

but it will be realized that when a cube is viewed along a body diagonal—or a (111)-axis—the resultant figure is a hexagon, so that the corners can be alternately black and white.

Since we have 48 operations and eight classes, the dimensionalities of the irreducible representations are given by the relation

$$1^2 + 1^2 + 2^2 + 2^2 + 2^2 + 3^2 + 3^2 + 4^2 = 48 \qquad \textbf{(5-247)}$$

so that the three additional representations are of degree 2, 2, and 4. Again using Bethe's method, we obtain Table (5-40) for the double group.

TABLE 5-40 ***Character Table for Double Group $E \times T_d$***

	E	$\bar{E}$	$6C_4^2$	$8C_3$	$8\bar{E}C_3$	$6JC_4$	$6\bar{E}C_4$	$12JC_2$
Γ_1	1	1	1	1	1	1	1	1
Γ_2	1	1	1	1	1	-1	-1	-1
Γ_3	2	2	2	-1	-1	0	0	0
Γ_4	3	3	-1	0	0	-1	-1	1
Γ_5	3	3	-1	0	0	1	1	-1
Γ_6	2	-2	0	1	-1	$\sqrt{2}$	$-\sqrt{2}$	0
Γ_7	2	-2	0	1	-1	$-\sqrt{2}$	$\sqrt{2}$	0
Γ_8	4	-4	0	-1	1	0	0	0

Proceeding as before, character tables corresponding to the Δ-, Λ-, and Σ-axes may also be calculated and compatibility relations determined from them. Dresselhaus[41] has pointed out that this table predicts the nature of the *spin-orbit splitting* in crystals. That is, the Schroedinger equation goes over into the Dirac equation and the Hamiltonian contains a term $(e^2/4m^2c^2)$ **grad** $V \times \mathbf{p} \cdot \boldsymbol{\sigma}$, where the components of $\boldsymbol{\sigma}$ are the Pauli spin matrices. Table

[41] G. Dresselhans, *Phys. Rev.*, **100**, 580 (1955).

5-41 indicates that the spin-orbit energy splits the levels which are highly degenerate at Γ into single or doubly-degenerate levels along the axes. This is similar to the situation shown in Herman's calculations for the valence bands of silicon and germanium [Figs. 5-65(b) and (d)]. Hence, spin-orbital splitting is responsible for the existence of several kinds of holes in these materials.

TABLE 5-41 ***Compatibility Relations for Double Cubic Group***

Γ_6	Γ_7	Γ_8
Δ_5	Δ_5	$\Delta_5\Delta_6$
Λ_6	Λ_6	$\Lambda_4\Lambda_5\Lambda_6$
$\Sigma_3\Sigma_4$	$\Sigma_3\Sigma_4$	$\Sigma_3\Sigma_3\Sigma_4\Sigma_4$

This discussion of double groups has been very brief. In particular, we have considered their application to energy band theory, but have ignored their use in understanding magnetic materials. A review by Cracknell,[42] to which we refer the reader, considers symmetry and its relation to the various kinds of magnetism: paramagnetism, ferromagnetism, and so forth.

[42] A. P. Cracknell, *Rep. Prog. Phys.*, **32**, 633 (1969).

AUTHOR INDEX

SUBJECT INDEX

T